THE GOSHAWK

For Bridget, for years of patience and encouragement.

THE GOSHAWK

ROBERT KENWARD

Illustrated by
ALAN HARRIS

T & A D POYSER
London

Published 2006 by T & A D Poyser, an imprint of A&C Black Publishers Ltd., 38 Soho Square, London W1D 3HB

www.acblack.com

Reprinted 2007

Copyright © 2006 text by Robert Kenward
Copyright © 2006 illustrations by Alan Harris

The right of Robert Kenward to be identified as the author of this work has been asserted by him in accordance with the Copyright, Design and Patents Act 1988.

ISBN: 978–0–7136–6565–9

A CIP catalogue record for this book is available from the British Library

All rights reserved. No part of this publication may be reproduced or used in any form or by any means – photographic, electronic or mechanical, including photocopying, recording, taping or information storage or retrieval systems – without permission of the publishers.

This book is produced using paper that is made from wood grown in managed sustainable forests. It is natural, renewable and recyclable. The logging and manufacturing processes conform to the environmental regulations of the country of origin.

Commissioning Editor: Nigel Redman
Project Editor: Jim Martin

Typeset by RefineCatch Limited, Bungay, Suffolk

Printed and bound in China

10 9 8 7 6 5 4 3 2

Contents

Prologue

Foreword

1	NAMES, RACES AND RELATIVES	17
	What's in a name	18
	Plumage	19
	Racial categories	24
	Evolution within *Accipiter gentilis*	29
	Goshawk relatives	30
	Goshawk origins	33
	Implications for conservation and management	33
	Conclusions	34

2	WEIGHTS AND MEASURES	36
	Size matters	37
	Variation between the sexes	37
	Variation in body composition	42
	Variation through the seasons	46
	The benefit of high mass in winter	48
	Causes and consequences of seasonal variation in body-mass	49
	Variation in mass with latitude	51
	Possible origins of reversed size dimorphism	52
	Size rules	54
	Implications for conservation and management	54
	Conclusions	55

3	NESTING AND LAYING	57
	Tamed and wild	59
	Nest habitats	60
	The nest	60
	The nest site	61
	The next area	64
	Nest and area occupancy	68
	Courtship behaviour	70
	Flight behaviours	71
	Calling	73

	Territoriality	75
	Nest-building	75
	Provisioning	76
	Copulation	77
	Laying	80
	Laying date	80
	Clutch size	82
	Repeat clutches	84
	Implications for conservation and management	84
	Conclusions	88
4	INCUBATION AND REARING	90
	Incubation	91
	Hatching	93
	Feeding the young	95
	Parenting	97
	Development of the young	100
	Sex ratios	104
	Feather growth and moulting	107
	Implications for conservation and management	112
	Conclusions	113
5	MARKERS AND MOVEMENTS	115
	Marking and trapping	116
	Traps	116
	Identification by natural characters: feathers and DNA	119
	Artificial markers for visual detection: rings (bands) and dyes	120
	Markers for electronic detection: micro-transponders and radio-tags	121
	Movements in the post-fledging dependence period	123
	Dispersal	125
	Post-fledging brood-switching	128
	First-winter movements	130
	How dispersal movements occur	131
	Homing	137
	Migration	138
	Irruptions	140
	To move or not to move?	142
	Pre-nuptial movements and site fidelity	143
	Implications for conservation and management	145
	Conclusions	146

6	DIET AND FORAGING	148
	Studying diet at nests	149
	Studying diet in winter	151
	Foraging behaviour	153
	Studing Goshawk behaviour in lowland Britain	154
	Foraging in lowland Britain	155
	Attacking, feeding, loafing and bathing in lowland Britain	159
	Foraging in Fennoscandia	160
	Foraging in towns	163
	Foraging in North America	165
	Home-ranges and territoriality	166
	Variation in diet with latitude and altitude in Europe	170
	Variation in diet with time	171
	Variation in diet between sexes and ages	175
	Carrion	177
	Comparing diet in Eurasia and North America	177
	Implications for conservation and management	180
	Conclusions	182
7	PREY SELECTION AND PREDATION PRESSURES	184
	Prey vulnerability	185
	Attack success	187
	Arranged attacks	188
	Selection	190
	How Goshawks choose prey	193
	Specialisation	194
	Food requirements	195
	Estimating the impacts of predation	196
	Goshawks and pheasants	199
	Goshawks and woodland grouse	202
	Goshawks and corvids	202
	Goshawks and other uncommon species	204
	How Goshawk predation relates to change in prey densities	206
	Implications for conservation and management	209
	Conclusions	215
8	DEATH AND DEMOGRAPHY	217
	Population dynamics	218
	Egg-laying and clutch size	220
	Total and partial brood failures	220

	Spatial and temporal variation in productivity	224
	Productivity of individuals and at single sites	227
	Occupancy and breeding rates	229
	Mortality and survival rates	233
	Causes of death	234
	Juvenile and adult mortality rates	237
	Temporal and sex-linked variation in mortality	239
	Life-time reproductive success	240
	Population models	241
	Dynamics and density	243
	Population regulation	244
	Dynamics and evolution	248
	Implications for conservation and management	248
	Conclusions	250
9	FALCONRY AND MANAGEMENT METHODS	252
	History of falconry	253
	Training Goshawks	257
	Location aids	263
	What happened to trained Goshawks?	264
	Domestic breeding	266
	Hacking	270
	Predation management	270
	Implications for conservation	273
	Conclusions	276
10	CONSERVATION THROUGH PROTECTION AND USE	277
	Cooperation or conflict?	278
	Pesticides	279
	Past persecution and present illegal killing	284
	Predation	287
	Prey deficits	289
	Goshawk habitats	293
	Goshawks remain widespread	295
	The application of some conventions	296
	Cooperative conservation	299
	Conclusions	302

11 *Appendices*
    1 Scientific names of vertebrates mentioned in the text   304
    2 Sources for figures that used data from more than 10 publications 307

References   312

Index   358

Prologue

It's one of those dark December nights that are damp-cold rather than frosty, because a cloud layer obscures the moon and open sky. I'm six metres up a pine tree on the edge of a firing range, a short way from home. These mid-life Scots pines are nice trees to climb, with branches that still come down to hand-reach and don't break easily if you keep your weight on the junction with the trunk. I can climb easily, without shaking the tree or getting twigs in my face.

I've climbed very many trees to collect Goshawks, but not at night. The torch held by my son Ben is fixed on the juvenile female still sitting with one leg tucked up, some two metres above me. I start talking to her as I would in the mews, where she'd normally be roosting on her choice of perch.

The first time I climbed for 'Miss Piggy' was in early July, on the Baltic island of Gotland. An early morning ascent of a substantial pine was followed by prolonged discussion at Visby airport, and finally by flights to Stockholm and Heathrow. The airline staff and customs found it hard to understand how the Swedish agriculture ministry had given a Brit a licence to raid their wildlife.

Miss Piggy got her name after being tamed and taught rapidly to fly to the fist, so that she could be released to study her early hunting and settling behaviour at a natural age. The name came from the killing of 17 pheasants in 22 days. Radio-tracking and filming (for 'Phantom of the Forest') showed how she found a pheasant pen two miles from her release site. Happily, a price for each possible depredation had been agreed in advance, but the TV firm raised eyebrows at the bill. The predation rate was an eye-opener for all concerned, especially the radio-tracker on study leave from the Department of the Environment.

After being trapped on one of her kills (because the UK licence was only for experimental release) Miss Piggy is destined to hunt squirrels with me and then for study of her breeding behaviour. However, such is my overconfidence that yesterday evening I flew her free without a bell, and then failed to find the right place to call her down because I couldn't see her in the dusk. She's also plump enough to have hunted today without conviction, growing too wild for much interest in my offerings. An important meeting tomorrow gives no time to tempt her in the morning.

So I gently negotiate three more branches and reach up to slip one hand round both her legs, just as she tenses for flight. Of course, one leg eludes my grasp and completes our connection by seizing my hand instead. After folding her wings and tucking her under my jacket, the one-handed climb down is awkward and messy. 'You're bleeding, Dad', says Ben.

Foreword

If you share my tendency to skip prologues, please go back a page and read a short tale. It sets the tone for a book in which each chapter starts with an anecdote to leaven the weightier matters that follow. It also starts to illustrate a conservation perspective that will be developed at the end of each chapter. In other respects, the format follows that agreed with Trevor Poyser 25 years ago. Thus, the first two chapters consider the taxonomy and measurements of hawks, followed by two on breeding and a fifth on movements, starting as the young hawks fledge and then disperse. The next two chapters are on diet and predatory impacts, followed by an eighth on how Goshawk populations function. The ninth chapter considers falconry and other management techniques, with a final chapter that brings together the many conservation issues and looks to the future.

The anecdotes in each chapter will help show more about the Northern Goshawk than mere biology, because this is a bird to be viewed from many sides. It is cherished by some, such as birdwatchers and falconers. It is an intriguing subject for rehabbers, vets and scientists. Conversely, the Goshawk can be loathed by pigeon fanciers and keepers of poultry and pheasants. It has joined with owls to irritate lumberjacks. Goshawks can be bread-and-butter for some, yet occasionally threaten the livelihoods of others. As a result of all this interest, the Northern Goshawk vies with the Golden Eagle (scientific names are in Appendix I) for the bronze medal for abundance of scientific publications, after the American Bald Eagle and the Peregrine Falcon. Unlike those species, it has been studied to a similar extent on both sides of the Atlantic. Perhaps, due to the variety of research year-round on tamed as well as free-living Northern Goshawks, this is the most comprehensively studied wild raptor.

My own connections with the Goshawk are varied and long. I grew up on a farm in Hampshire, in a post-war Britain that lacked wild Goshawks but was otherwise rich in the wildlife of well-loved land that still pensioned working horses. In common with the other children of Upham, I collected wild flowers and bird's eggs (only one egg from each species: the first sight of multiple clutches from a 'real' collector was a shock). My mother introduced me to the huge caterpillars of hawk moths, and in her family home at Dumbleton I kept a museum, encouraged greatly by gifts of fossils and other faunal mementos from the travels of her sister Joan. My pre-school teacher Rose Wright loved nature rambles, and my father taught me to shoot and fish.

My serious tree-climbing began with nests of the Rook, and was not discouraged by my uncle Esmond Giles, a recorder of birds at Romney Marsh. He knew an oologist who could explain about unusual eggs. In days when birders and collectors could be friends, Esmond taught me to recognise many avian species.

The first Goshawk in my life came from central Europe, an eyass tiercel (nestling male) sent by an ex-army sergeant, Bill Ruddock, who was also starting to release older hawks in northern England. It cost £18, and shared a barn with its sister until it was hard-penned (flight feather growth completed) and ready for training. Since then, thanks to a tolerant school (Uppingham) and university college (Christ Church) I have worked with about a dozen trained Goshawks. With so much Goshawk biology written in German at that time, it seemed sensible to learn that language and hitch-hike to visit falconers with special biological skills, including Walter Bednarek, Heinz Brüll, Gustl Eutermoser, Helmut Link, Horst Niesters and Renz Waller.

As a student at Oxford University, I was too much in awe of lecturers and demonstrators like Richard and Marian Stamp Dawkins, David Lack, John Lawton, Tony Sinclair and Niko Tinbergen to be confident of a career in zoology. So I sought to assuage a fascination with flight by joining the RAF. However, my entomology tutor George Gradwell had other ideas and persuaded me to go with him to visit Chris Perrins in the Edward Grey Institute of Field Ornithology. The result was an invitation to do a doctorate, studying whether Goshawks (then re-colonising Britain) could help to reduce crop damage by Woodpigeons.

Chris Perrins provided an introduction to another friend, who was just starting work for the Nature Conservancy on the Eurasian Sparrowhawk. Ian Newton suggested that, if I was going to import Goshawks for training and release in the Woodpigeon study, perhaps he could have some to release by fostering with Sparrowhawks. I had joined the committee of the British Falconers' Club, whose members Gordon Jolly and Jack Mavrogordato agreed to try loaning Goshawks to falconers and then releasing them.

So I travelled in 1971 to Finland, where Seppo Sulkava and Kaukko Huhtala gave the first research hawks for release in Scotland during the next two years. Teppo Lampio then arranged in the Central Hunters' Organisation for a voluntary scheme that ran for another five years through Svante Andersson and Seemi Pihläjämäki. There was help in Scotland from David Kent (who had already arranged hawks from Sweden for release in the Welsh borders with Russell Coope), from Nick Fox and Ian Newton's assistants Mick Marquiss and Herman Ostroznick, with welcome hospitality from Halina Newton. The first release area for trained hawks was organised by Doug Weir, not far from my father's and stepmother Bridget's cottage in Scotland. Much help with the falconry side of these voluntary efforts also came from Henry Clamp, Pat and Patricia Coles, Ted Davis, Pat Fields, Caroline Hunt, Tony Jack, Josephine Mitchell, John Murray, Brian Simpson, David Stoodley and Tony Walker.

There were many results from the affair with Finnish Goshawks. One was a successful thesis, thanks to help and comment from Mike Cullen, John Krebs, Hans Kruuk, Ron Murton, Ian Newton, David Macdonald, Robert Prys-Jones, Luke Schifferli, Richard Sibley and Mike Webber, as well as Chris Perrins, with further stimulation from flat-mates Nick Davies and Tim Birkhead. After following Goshawks released with radio tags through many cold winter days in northern Oxfordshire, an impoverished student also owed much for hot meals, coffee and empathy from Jim and Kaye Wiggins.

Another outcome was a series of Oxford conferences that were run with John Cooper, Tim Geer, Ian Lindsay and Gavin Wakley, on raptor health and breeding (1975), on raptor management (1977) and finally on 'Understanding the Goshawk'

(1981). A third outcome was the establishment of wild Goshawks breeding near Oxford, albeit in tiny numbers compared with the population introduced into the Welsh borders (with which they now merge). A fourth outcome, after surveys on survival and breeding in trained Goshawks and other raptors, was an invitation to begin what turned out to be more than 30 years of service with the International Association for Falconry and Conservation of Birds of Prey.

For this book, the crucial outcome was a continuation of links with the Fennoscandian countries of Finland, Norway and Sweden for more work on Goshawks. Travels and meetings had already brought friendship with northern raptor biologists, including Per Bro, Björn Helander, Peter Lindberg, Pertti Saurola and Marcus Wikman, but it was not until 1975 that I went to Uppsala and met Vidar Marcström. Vidar, who at first doubted whether one really could radio-tag and stalk Goshawks to record their kills, nevertheless invited me to begin a post-doctorate in Sweden. Radio-tagging seemed in those days unbelievably 'James Bond'. Ian Newton and Nick Picozzi sought a demonstration in Wytham woods near Oxford before trusting the technique.

In contrast to the farming question 'can Goshawks be useful against damage by pigeons', Swedish hunters were asking 'how much of a problem are Goshawks for pheasants'? Vidar's research group had doubts about whether a falconer could look squarely at this issue: a first enquiry by one of his post-graduates, Rolf Brittas, was 'have you eaten Goshawk'? Another post-grad at Uppsala was Per Widén, who came to Oskar Bernadotte's estate (Frötuna) to learn radio-tracking. There was also Mats Karlbom, working for Lasse Lans as an apprentice in charge of the released pheasants. Mats joined the work for a second year in a wild pheasant area, and later brilliantly ran our seven-year study of the Goshawk population on the Baltic island of Gotland.

In 1979, one of many relaxing stays with Vidar and Barbro Marcström was followed by an exploration of Gotland with its hunting supervisor, Ralf Beinert. Swedes were still allowed to kill Goshawks that attacked poultry, and to trap them under licence for protecting game. The Swedish Hunters' Association wanted to know whether this killing reduced the size of a relatively isolated Goshawk population, on an island rich in smallholdings and wild pheasants. I was now working mainly on woodland damage caused by Grey Squirrels, after joining the Institute of Terrestrial Ecology at Monks Wood in 1978, but supervisors David Jenkins and Jack Dempster kindly agreed for me to spend a month each year on Gotland, which was extended by sharing happy family holidays there with my wife Bridget and our children Ben and James.

The need for hundreds of radio tags on Gotland, for squirrels as well as for Goshawks, gave rise to the firm Biotrack, which has grown in Bridget's care for 25 years. Mike Dolan at Oxford had taught me to build tags for the first work in Sweden. Jessica Holm was Biotrack's first employee, building radios and working on Grey Squirrels before moving to the Isle of Wight for a thesis on the Red Squirrels so loved by Sweden's Goshawks. Tags were also made for the early Goshawk radio-tracking of Veronique Herrenschmidt, Fridtjof Ziesemer and Pat Kennedy. Andy Village and Brian Cresswell (who joined Biotrack for its 1984 move to Dorset) helped with the summer fieldwork on Gotland, as did Frank Doyle before he went in 1989 to start research on Yukon Goshawks at Kluane Lake with Sue Hannon, Charley Krebs and Jamie Smith.

My links with Goshawk work in North America started long before helping at Kluane. A letter in 1972 from Mark Fuller, a student with the pioneering automatic tracking at Cedar Creek, gave me the first advice and confidence to radio-tag hawks. That letter started many kindnesses from friends across the Atlantic. Dan Brimm, an aerospace engineer who visited one of the Oxford conferences, funded me to a Raptor Research Foundation (RRF) meeting in Arizona. Joe Murphy and Clay White then hosted me in Utah. At a later RRF meeting in Salt Lake City, there were many questions from Pat Kennedy as she started radio-tracking Goshawks, and opportunities to talk with Richard Reynolds, Tom Erdman and Dan Brinker. Mark Fuller nominated me to work with RRF, and thus to make friends with Marc Bechard, Pete Bloom, Sandy Boyce, Tom Cade, Cole Crocker-Bedford, Phil Detrich, Pat Hall, Fred and Fran Hamerstrom, Chuck Henny, Tim Kimmel, Mike Kochert, Jeff Lincer, Brian Millsap, Jim Mosher, Sergei Postupalsky, Pat Redig, Jim Ruos, Helen Snyder, Karen Steenhof, Kim Titus, Jim Younk and many other raptor biologists and falconers in North America.

The 1990s brought 'Miss Piggy', for teaching the diploma students of Mike Nicholls and Mima Parry-Jones. Sean Walls trained her, because we were eager to try new software to analyse settlement. Nick Williams tracked her while Hugh Miles filmed. Christian Saar taught me about Goshawk breeding, after working with hawks sent from Finland in the 1970s (as well as breeding Peregrines to replace the DDT-ravaged population in northern Germany). Family holidays in Fennoscania had welcome hospitality from Torgeir Nygård and Risto Tornberg, who started projects with radio-tagged hawks in northern forests. Meetings in Europe, especially those organised by Bernd Meyburg and Robin Chancellor of the World Working Group on Birds of Prey, provided opportunities to talk with Patrick Byholm, Jörg Dietrich, Herman Ellenberg, Vladimir Galushin, Anita Gamauf, Achim Kostrzewa, Volkher Looft, Santi Mañosa, Jan Nielsen, Thérése Nore, Ludmilla Olech, Paul Opdam, Steve Petty, Zygmund Pielowski, Roland Schulz, Peter Sunde, Jean-Marc Thiollay, Günther Trommer, Paul Toyne, Jan Wattel and Michael Wink.

The perspectives brought by so many kind people came together during April 2004 in a courtroom in Ludlow, a pretty town on the border of England and Wales. A law firm in Sussex had invited me to help defend a young man accused of disturbing Goshawks at a nest nearby. It quickly became clear not only that the birds had probably been disturbed by those seeking to trap the defendant, but also that successful breeding by hawks nesting close to a footpath indicated tolerance of passing humans. The case was dismissed, but it was shocking that someone thrilled by the birds and wishing no harm to them had been prosecuted at all. Afterwards, the young man got advice from his defence team about the importance of working to restore relationships that others had damaged. How far we have come from tolerance between different types of naturalist!

I am indebted for the continued confidence of Andy Richford, and Jim Martin and Nigel Redman at A&C Black, to editors Ernest Garcia and Simon Papps, to Anne Horsfall and James Kenward for transcribing original longhand to electronic format, to Katrina Cook, Frank Doyle, Rory Hill, Mats Karlbom, Tomi Muukkonen, Jari Peltomäki, Vencenzo Penteriani, Ted Swem, Markus Varesvuo and Doug Weir for help with photos and to Alan Harris for his beautiful artwork. Christian Rutz gets unreserved credit and thanks for somehow finding the time and tact to comment on every chapter while completing a brilliant thesis. My thanks for further advice or materials for the book go to Walter Bednarek, Rob Bijlsma, Frank Doyle, Graham Fairhurst, John Keane, Pat Kennedy, Tim Kimmel, Helen Macdonald, Santi Mañosa, Mick Marquiss, Torgeir Nygård, Mihaela Pavlicev, Remo Probst, Sarah Sonsthagen, Peter Sunde, Risto Tornberg, Per Widén, Michael Wink and others who kindly gave permission to recreate figures from their work. Work with the World Conservation Union (IUCN) has opened my eyes to socio-economic issues, and especially to the importance of rejecting the blame-culture and tribalism between groups that share an interest in conservation but differ in their approach. Goshawks and other wildlife need humans to combine diverse skills and resources for them, not to waste time fighting.

I am still enjoying research on the Northern Goshawk, thanks to invitations from Mike Morrison and Christian Rutz, working with Rob Bijlsma and Mick Marquiss, to provide material for a book and other papers. The completion of Mike's volume of research papers and reviews, together with the proceedings of RRF's 2003 conference in Alaska, make this a very good time to pull together all the new material in a form digestible for birders and all others seriously interested in the Northern Goshawk, including falconers. My hope is to work again with tamed and wild Goshawks, to help improve their coexistence with humans. There is still a debt to repay for many happy memories of hawks and the human lives they entwine. My heartfelt thanks to you all.

CHAPTER 1

Names, races and relatives

It is the summer of 1982, in Moscow. The metro is admirable, with other architecture a blend of function and stark grandeur. The display of a mere half-dozen products in neighbourhood food shops is a shock. Vladimir Galushin, a host at the 18th International Ornithological Congress, has kindly arranged for me to visit the museum, where large wooden chests carry magical names: '*buteoides*', '*fujiyamae*', '*schvedowi*', '*albidus*'.

Fascinated by the concept of a white Goshawk, I open the '*albidus*' box and start laying out birds to photograph (Plate 7). The first bird, with label neatly written in Cyrillic on a flattened and stiffly crossed leg, is the pale colour I had expected, perhaps helped a bit by bleaching in the years since its death. The second bird is pale too.

However, as I work down through the box, taking measurements and scoring plumage characteristics, some of the birds start to look remarkably like the adults and young I've known from Fennoscandia. It is the same in other boxes. The birds at the top make fine photographs of what is considered typical for the appropriate race, but those lower down could have come from any of the boxes. Some birds even have dark central streaks on their breast feathers, reminiscent of Goshawks from North America.

Of course, these stiff skins that smell of moth-proofing naptha lack eyes, which also change colour geographically. However, one thing is clear. If there ever were truly distinct races of Goshawk across the Eurasian land-mass, perhaps trapped by ice-ages in isolated pockets, interbreeding has long since softened the distinctions between them. Only the most typical of these morphs, the sort that are provided to museums as outstanding specimens, can be distinguished by physical characteristics alone.

Intrigued, I start measuring wing-lengths and middle toes, and scoring the colours and patterns of breast feathers and backs, in order to seek trends. The helpful museum staff read the place names in Russian. It worries me that the hawks collected in winter may have travelled some distance from their nest sites. Across the 180° of Eurasia this should be a relatively small latitudinal error, but some northern hawks may have been collected far to the south. Ideally one should examine live hawks sampled at nests at regular intervals of longitude and latitude. It is also worrying that some hawks may have been sent to the museum specifically because they were atypical: one hopes that these were few among the 209 skins. After entering the museum with a view of taxonomy as rather mundane, I leave with more appreciation of the fascinating challenges it contains, and make trips in the autumn to measure more museum skins, in Stockholm and Chicago.

WHAT'S IN A NAME?

Raptors are named from their prey in many languages. In English we have mouse-hawk, duck-hawk, pigeon-hawk, chicken-hawk, sparrow-hawk, for species which made themselves noticeable by what they ate. The Northern Goshawk is a pigeon-hawk in modern Danish (*Duvhøk*), French (*Autour des palombes*), Hungarian (*Galambasz heja*) and Swedish (*Duvhök*), while the Finnish (*Kanahaukka*) and Norwegian (*Hönsehök*) names allude to the theft of chickens. The Germans, in the spirit of re-naming raptors *Greifvögel* (birds which seize) from the earlier *Raubvögel* (birds which steal), have dropped an earlier 'chicken' prefix and just use the word 'hawk' (*Habicht*), like the Dutch (*Havik*). The Russian is *Yastreb Teterevyatnik*, literally 'black-grouse hawk'. The Slavic word *yastreb* for hawk may be related to the original Latin generic name *Astur*, which has connotations of speed, as in *strela* (Russian for 'arrow'). The Turkmen name of *Garynchykar* implies both speed and powerful seizure, while the Ottoman name of *Çakýrkupu* (chakirkushu) means a greyish-blue bird, although Turkish villagers now often use *Tavukkapan*, meaning 'chicken-snatcher' (O. Borovali, pers. comm.).

In their book *Pedigree: Words from Nature*, Potter & Sargent (1973) had no doubt that the prefix 'gos' is Old English for goose, as in Gosport and gosling. An adaptation of the word 'gross (groß)' for large, as in grosbeak, is also conceivable. The large size of Goshawks is recognised in a German local name *Doppelsperber* (double sparrow-hawk) and the Turkish *Atmaca pahini* (greater sparrow-hawk). However,

Germans have also used the word *Gänsehabicht* (goose-hawk) in the past (Fischer 1980). Goshawks seldom take wild or farmyard geese nowadays, but perhaps domestic geese were smaller where Goshawks and Gänsehabichte were named.

In the scientific name, *Accipiter* is literally hawk and *gentilis* means 'belonging to a clan or species' (Cooper 1981). In the 12th century Latin used by Linné in 1758 to name the species, *gentilis* also implied 'of good birth' (hence the term 'gentlemen'). The original generic name *Astur* links to names in Romance languages of *Astor* in Catalan, *Autour* in French and *Azor* in Portuguese and Spanish. The Azores are 'hawk-islands', although the hawks there are buzzards.

The Northern Goshawk is one of those few species with a Holarctic distribution, i.e. extending across the Eurasian Palaearctic zone as well as the North American Nearctic. If we travel a third of the circumference of the world to the west from Europe, this species continues to be named with respect and irritation. Thus, the totem pole of St'aawaas X̱aaydagaay, the ruling family of the Cumshewa First Nation on Haida Gwaii (in the Queen Charlotte Islands) is topped by a blue hawk with red eyes, which is most likely a Goshawk (F. Doyle, pers. comm.); the bird is known colloquially as a chicken hawk. If we travel a similar distance east, to Manchuria, the name is *Cang Ying*, meaning 'a hawk with pale plumage' (Zhang Zhengwang, pers. comm.). Other Chinese names include *Huang Ying*, meaning 'yellow hawk' (perhaps named from the juvenile) and *Ji Ying* (gamebird hawk). In Japanese the Goshawk is *Ootaka*, or 'blue hawk'.

The names for this species reflect human attitudes. Europeans name the Northern Goshawk from observations of predation, with northerly 'grouse hawk' and southern 'pigeon hawk' being rather accurate according to Chapter 8, and the older German 'chicken-hawk' less sympathetic. Moving east from Slavic to Turkic languages, ideas of speed, strength and colour became prevalent. Names based not on imputation but colour alone are used in the prolonged eastern civilisations of China and Japan, where Goshawks remained esteemed through a long history of falconry (Chapter 9).

PLUMAGE

The plumage of nestling Goshawks has been described by many authors (e.g. Siewert 1933, Bond 1942, Boal 1994a). Newly-hatched hawks have a short coat of snow-white down, grey-black eyes, and legs which turn within a few days from pinkish to pale yellow. The beaks are jet black and relatively smaller in relation to the greenish-yellow cere than at fledging.

The second coat of down, which develops at 7–10 days old, is longer and silkier, with a slight grey tinge on the back. Down always remains shorter and sparser on the chick's underside. The main tail feathers start at 14–16 days, followed by the contour feathers on the back. By the time the birds fledge, at 35–42 days, the main flight feathers are two-thirds grown, the eyes have faded through slate-grey to blue-grey, and the cere and legs are becoming darker yellow. Feather growth is complete two weeks after leaving the nest.

There is considerable variation in the juvenile first-year plumage. Northern European birds, ascribed to the nominate race *A. g. gentilis*, have chocolate-brown to mid-tan crowns, body and upper wing surfaces, the feathers being edged with a paler shade which sometimes contrasts sufficiently to give a gentle speckling. The underparts have a lighter base colour (Plates 3, 5 & 7), which varies from pale cream (var. *fulviana*) through buff to a pale rusty colour (var. *rufina*). These may be distinct phenotypes, because young occasionally show very different colours in the same nest. Among museum specimens, the darker type was more common in skins from central Siberia and China, but uncommon again in the far east of Siberia and was not attributed to North American skins (Figure 1).

Most ventral feathers also have a chestnut to deep-chocolate central marking along the shaft, although the undertail feathers tend to be unmarked. This marking may be a relatively narrow streak of fairly constant width, or may vary along its length for up to half the width of the feather, or become a broader, often heart-shaped character (Figure 2). The heart-shaped markings are usual under the wings, and often to be seen on the flanks, while on some birds they are also noticeable on the breast (var. *cordata*). The juveniles that remain streaked (var. *striata*), have these contrast-rich markings widest and most prominent on the chest, and finer on throat and flanks (Fischer 1980). Streaked juveniles become more frequent to the east of Eurasia, but are rare in North America (Figure 3).

The pale ventral colouration is also noticeable on the eyebrows (Plate 5), although the superciliary stripe is by no means as striking as in adult hawks (Plates 1 & 4), and towards the base of the nape feathers. These nape feathers are quite white at the base, another feature which is more prominent in adult hawks and probably has a signalling function (Brüll 1964).

As in most other accipiters, the seventh of the ten primaries (fourth from the front) is normally the longest. The 12 main tail feathers (very rarely 13 or 14, Slijper 1978) have distinct dark-brown to black transverse bands, which are as wide as the paler base

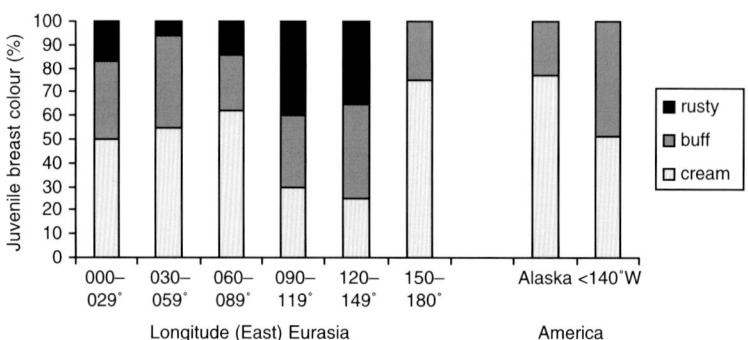

Figure 1. Breast colour of 161 juvenile Goshawks in musem collections.

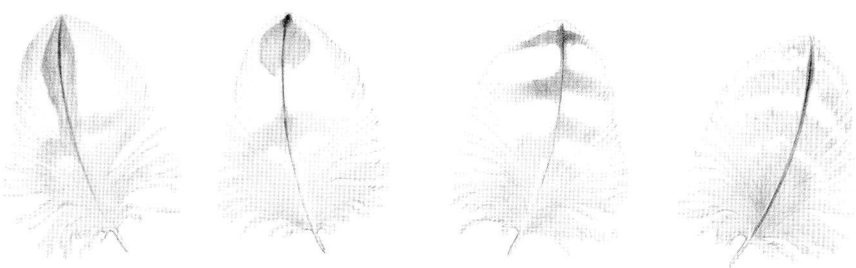

Figure 2. Breast feathers, showing (left to right) juvenile streaks, juvenile hearts and adult bars typical for Europe and North America. Relatively pale bars on American adults make shaft streaks especially prominent on the breasts (Plates 7 & 13).

colour and more clearly fringed with white in some hawks than in others. There are typically five bands on the central tail feathers, six or seven on the outermost pair. The dark terminal band has an especially broad white fringe at its outer end, providing a pale tip to the tail which stands out in nestlings and first-autumn hawks, but can wear down later in the first year. The black beak and claws of nestlings become more blue-black after the birds have fledged, while the cere and legs become darker yellow. The claws on the rear and inner front toes are much larger than the others, for delivering the formidable killing grip, while the other toes are finer and usually have better developed ventral pads. The eyes have become pale yellow within a month or two of fledging (Plate 5), sometimes even before hawks leave the nest.

The plumage of adult and first-year Goshawks is so different that Linné considered them two species, *Falco* (=*Astur*, =*Accipiter*) *gentilis* for the young hawks and *Falco palumbarius* for the adults. At the first moult, when the hawks are a year old, the upper parts change from brown to a dark grey, typically with a slightly brownish tinge in second-year birds but becoming grey in subsequent moults, in shades ranging from black to an attractive blue-grey. The crown and sides of the head usually become darker than the back in later moults, and may be completely black. Contrast between back and head is enhanced by the white inner parts on the nape feathers, which give a slightly speckled impression with the feathers flat, and are very conspicuous when the feathers are raised at the back of the crown by an aggressive hawk. The superciliary stripe on each eyebrow becomes most conspicuous after the second moult, the white being flecked with black on some of the tiny feather shafts (Plate 4). Adults' back colours were palest for museum hawks from central Eurasia and North America (Figure 4).

The main tail and wing feathers are greyer in adults than juveniles, but retain much of the first-year markings at least in the second year. The bands usually become less distinct in later moults (Brüll 1964). The detailed pattern of these feathers is consistent enough between moults for the recognition of individual hawks at nest sites (Kollinger 1964, Opdam & Müskens 1976, Ziesemer 1983, Rust & Kechele 1996, see Chapter 5). The snow-white undertail coverts are particularly striking when spread during courtship (Plate 12).

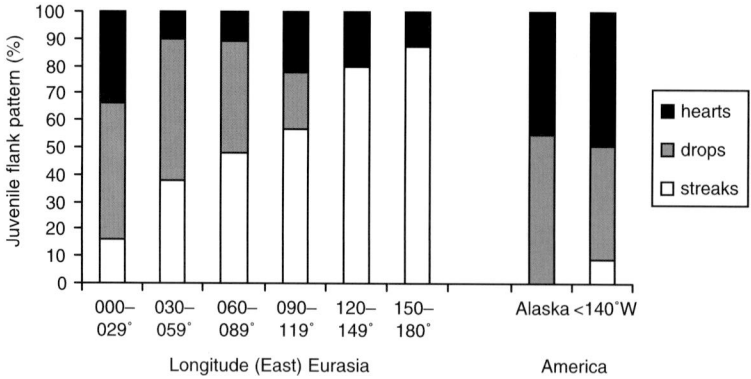

Figure 3. Pattern on flank feathers of juvenile Goshawks in museum collections.

The rest of the underparts are barred black on white (Plates 1 & 2), although the feathers often have a brownish tinge in the second year and this does not always disappear with increasing age. The bars are most prominent on the chest, and finer at the throat and neck, where absence from some feathers gives a speckled appearance. Fine shaft streaks often link the 3–5 bars on each breast feather, to form a row of 'bird flight silhouettes' (Fischer 1980).

The best way to distinguish second year birds from older hawks is the presence of any unmoulted brown feathers in the wing coverts or (more rarely) of any juvenile feathers on the breast. This gives about 95% accuracy, because ringed second-year birds sometimes have no noticeable juvenile feathers left and third-years very occasionally have such feathers.

The shaft-streak and bars are also thickest on second-year birds (Bond & Stabler 1941), and often do not overlap regularly between adjoining feathers, but become finer with progressing age to give lines right across the breast of older birds. With bars scored from 1 for the thinnest to 4 for the most pronounced, 15 Eurasian adults classed as second-year by residual brown feathers scored an average 2.8, whereas 61 older hawks scored 1.9. Barring was appreciably finer on 36 American adults (and virtually invisible on some), which scored 1.3.

Despite the tendency for barring on the breast of adults to become finer with age, a few second-year hawks in Eurasia had very fine bars and some older birds retained thick bars. So the width and regularity of barring is not a reliable age-guide. Moreover, although a few barred feathers sometimes occur on the lower legs or underwings of a hawk in first-year plumage, these are not second-year birds. They have probably regrown feathers in adult pattern after losing the originals in a struggle with nest-mates or prey.

Another unreliable age-guide is eye-colour. In the nominate race *A. g. gentilis*, the pale yellow eyes of first autumn hawks gradually deepen in hue, eventually becoming orange or even orange-red. However, the rate of deepening must vary between birds,

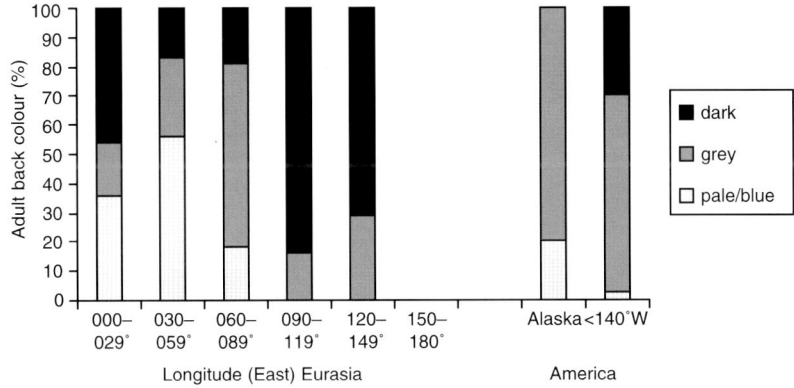

Figure 4. Back colour of 111 adult Goshawks in museum collections.

because wild birds known to be second-year or third-year differ considerably in eye colour. Trained hawks fed on food rich in yellow pigments (e.g. day-old chicks containing their residual yolk) have much deeper yellow eyes, ceres and legs than birds without the food. Exposure to light may also be important. Heidenreich (1996) describes a hawk with a fused eyelid on one side which, when opened surgically, revealed a bright yellow iris, contrasting with the dark orange/red-coloured iris of the other eye. However, genetic variables may be more important than diet in the wild.

A number of trained Goshawks have now reached ages up to 27 years (Bednarek in press). With great age, especially after egg laying ceases in females (by about the 17th year), further plumage changes occur. The banding that tends already to have vanished on main flight feathers is also lost progressively on very old hawks from the breast feathers, which may become white, grey, and even show a white-on-grey spotting or marbling. The development of these changes at an early stage in a hawk with a failed ovary shows that the process is influenced by sex hormones.

Seen in flight at a distance, the usual impression of a Goshawk is of a large dull-coloured bird, with wings more rounded than a falcon's and a long tail (Plates 1–3). A few forceful wingbeats alternate with short glides, in low as in soaring flight. The wingbeats are a trifle slower, giving a slower 'flicker' than for sparrowhawks when Goshawks display at a distance. When soaring overhead, the relatively rounder tail and larger head distinguish them from the smaller accipiters in Europe. If the hawk is not too far away, those used to Goshawks will notice whether the bird is juvenile (brownish), or adult (greyish) and whether it is a female or the smaller, more agile male. Seen perched, Goshawks are appreciably bulkier than the smaller northern accipiters, with an appreciably flatter head and more pronounced brow ridges (except when the feathers are puffed out) than smaller accipiters except for Cooper's Hawk. This makes their eyes appear much less rounded and pronounced than in the Eurasian Sparrowhawk. Unlike most of the smaller northern accipiters, Goshawks never have a reddish tinge to their barring as adults.

RACIAL CATEGORIES

Goshawks may be found from the Arctic Circle to the Tropic of Cancer, wherever there is suitable woodland. In Eurasia they breed from the northern edge of the forest-tundra south to Morocco, northern Iran and the southwest foothills of the Himalayas (Fischer 1980). In winter, they have been recorded as far south as Taif in Saudi Arabia (P. Paillat, pers. comm.) and perhaps Tonkin, Vietnam, in the east (Chicago museum specimen). In North America, they occur in mountains into the south of Mexico in the west (Figure 5), and in the east have been recorded in central Florida in winter (American Ornithologists' Union 1983).

The main racial categories of Goshawk to be considered here were named 60–180 years after Linné established the species in 1758, by Wilson (*atricapillus*) in 1812, Menzbir (*buteoides, albidus* and *schvedowi*) in 1882, Kleinschmidt (*arrigoni*) in 1903, Swann and Hartert (*fujiyamae*) in 1923, van Rossem (*apache*) in 1938 and Taverner (*laingi*) in 1940. A number of other races have been claimed, including *balcanicus, caucasicus, dubius, gallinarum, khamensis, koeneni, marginatus, moscoviae, peocilopterus, striatulus, suschkini* and *tischleri*.

The concept of Goshawk races needs treating with caution. Looking at a crowd of people in a western metropolis, a majority could probably be categorised from faces alone into caucasian, negroid and oriental. A similar collection of Goshawks could be divided fairly easily, on looks alone, into Nearctic and Palaearctic races. After that, however, only a minority of the juveniles, and even fewer adults, could be put confidently in one of the Old World or New World 'races', at least without taking size measurements. Among the Eurasian Goshawks in the Moscow Museum, perhaps 10% of the juveniles in each box could be considered 'classic' examples of each race, with another 20% having enough 'typical' characteristics to be ascribed to the race on looks alone. The adults could scarcely be categorised by their plumage at all. Most of the hawks had probably been assigned to a race on account of the area in which they were collected.

Nevertheless, the results of measuring and colour-coding these museum skins showed trends in size and plumage patterns that generally confirm the qualitative observations of authors such as Menzbir (1895), Dementiev (1955), Vaurie (1965) and Fischer (1980). For example, it is well known that Goshawks in Europe show a latitudinal cline in size. Goshawk wings in Europe get about 4% longer, on average, for every 10° of latitude towards the north, with samples from the north of Fennoscandia having wings about 10% longer than hawks on islands in the Mediterranean (Figure 6).

This trend of increase in size to the north occurred in the museum skins right across Eurasia, with averages for both males and females south of 50°N being smaller than for northern samples in every 30° band of latitude (Figure 7). Other tendencies were for the very largest birds to be those from the north-east of Siberia, and for southern birds to become large towards the centre of their distribution, but to be small on Mediterranean Islands and smallest of all at 90°E–120°E, in Tibet, China and

Names, races and relatives 25

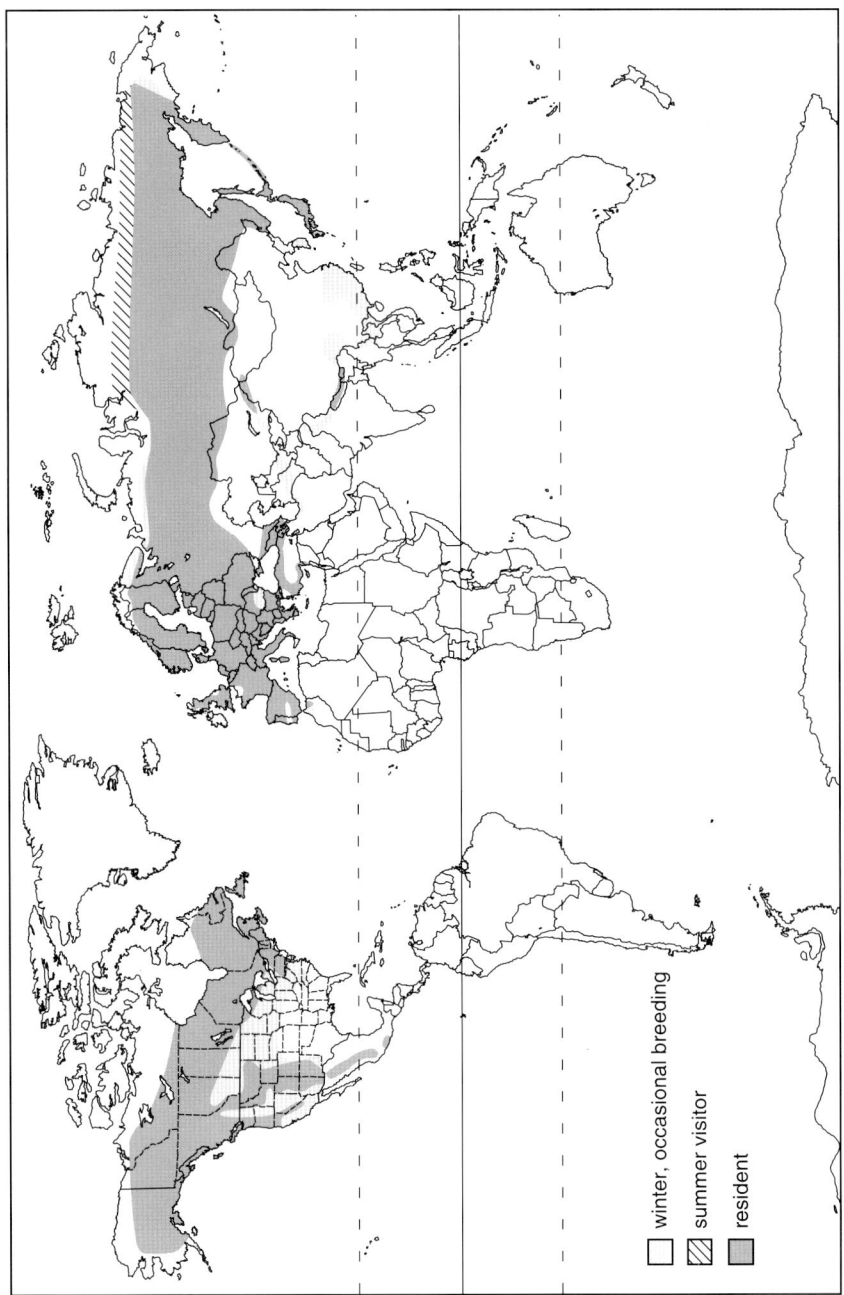

Figure 5. Global distribution of the Goshawk, *Accipiter gentilis*. Revised from Ferguson-Lees and Christie 2005.

Japan. Another commonality throughout the Eurasian distribution was for female Goshawks to have wings about 13% longer than those of males (see Chapter 2).

Wayne Whaley and Clayton White (1994) took many measurements of Northern Goshawks in museums throughout North America. In the parts of the species' range equivalent to the distribution in Europe (roughly 35–65°N), they found a similar decrease in size to the south. Thus, Goshawks in Alaska had wings averaging 3–4% longer than those on the mainland further south and were similar in size east and west of the 100° meridian, while those on islands off the west coast were 3–4% smaller than those on the mainland at similar latitude. However, the very largest Goshawks in North America were from the south-west, where the relatively cool and moister conditions of the southern Rocky Mountains and Sierra Madre ranges provide suitable conditions for coniferous woodland in the southwesternmost USA and Mexico (Figure 7). This reversal of the size trend gave some justification for the separation of these hawks into the *apache* race.

A discontinuity apparent in Figures 1 and 4 was for both juvenile breast colour and adult back colour to be especially dark in Eurasia in the region from 90°E–150°E. This is an area that includes the Siberian uplands between the Yenisei and Kolyma rivers in the north, and from the Altai to Magadan in the south, plus most of China. Then, in the far north-east and on through Alaska into North America, the juveniles become pale again and the paler back colour of adults contrasts with their dark heads. The dark, contrasting heads give the name *atricapillus* to the principal American subspecies.

Across Eurasia, the proportion of pale birds is highest in the north, although some dark birds occur in northern populations and *vice versa*. Among southern hawks, there tends to be a higher proportion of pale birds in central Asia than in Europe or the Far East. In the north there is a marked tendency for birds to become paler to the

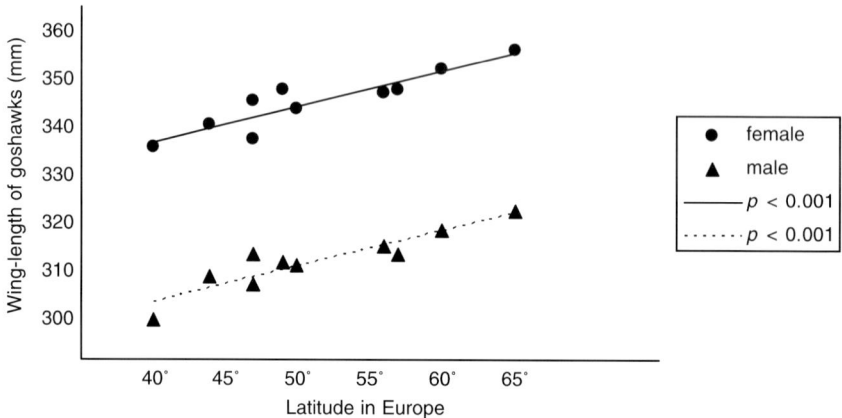

Figure 6. The wing-length (maximum chord) of Goshawks in Europe increases south to north. Data from Fischer (1980), Tornberg *et al.* (1999), Marcström & Kenward (1981a) and Kenward, Marcström & Karlbom unpublished.

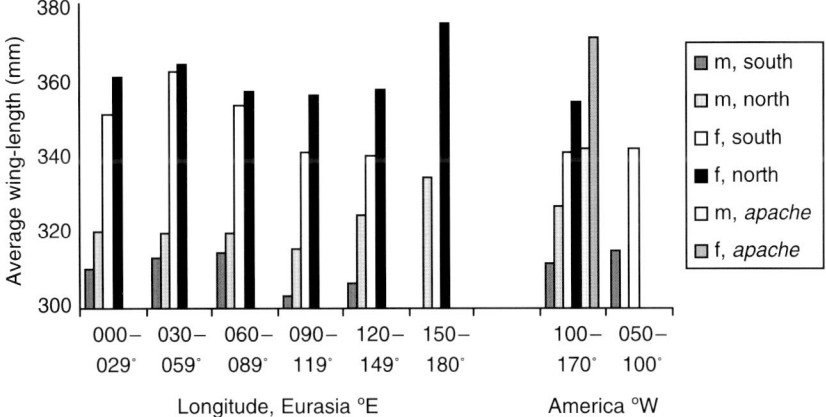

Figure 7. Goshawks tend to be smallest at the southern fringes of Eurasia and largest in the north-east; the *apache* type is the largest in North America and occurs in the south-west.

East, while in the far north-east can be found an almost completely white phase, which Menzbir (1882) typified as '*albidus*'. This is a distinct pale colour phase that accounts for up to 50% of the Goshawks in certain populations (Fischer 1980), especially in the Beringian forest tundra such as the Anadyr basin (Krechmar & Probst 2003). The young of both pale and normal phases have been seen in at least two nests (Dementiev & Böhme 1970, V. Pererva, pers. comm.). These young are pale enough to be hard to tell from adults, especially in flight (Krechmar & Probst 2003). Even the alternative darker phase tends to have unusually pale grey and finely barred adults (Krechmar & Probst 2003) and with relatively long wings (Figure 7), which may justify a separate '*albidus*' race for east Siberian hawks in general.

A white phase has also been described (Menzbir 1882) for the otherwise darkish birds found in Tibet and pale 'isabelline' versions also occur of the *gentilis* plumage (Fischer 1980). Pale morphs may account also for as many as 10% of hawks in the areas typical for *buteoides* (Ferguson-Lees and Christie 2001). Rutz *et al.* (2004) found plumage dilution as a result of a simple genetic defect among Goshawks in Hamburg.

One other plumage pattern shows interesting discontinuities across Eurasia. This is for contour feathers on the nape, back and upper wing surfaces to have pale bases expanded beyond the cover of overlying feathers, and wide pale margins, giving a pale, flecked upper surface, which in extreme cases appears barred (Kleinschmidt 1934). This flecking occurs mostly on birds in the north, and was most common in museum skins from the west Siberian plain, just east of the Urals (at 60°E), and in the far north-east (Figure 8). It is considered characteristic of the *buteoides* race (Plate 7. Another *buteoides* trait is a tendency for the eyes of adults to become tan to pale brown. Unfortunately, the eye colour of Eurasian Goshawks could not be assessed from skins.

In the Chicago museum, Goshawks from Alaska showed more flecking and speckling than those from elsewhere in North America (Figure 8), just as they also tended to have paler plumage (Figure 1). Adult birds from easternmost Asia also show an increasing prominence of dark shaft streaks on the adult breast feathers, and streaks are usually present, albeit palely, even on '*albidus*' hawks. Shaft streaks are a strong characteristic of American Goshawks due to less substantial barring (Figure 2). The less substantial bars (and hence prominence of shaft streaks) in America may indicate a 'founder effect' during colonisation from east Eurasia.

Since some Goshawks in Eurasia possess the black head for which the main New World Goshawk race, *atricapillus*, is named, the skins of adult American Goshawks may best be distinguished by the irregular barring of their undersides. These bars tend to form zigzags or even to be broken up as dark flecks, unlike the neater horizontal lines of most adult Eurasian hawks, and dark vertical shaft streaks are very prominent on some breast feathers (Plate 13) and often on the nape. Another contrast with the Old World is that the eyes of the American Goshawks become a fiery wine-red at an age when *gentilis* eyes would still only be yellow. However, juveniles from Eurasia and America are not easily distinguished, although breast markings tend to be particularly wide and prominent on *atricapillus* juveniles, which are thus more like young *gentilis* than their east Eurasian relatives (Figure 3). As in the Mediterranean and Japan, there is a darker, more heavily barred form *laingi* in the rainforests and islands of the west coast (Taverner 1940).

In mainland Europe, including European Russia, 12 Goshawk races were named between 1758 and 1950. Since there is clinal variation in both size and colouration with latitude and longitude, it is easy to see how this arose: almost any local population can be shown to be different in size or colour from those in other parts of the range. Bährmann (1965) drew attention to these clines and started a process of simplification, by separating a small, dark *marginatus* race around the Mediterranean from the larger *gentilis* in the north, and the paler, more speckled *buteoides* to the

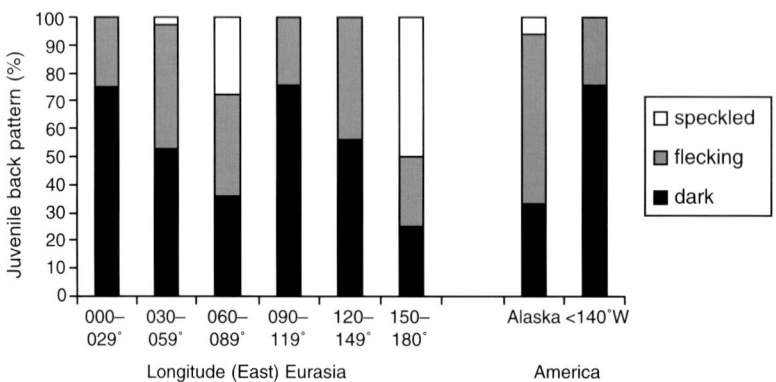

Figure 8. Juvenile backs are most speckled in the west of Siberia and far north-east.

east, with wide zones of hybridisation between. The same line was taken by Fischer (1980), who also contrasted the larger pale *buteoides* and *albidus* with smaller, darker *schvedowi* and *fujiyamae* in the south. Others (Brown & Amadon 1968, Wattel 1973, Cramp & Simmons 1980) have not separated *marginatus* from *gentilis*, but have retained the island race *arrigoni* on Corsica and Sardinia.

EVOLUTION WITHIN *ACCIPITER GENTILIS*

It seems sensible for descriptive purposes to maintain the present races, or types, *gentilis, buteoides, schvedowi* and *atricapillus*, which at the least represent regional differences in plumage and eye-colour. Among these four main types, the American *atricapillus* is most distinct and also shows less sexual dimorphism than the others. Studies of genetic relatedness are starting to be used for Goshawks, and may show differences between *gentilis* and *atricapillus* to be as great as in races recognised among other species. It remains to be seen whether the island types *arrigoni, fujiamae*, and *laingi* are really sufficiently different from adjacent mainland birds to be treated separately. Preliminary analyses suggest that hawks on the outlying west Canadian islands may be genetically distinct (F. Doyle, pers. comm.). Genetic research should also show whether the unusually large *apache* really deserves separation from the northern forms, and whether *albidus* is more than increased frequency of a pale morph.

Genetic studies could also help explain why Goshawks show so much variation across Eurasia, whereas the Eurasian Sparrowhawk remains relatively uniform (Newton 1986). Two processes could have occurred, and they are not exclusive. One possibility is that isolation in the past enabled gene combinations to assort as distinct morphs that suited conditions in different geographic areas, followed by remixing of these genotypes to result in clines. This scenario might have involved ice-age refuges in southern Europe for *gentilis*, in Asia Minor for *buteoides* and in south-east Asia for *schvedowi*, with subsequent northward expansion and interbreeding for so long that most pure forms have been lost in a sea of intermediate types.

This 'past isolation' hypothesis may seem unlikely, because Goshawks disperse over relatively long distances, so that any previously isolated populations have been able to interbreed for several thousand generations since the last ice-age. A second explanation is that subtle variation in modern selection pressures maintains a diversity of hues and patterns. In this context, it is worth noting the marked contrast between adult and juvenile plumages, which shows that selection pressures on Goshawk plumage can be strong.

When searching for trained hawks that have a kill on snow-free ground, juveniles are much more difficult to spot than adults. The brown and paler hues of juveniles provide excellent camouflage against earth and vegetation, but from what are they trying to hide? Concealment from predators (e.g. Eagle Owls) may be more crucial during the first few months after fledging for inexperienced juveniles than for adults, and first-year birds may sometimes need to avoid the attention of other Goshawks.

The paler plumage of adults, especially when seen from below, would be less conspicuous than juvenile browns against the sky, which may be especially important when hunting. The grey-white and grey-black contrasts of adults, when seen from the sides or above, may help them to be seen and avoided by other hawks (and bars increase the apparent bulk of the wearer). From this line of reasoning, juvenile plumage may vary as an adaptation to geographic differences in habitat and especially ground-cover. Adult colours may be a compromise between the needs of concealment from prey and advertisement to other Goshawks.

Plumages of Goshawk adults and juveniles certainly become paler where there is most snow in Eurasia, but no one has investigated whether the speckling of *buteoides* or the extreme pale morphs that typify *albidus* might be especially adaptive in Siberia and the Russian far-east. However, these north-east Eurasian Goshawks live in larch, birch and willow-alder bordering the tundra and in poplar woods elsewhere. These woodlands are lighter and more open in winter than in most other parts of the Goshawk distribution, and are so poor in the forest tundra that hawks may hunt in the open much of the time, with ptarmigan and varying hares as their (also pale) main winter prey (R. Probst pers. com.).

A combination of 'past isolation' and 'continuous selection' hypotheses is also possible, with dark *gentilis* and *schvedowi* types having evolved at the edge of the continents, where denser forests with less snow-cover still exist, and the paler *buteoides* and *albidus* types in a central pocket with more open woods and more snow. In this case one wonders whether the presence of darker birds between the pale types in central Siberia and the far north-east (Figures 1, 4, 8) is evidence (a) of a *schvedowi* push north after pale types had become established, or (b) of a fourth isolation pocket linked to Beringia, or (c) of a sampling artefact among the museum skins. There are some interesting questions for geneticists.

GOSHAWK RELATIVES

The northern Goshawk is the most widespread member of the genus *Accipiter*, which contains nearly 50 species and is thus the largest genus in the family *Accipitridae*. This broad family, which also includes the various eagle and buzzard genera, the harriers and old world vultures, is generally considered to have evolved from kite-like scavenging birds (Brown & Amadon 1968, Jollie 1977). A number of lines in the family have developed as active predators rather than carrion-eaters, with the accipiters being the most specialised for catching agile prey. Nevertheless, the Goshawk is not averse to carrion, and will eat the fresh kills of other predators or even feed from carcasses which have started to decay. (Plate 21).

The exact lines of evolution within the *Accipitridae* were the subject of much debate, as were the relationships between this family and others in the present order *Falconiformes*. The original classification of bird groups was based mainly on morphology. Outward appearance is often reflected in skeletal features that can also be studied in fossils. However, some features evolve convergently, to adapt for similar conditions in otherwise unrelated birds. For instance, hooked beaks have been developed for tearing flesh by several groups of birds which are very different in most other respects.

Many questions about raptor evolution have now been resolved, at first by analysis of egg-white proteins (Sibley & Monroe 1990) and latterly by studying DNA. Comparing genetic material is the best approach, because it is sensitive not only to all physiological differences (because DNA codes for the proteins that produce those differences) but also to small changes in DNA which occur without altering protein function. These small changes accumulate with time since separation of populations and provide a 'molecular clock' that is unaffected by recombination processes in the case of mitochondrial DNA (mt-DNA). This mt-DNA can therefore indicate not only which species are closely related but also, with assumptions about the rate of change of the DNA, how recently they diverged. On this basis, Michael Wink and colleagues at Heidelberg have been building a picture of molecular relationships within the Falconiformes (Wink 1998, Wink *et al.* 1998, Wink & Sauer-Gürth 2004). DNA from cell nuclei is more variable, and is also now being used for studies of relationships within families and between current populations (Gavin *et al.* 1998, de Volo *et al.* 2005).

It is clear from these analyses that the accipiters are much more closely related to the harriers (*Circus*) than to the buzzards (*Buteo, Busarellus, Buteogallus, Parabuteo*), kites (*Milvus, Haliastur*), sea-eagles (*Haliaeetus*) and chanting Goshawks (*Melierax*), with an even greater evolutionary distance to the other eagles, vultures and other members of the *Accipitridae*. Noting the close link to harriers, one thinks immediately of the similar grey adult and brown juvenile plumages in several species, including the Hen Harrier. Within the accipiters, the Northern Goshawk is more closely related to Cooper's Hawk than to the Eurasian Sparrowhawk and its close relative, the Sharp-shinned Hawk. The Australian Brown Goshawk is an even more distant relative (Figure 9), as is presumably the case for the Grey Goshawk with which the Brown Goshawk has been recorded interbreeding in the wild (Cupper & Cupper 1981). Unfortunately, more DNA analysis is required to define all the relationships within this large genus, so we must return to morphological comparisons to try to identify the Northern Goshawk's nearest relatives.

The Northern Goshawk has long been included in a superspecies with three other large accipiters: the near-threatened Henst's Goshawk of Madagascar, Meyer's Goshawk of Melanesia, and the Black Goshawk (Black Sparrowhawk) from Africa (Kleinschmidt 1922–23). Jan Wattel (1981) has suggested that Meyer's Goshawk is the nearest living relative of the Northern Goshawk, meaning that they share an ancestor not shared by any other species. The Northern Goshawk is close to Meyer's Goshawk in proportions, although it is slightly closer to Henst's Goshawk in undersurface and cere colouration and in absence of a dark phase. Meyer's Goshawk is found on islands in the southwest Pacific, including the Moluccas, Japan, the Bismarck Archipelago and Solomon Islands, with a record also from the Kraethe mountains of New Guinea (Amadon 1964). This is distant from the nearest Northern Goshawks, but Wattel points out that other Palearctic raptors, such as Bonelli's Eagle and Short-toed Eagle, have populations on the islands east of Java.

The mainland ranges of these species may have extended further south in the ice ages than today, with the islands offering dispersing stragglers an empty niche to colonise.

The Black Goshawk has much longer and finer legs and middle toes than the Northern Goshawk, and was therefore considered by Brown & Brown (1979) to be a sparrowhawk. However, Fischer (1980) noted that the size and plumage are much more Goshawk-like. The juvenile plumage very closely resembles *A. gentilis*. Moreover, at least one Northern Goshawk has been recorded with the dark upper surface and white, unbarred lower parts characteristic of Black Goshawks (Voous & Wattel 1972), so their black and white plumage may be only a short step genetically from *gentilis*. Nevertheless, this contrasting plumage pattern is also found in the less-closely related Grey-bellied Goshawk from the forests of South America, and in several sparrowhawks which inhabit tropical forests, in which it is presumably adaptive. Wattel (1981) therefore considered that Black and Henst's Goshawks are descended from a common *gentilis*-coloured ancestor in Africa, with the Black Goshawk then evolving plumage most suitable for its habitat, and that Meyer's Goshawk is the closest relative of the Northern Goshawk.

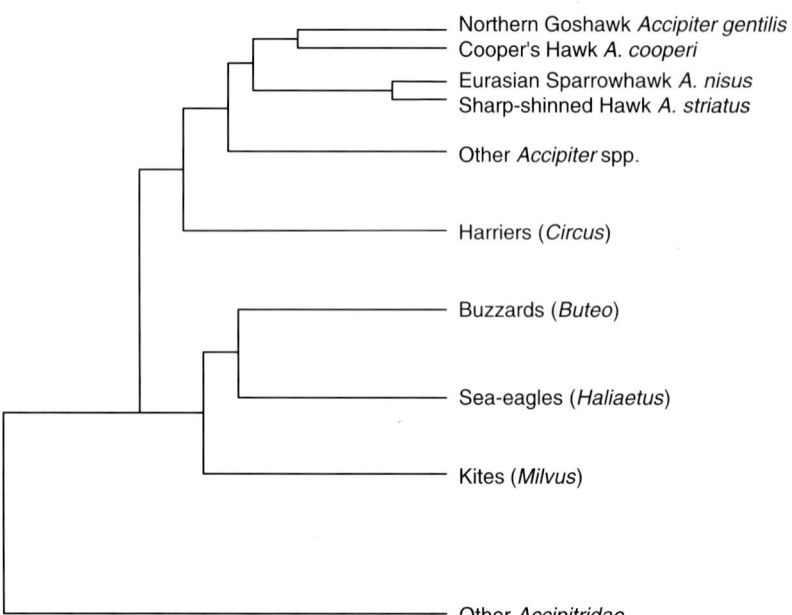

Figure 9. The relationships of cytochrome b from the Northern Goshawk and other accipiters and members of the Accipitridae. Data from Wink & Sauer-Gürth 2004.

GOSHAWK ORIGINS

The Northern Goshawk is not only the most widespread accipiter but also the largest. These facts may well be related. Since there are fewer large species than small ones and large animals tend to occur less densely than small ones, Goshawks are obliged to hunt over wider areas than small accipiters and to take whatever prey is available. This would tend to inhibit specialisation in habitat or prey choice. Without the specialisation that can lead isolated populations to diverge sufficiently for speciation, Northern Goshawks have remained generalists of the whole temperate zone where there is adequate woodland.

In contrast to Goshawks, smaller accipiters differ quite markedly between Nearctic and Palaearctic. The Sharp-shinned Hawk is smaller than the Eurasian Sparrowhawk, which is replaced by other species to the south. The Nearctic also differs from the Palaearctic in having a third northern accipiter, the intermediate-sized Cooper's Hawk.

Although the evolutionary process is not yet clear, the genetic analysis (in Figure 9) indicates an initial separation of the ancestor for the two large accipiters from that of the two small ones, followed by a separation of Northern Goshawk and Cooper's Hawk, with more recent separation of sharp-shins and sparrowhawks. This would be consistent with the separation as Cooper's Hawk of an early ancestral Goshawk in North America, in a niche that could be maintained (albeit with an emphasis to the south) despite (a) later invasion by the large Northern Goshawk from Eurasia and (b) even later colonisation by a small sparrowhawk ancestor. The Cooper's Hawk niche reflects a greater variety of nut and mast bearing trees, and hence of jay-sized seed-eating birds (and of squirrels) in the Nearctic than in the Palaearctic. A relatively recent Northern Goshawk colonisation of the Nearctic is also consistent with the relatively low genetic diversity found in North America (Gavin & May 1996).

IMPLICATIONS FOR CONSERVATION AND MANAGEMENT

This is a time of accelerated extinction of wildlife populations, as environments are changed to produce more of what humans want. In developing countries, where rural people are initially close to subsistence level and later adopt agriculture that feeds the cities, the use of land is changed primarily to produce food and essential materials. Wealthy countries change land increasingly through mechanised farming and construction and for recreation, although areas may be protected for science, aesthetics and to preserve ecological services. Increasing production of fuel-crops and other renewable energy sources may now join production of food and timber as dominant uses.

Raptor populations are impacted by land-use, pollutants, accidental effects of human activities and direct actions, including killing and collection. Most of these impacts are reversible, albeit relatively slowly if land-use has changed. Moreover, raptors are globally less threatened than many other bird groups, with fewer than 15% of Palearctic and Nearctic resident species (and about 20% of migrants) being in the categories of conservation concern (endangered, vulnerable and near-threatened) defined by the World Conservation Union (IUCN). As a species, the Northern Goshawk is in the 'least concern' category.

However, as we will see in later chapters, the Northern Goshawk can be impacted by change in land-use that reduces food supplies, by pollutants and by direct human actions. Alleviation of impacts has economic consequences and, with other species often more threatened than Goshawks, raises issues of priority. So, how much does it matter if a particular Goshawk race or type comes under pressure?

The answer will depend on how much genetic adaptation is associated with the type concerned. As we will see in Chapter 2, size may not matter greatly because it can change quite rapidly under selective pressure. Plumage may either not be subject to strong selective pressure (under the 'past isolation' hypothesis), or be flexible too, in which case plumage variation may be of interest more in terms of aesthetics than for conservation. However, it is also possible that variation in plumage is linked to other genetic traits that are useful for living in particular circumstances. Such traits might include the timing of egg-laying in response to day-length (see Chapter 4), or variation in shell structure to permit optimum development in different ambient temperatures.

As humans modify habitats and global climate changes, research on conserving genetic traits is important. It is probably important to preserve local traits as well as overall variability in order to help populations adapt as environments change. Populations at the end of clines, such as the small and dark hawks on islands, pale northern forms and *apache*, may be the most important for maintaining particular gene complexes that have evolved in isolation. Alternatively, Goshawks may be genetically diverse throughout their range but with particular alleles concentrated at the extremes of their distribution.

CONCLUSIONS

1. Names of the Northern Goshawk are in some languages descriptive of colouring, but often reflect their predation on species of socio-economic value.
2. The blue-to-black backs and dark-on-white ventral barring of adults is so different from the brown plumage of juveniles, streaked ventrally and speckled dorsally, that these age classes were originally named as separate species. Very old female hawks (which have lived to 27 years) lose the bands and bars on feathers after laying ceases.
3. It seems useful to recognise racial types across Eurasia that include *gentilis, buteoides, schvedowi* and *albidus,* despite clines in size and plumage between them. It is unclear whether the dark *arrigoni* and *fujiyamae* on islands are genetically distinct from Goshawks on the nearby mainland. The North American *atricapillus*

probably qualifies as a separate race from the Eurasian types, but questions remain about the *laingi* on outlying west-coast islands and the large *apache* hawks that use upland forests further south than Goshawks in Eurasia.
4. Genetic studies show the Northern Goshawk, the largest and most widely distributed accipiter, to be closer to the sympatric American Cooper's Hawk than to the Eurasian Sparrowhawk, and still less closely related to Australian and chanting Goshawks. Morphology suggests even closer relationships with Meyer's Goshawk from the south-west Pacific and perhaps to Henst's Goshawk of Madagascar and the African Black Sparrowhawk.
5. It is uncertain whether Northern Goshawk types represent past isolation, in which case they may reflect specific gene complexes, or whether they are products of allele concentration by continuous selection in different habitats. Genetic studies to resolve the question of whether types at the extremes of the distribution contain unique gene complexes would aid conservation strategies in a changing world.

CHAPTER 2

Weights and measures

The young hawk lies on her back on the kitchen table. A soft towel covers her head and her feet are tied loosely together, but she is otherwise unrestrained. Goshawks lie quiet if placed on their back, sometimes even if they can see, but are liable to lash out at any target if their feet are free, switching from supine tranquillity to upright fury.

It is February 1977, in the kitchen of a house at Gäddeholm ('pike-place'), on the eastern fringes of Lake Mälaren's extent from Stockholm far into central Sweden. The ground carries a half metre of snow and temperatures have reached −20°C, in what is the hardest winter for a decade. This hawk was hungry enough to try her luck at two pigeons in a closed compartment. At some point between my trap rounds at midday and dusk, she brushed a fine wire across the ground on one side of the pigeons' inner sanctum. That was enough to trigger a spring-loaded peg, which released a door to drop behind her.

I gently slide the wing-rule to butt on one carpal joint and flatten the primary feathers along it, without moving the wing from her side. Her 363mm wing-length equals the mean value for mainland Sweden. She continues to lie still as I slip one end of the metal callipers into the clavicular fork at the front of her sternum, and gently fit the other end tight into the slight notch at the far end of her sternum. The 92mm keel-length is again close to average. She lets me gently adjust her legs to fit a ring and even remains placid as I slip the hook of a Pesola spring balance under her foot-tie and lift, to hang her a few moments inverted. Her condition is good, at 1,440g.

It seems unjust to roll such a calm lady in a towel. However, it takes 20 minutes to fit a radio tag, and hawks seem least stressed if they cannot struggle. She also deserves recompense for her loss of precious hunting time, in winter days that are just seven hours long. She gets 100g of lean beef, water-moistened so that it pushes smoothly into her crop. Like the pigeons, she'll be in a box until dawn. As I shut up the Land Rover, the stars are bright and the snow squeaks crisply underfoot. Tonight will be cold.

SIZE MATTERS

When biologists have an animal briefly in hand, they routinely record measures of size and mass. The way in which these measures change with time and geography provides evidence about how animals have adapted to their environment, and can give warning of impending problems for individuals and populations. Thus, mass may be found by costly research to relate to survival or breeding success, and the relationship then be used by volunteers to indicate when and where there are problems with food shortage. Size too may relate to individual performance, in ways that indicate evolutionary pressures either between global regions or through changing environments. Later chapters will also consider measures that can be recorded remotely, for example with radio tags.

Chapter 1 showed that the plumage of Northern Goshawks is highly variable in Eurasia, but that adults at least can generally be distinguished from the same species in America. However, the variation of Goshawk size in Eurasia was great enough to encompass that of American Goshawks (Figure 7). This Chapter looks in detail at variation in linear dimensions, mass and body composition between the sexes, in different age classes and through the seasons. It also starts to look at relationships between the size and performance of hawks.

VARIATION BETWEEN THE SEXES

Like other accipiters, the Northern Goshawk shows strong sexual dimorphism. Unlike the size difference in many mammals and other bird species, the female is larger than the male, which is known as 'reversed size dimorphism' (RSD). The reasons for this size difference between males and females have fascinated biologists, and Goshawks have provided much information on the subject.

During the 1970s, wardens were permitted to remove Goshawks at game release sites in Sweden, which provided important material for studies of live and dead hawks that were arranged by Vidar Marcström. Many of the hawks were live-trapped, ringed and released to study their movements, but some were shot and some trapped birds were used for detailed studies of disease and physiology. These hawks could be sexed by gonad examination, at a time before modern genetic sexing methods were available.

Among these hawks, which were almost all from central and southern Sweden, none of the 365 males had wings longer than 340mm, and none of the 269 females had wings shorter than 340mm, although two males and one female (0.5% of the 634 hawks) had wing-lengths of exactly 340mm (Marcström & Kenward 1981a). The mean female wing-length was slightly but statistically significantly greater in 69 adults (366mm) than in 197 juveniles (363mm), but the mean wing-lengths of 37 adults and 308 juvenile males were the same (323mm). Among 234 hawks live-trapped on the island of Gotland, where regional variation in size would have been minimal, there were no wing-lengths at all between 335–340mm (Figure 10). The wing-lengths of 112 juveniles on Gotland averaged 5mm less than 124 adults for both male and female hawks.

The standard way of expressing size dimorphism is the difference in mean dimension between the two sexes as a percentage of half the sum of those means (Storer 1966). The dimorphism index for adult hawks was $100 \times (366 - 323)/(0.5 \times (366 + 323)) = 12.5\%$ on the Swedish mainland. On Gotland, the index was 12.3% for both adults and juveniles, almost exactly the same as for Northern Goshawk studies throughout Europe (Figure 11). If the skins in museums had been sexed with

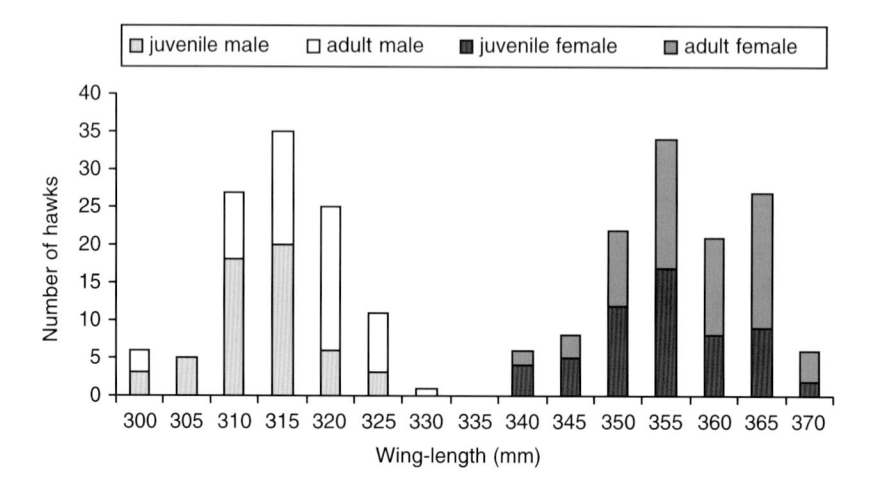

Figure 10. The wing-length of 234 Goshawks trapped at >65 days old on Gotland.

similar care, it appears that the dimorphism of Goshawks in Europe is greater than that of those from east of the Urals, i.e. beyond 60°E. Throughout the north, south, east and west of the Nearctic, Goshawk sexual size dimorphism is close to 8%, uniformly lower than in Eurasia.

A dimorphism of 12.5% in wing-length is equivalent to a much greater difference in mass between males and females. On the Swedish mainland, the average mass of females exceeded the average of males by 54% for juveniles and 61% for adults. Mass alone was less good than wing-length for estimating an individual's sex, because 1.9% of males exceeded 1,050g while 2.8% of females were less. However, hawks could be sexed with complete confidence by combining these measures. All females and no males had a wing-length of more than 340mm and mass more than 1,050g. This was because there was great variability in body mass, with the longest-winged males not being the very heaviest, and the shortest-winged females not the lightest (Marcström & Kenward 1981a).

Nevertheless, there was a tendency within each sex for long-winged hawks to weigh more than short-winged ones. There was considerable variation in body mass, with rather little due to change in the linear dimensions of hawks. Change in wing-length, as one size measure, accounted for 4% of change in body mass of males and 8% for females on the Swedish mainland. The 10% of females with the shortest

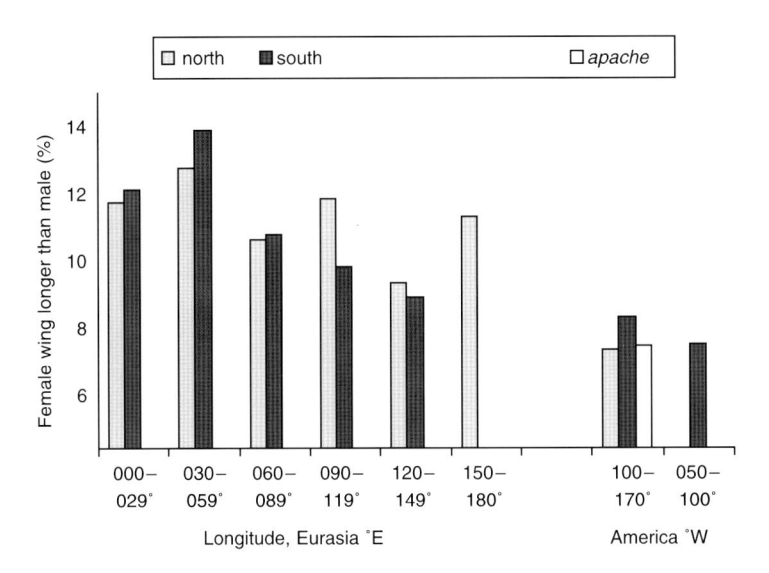

Figure 11. The size dimorphism of goshawks in European field studies and museum collections, shown by the excess of female wing-length as a percentage of the mean wing-length for both sexes. Goshawks in American museums were less dimorphic, and Eurasian goshawks seem to be least dimorphic in the southeast.

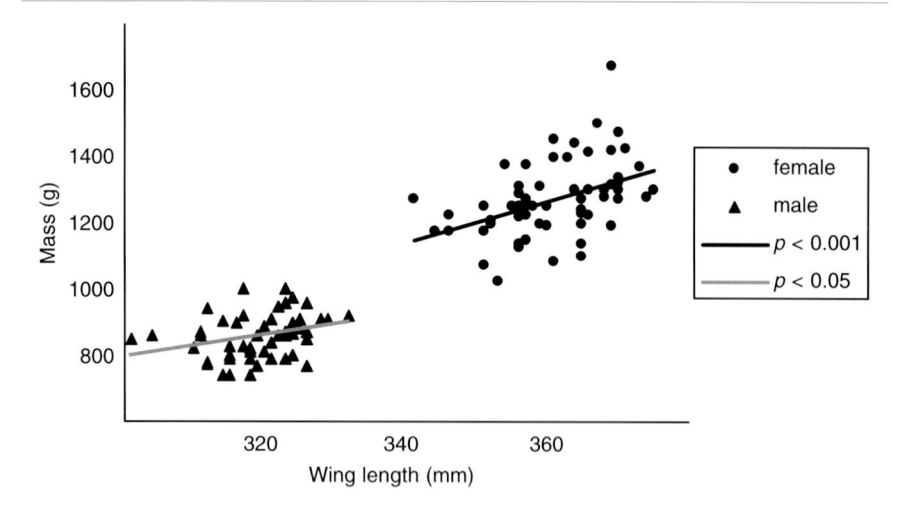

Figure 12. Relationships between body masses of adult Goshawks on Gotland and their wing-lengths: regression lines show the tendency for the longest-winged males, and especially females, to be the heavier birds.

wing-lengths averaged 159g less than the 10% with longest wings. A similar size change accounted for 75g in males (Marcström & Kenward 1981a). In the more consistent data from adults on Gotland, 9% of variation in male body mass and 19% in female body mass was due to change in wing-length (Figure 12).

Wing-length is much used by ornithologists as a size measure, but it has disadvantages. Feathers may be shortened by abrasion. They may also, perhaps through shortage of nutrients, not reach their full length, which may explain why Goshawk wing-lengths were shorter in juveniles than in adults. Moreover, although the growth of wings is convenient for ageing nestlings (Chapter 4), other measures are needed for sexing them.

Another size measure that can be obtained readily for full-grown Goshawks is the length of the sternum. For male and female hawks killed in mainland Sweden, 6–11% of variation in body mass was explained by variation in sternum-length, 2–3% more than in the case of wing-length. However, sternum-length was no better than wing-length as a measure of body-size for live hawks on Gotland. Moreover, it is not easily measured until the cartilage at the end of the sternum has hardened (which occurs after leaving the nest) and it is not usually available for museum skins.

Fortunately for Risto Tornberg, Mikko Mönkkönen and Maarit Pahkala, the University Museum of Oulu in northern Finland had kept skeletons as well as skins of 258 hawks, which enabled an especially thorough examination of dimensions, using multivariate techniques (Tornberg *et al.* 1999). Their measurements included wing-length, tail-length, beak-length, tarsus-length and total length of skins. Skeletal measurements included length, breadth and height of the sternum, length and breadth of the pelvis, and lengths of the coracoid, humerus and femur. The skin measurements that most strongly indicated body-size were length of wing, tail and body, while the most useful skeletal measurements were length of sternum and other bones, and breadth of pelvis. The strongest age-linked difference in each sex was for tails to be longer in juveniles, especially males.

Other measurements used for museum skins include indices of wing shape and the length of toes from the notch where they join the tarsus to the emergence of the claw (usually for the middle toe). Thus, Wattel (1973) proposed the distance from the tip of the first secondary to the longest primary as an index of wing breath, and from primary 10 (at the front of the wing) to primary 7 as an index of rounding at the front of the wing. Whaley & White (1994) used these wing-shape indices to show that American Goshawks from the north and inland had wings that were pointed and long relative to the tail, which could aid long-distance dispersal. Hawks from the dense coastal temperate forests had broader and more rounded wings, which would increase their ability to manoeuvre in woodland.

The length of raptor toes varies considerably in relation to other size measures, such as wing-length. Long toes are considered to be an adaptation for gripping birds, while short toes can exert greater pressure for driving talons through the thick skin of mammals. Toes of Eurasian Goshawks are longest near the Atlantic and Pacific coasts, and become shorter in the north than in the south as one travels east. The length of toes relative to wing-length seems to be less variable in North America (Figure 13).

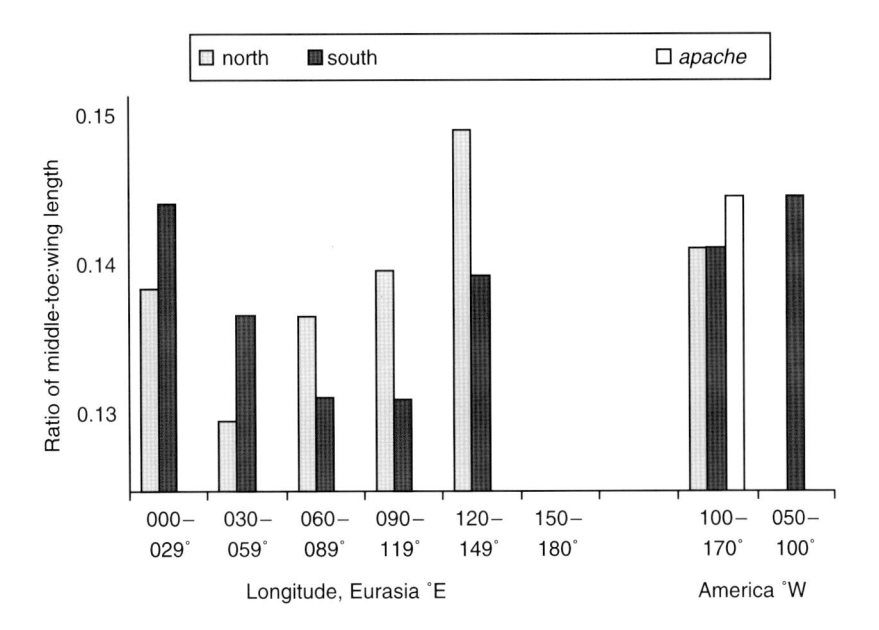

Figure 13. Mid-toe length of male Goshawks from museums in Chicago, Moscow and Stockholm, in relation to wing-length. Female toe:wing ratios have similar trends.

For young hawks in the nest close to fledging, or when trapped before leaving the nest area, a combination of foot measurements and mass were convenient for sexing. Like the toe-length, the minimal width of the tarsus (with callipers tightly closed) is available both from live hawks and museum skins. However, tarsus-width can be taken with one hand, and thus by one person at a nest, whereas it needs two hands to find the tarso-metatarsal junction and straighten a toe for measurement. Tarsus-width was 98% accurate for sexing fledglings that were recovered when full grown: only two of 87 males had a tarsal width greater than 6.5mm and only one of 82 females had a narrower tarsus (Kenward *et al.* 1999).

A comparison of plots of body mass and tarsus-width for juveniles (Figure 14) and adult hawks (Figure 15) on Gotland shows that tarsus-width did increase slightly with further growth, but that 6.5mm remained a useful length for discrimination between sexes. This was partly because small females, and large males, seemed less prevalent among adults.

However, tarsus-width is not as good a measure of body size as is wing-length or sternum-length. It did not correlate strongly with these or with body mass in adults. The span of the foot, from claw emergence on the central and back toes, was a better size measure to combine with mass for sexing nestling falcons (Kenward *et al.* 2001a). Unfortunately, foot-span cannot easily be measured in the nest or for museum skins

VARIATION IN BODY COMPOSITION

The Goshawks killed in central and southern Sweden to protect game also gave useful information on flight muscles and body reserves, on which hawks depend for survival.

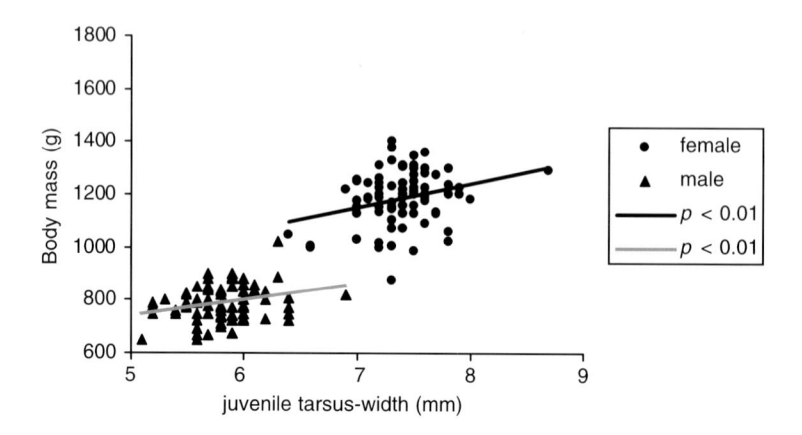

Figure 14. The mass of 166 fledging Goshawks on Gotland plotted against tarsus-width.

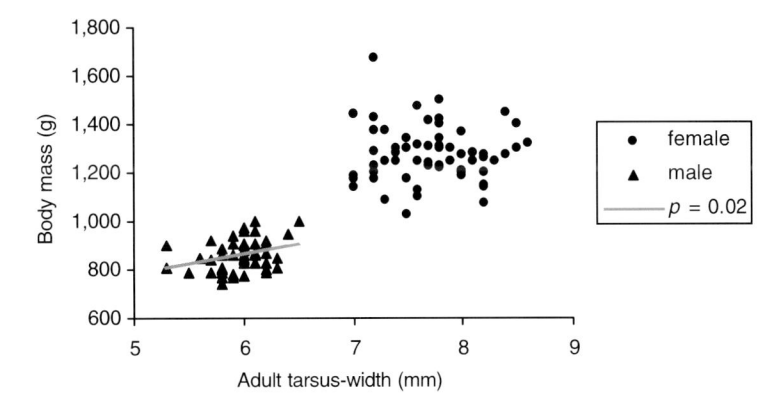

Figure 15. The mass of 110 adult Goshawks on Gotland plotted against tarsus-width.

Flight muscles were carefully dissected off, along with any overlying fat, dried to constant mass, and put in an apparatus which dissolved all the fat by repeatedly flushing the muscles with a 1:3 mixture of alcohol and liquid ether. The new dry weight was known by reference to other animals to be more than 90% muscle protein (Callow 1946), and its difference from the previous dry weight represented fat. Complete analysis of 23 hawk carcasses showed that the total fat could be predicted with 81–91% accuracy from the fat associated with these pectoral muscles, and total lean dry matter with 93–96% accuracy from the lean dry muscle mass. The pectoral muscles could therefore be used to estimate the body composition of all the carcasses, to show the way in which it changes with mass for each sex (Figure 16).

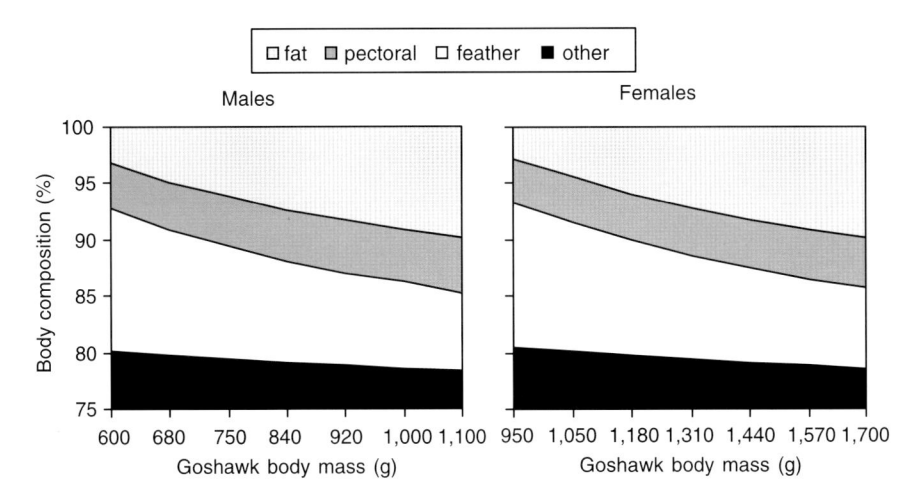

Figure 16. Models of fat, pectoral muscle protein and feather content as a percentage of total mass in adult Goshawks. Data from Marcström & Kenward 1981a.

At the mean body mass, hawks of each sex were about 9% feather, 7% fat, 24% lean dry matter and 60% water. Over the range of body mass in the analysed hawks, water content averaged 57%–61%. There was a compensatory decline in total lean dry matter (muscle, bone, internal organs) from 27% to 22%, despite a slight rise in the proportion that was pectoral muscle protein (Figure 16), which provided more power to keep the heavier birds airborne. As body mass increased, a decline in the percentage contributed by components with fixed mass (e.g. feathers) was offset by a considerable increase in fat.

The proportion of fat in individual hawks varied much more than indicated in Figure 16. A 600g male and a 950g female which had starved to death had only about 1% of their mass as fat. Indeed, birds seem to die when their fat is exhausted, since the starvation level for most species is less than about 1% of body mass. At the other extreme, maximum fat estimates were 14% in a 1,115g male and 17% in a 1,735g female. Fat therefore varied through 6–157g in males and 9–298g in females. These changes seem large, but they are less than the variation in fat reserves in smaller and more migratory species. There can be up to 42% fat in passerine and wader migrants (Odum 1960, Fuchs 1973), 31% in Tufted Ducks (Laughlin 1975) and even 26% in crows (Houston 1977).

The measured body composition of males and females was similar through a large variation in mass, except in one important respect. The flight muscles of females were slightly but significantly smaller, as a proportion of total body mass, than in males. Across a range of 20% above and below the mean mass of 840g for males and 1,310g for females, flight muscle of females averaged 44–51% greater in mass than in males, whereas total mass of females was 54–61% greater. Classic work on dynamics of bird flight (Pennycuick 1972, 1989) estimates that in birds with similar shape the power requirement for flight increases with mass to the power of 1.17, whereas the power available increases with the muscle cross-section (approximately muscle mass$^{0.67}$). On this basis the power:mass balance of females would be 79–81% that of males, if their muscles were the same size and equally efficient. Females would need relatively larger flight muscles to compensate for this difference, but in fact their muscles are relatively smaller, giving a power:mass balance of only 75–77% that of males, 5% less than with muscles of equivalent size.

The relative lack of female flight muscle, when compared with males, is not the only disproportion between them. Females are not the same shape as males. For a start, their wings are shorter relative to their mass. An increase of 54–61% in mass would increase wing-length at the same shape by 15.5–17.2% (proportional to mass$^{0.33}$). However, wings of female Goshawks were only 12.3% longer than in males on Gotland (Figure 17). Assuming that wing area in Goshawks rises with the square of wing length, the wing loading (mass per unit area) of female Goshawks is 22–25% greater than that of males. That is 6–7% greater than the 15.5–17.2% expected if their proportions were equal.

Since the flight muscles of females account for less of their body mass than in males, something else must be relatively larger. The next biggest muscle mass is in the legs. There are no comparable data on these muscles for Goshawks. However, female tarsus width on Gotland was 25% greater than that for males (Figure 17), double the increase expected from the difference between sexes in body mass, whereas the length of middle toes shows the 10–15% level of dimorphism found in most other body components by Tornberg *et al.* (1999). The same excess of tarsal dimorphism, albeit with less sexual size dimorphism overall, occurs in North American Goshawks. So at least some of the extra mass is in the legs. With more massive legs relative to their toes, a stronger grip might be expected from female Goshawks than from males. This too has yet to be measured, but as a falconer who has had bare hands gripped (in unguarded moments) by birds of each sex, I strongly suspect that the female grip is disproportionately strong.

Higher wing-loading means that female Eurasian Goshawks cannot turn as sharply as males and must fly faster to stay in the air. A lower power:mass balance also means that females cannot accelerate as rapidly as males. By timing the flights of trained hawks, Slijper (1980) has shown that female Goshawks do fly faster and accelerate more slowly than males.

The details of Goshawk dimorphism have not been examined to the same extent in North America, but data on wing-lengths, body mass and wing-loadings have been analysed for the smaller Cooper's and Sharp-shinned Hawks (Mueller *et al.* 1979, 1981). For sharp-shins, the wing loading increment in females was almost exactly that expected from their mass increment over males, because wing-length and area increased in proportion to mass. Thus, whereas female Sharp-shinned Hawks are larger versions of the males, female Goshawks are not simply 'super-males'. Compared with males, female Goshawks are under-winged and over-legged. This adapts females for a diet of mammals (Chapter 7).

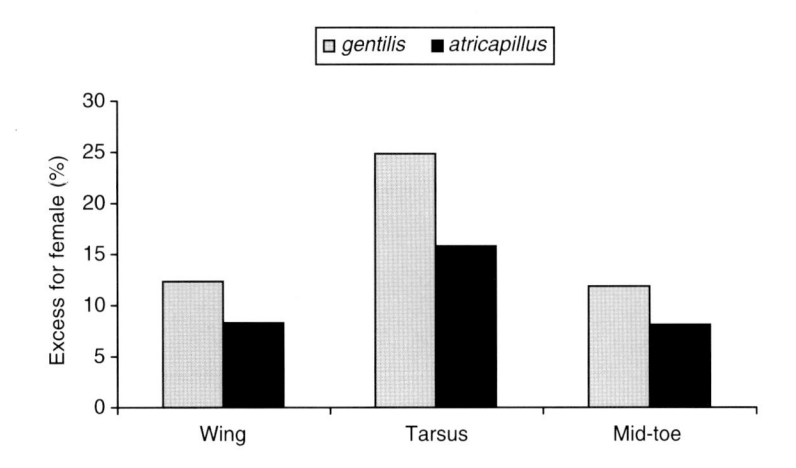

Figure 17. Size dimorphism of *A. g. gentilis* on Gotland (*n*=69) and *A. g. atricapillus* in a North American museum (*n*=61), shown by the excess of wing-length, tarsus width and middle-toe length in females as a percentage of the means for both sexes.

VARIATION THROUGH THE SEASONS

It has been known for more than 60 years that the mass of Scandinavian Goshawks varies considerably throughout the year. Yngvar Hagen (1942) collected measurements from 136 hawks shot and trapped between August and June in southern Norway. His material indicated that body mass peaks in late winter. The peak mass for females, in March, was 141% of their late autumn low, in November. The change was relatively less in males, whose mass peaked in February at only 114% of their October low. Hagen had only one or two hawks for the months of August, September and June, and too few data altogether to examine seasonal changes in adult and juvenile mass.

There were more extensive data from the 1,277 live and dead Goshawks in central and southern Sweden (Marcström & Kenward 1981a). Peak mass was in this case 32% greater than the August minima for both sexes (Figure 18) in the large sample of trapped birds, but with a tendency for mass to be somewhat variable during January to March.

Average mass of shot hawks was greater than that of live-trapped hawks in each sex for every month before the mid-winter peak (Figure 18). Shot hawks might often have been killed flying up from game or poultry, and might therefore have tended to greater mass due to food in their crops. Indeed, 47% of the shot hawks had more than a trace (0.5g) of food in their crops, compared with only 19% of 466 hawks that were killed after being live-trapped. However, the average food content of shot birds (23g), was just 14g more than the 9g of food in trapped hawks. Taking monthly variation into account, among dead hawks with crops and gizzards emptied, the

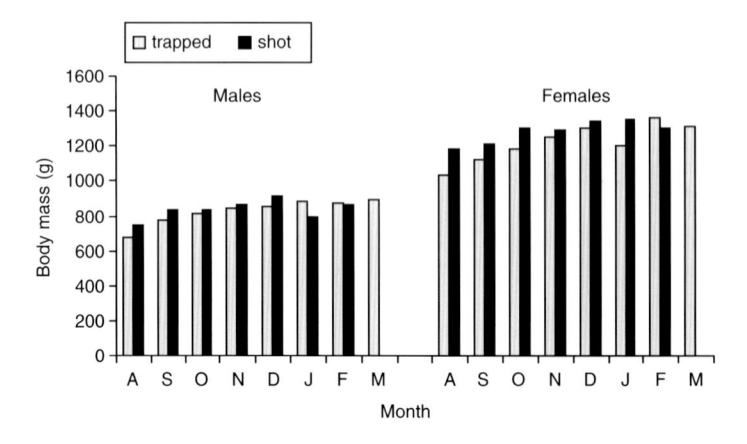

Figure 18. Variation in body mass of 1,113 trapped and 164 shot Goshawks in central and southern Sweden between August and March. Mass is combined for adults and juveniles because they show the same trends; mass of adult females (but not males) is slightly greater than for juveniles. Data from Marcström & Kenward 1981a.

average mass of shot juveniles was greater than for live-trapped birds by 23g for males and a (significant) 89g for females.

Most live-trapped birds had entered compartment traps baited with live domestic pigeons. It is probably the hungriest hawks which enter these strange contraptions in pursuit of prey that is unfamiliar for many hawks in Fennoscandia. Helmut Mueller and Dan Berger (1968) showed that Sharp-shinned Hawks caught in 'obvious' traps had lower body mass than those caught in fine aerial nets at the same sites. I have seen Goshawks approach traps without entering them, presumably because the hawks were not hungry enough.

The mean mass of live-trapped Goshawks is likely to be below the true population mean. The mass of shot hawks, on the other hand, may be biased upwards by being shot in areas with game rearing, where hawks tended to be heavier than elsewhere (Kenward *et al.* 1981a). A representative value for mean mass of male and female hawks in each month was probably between the values for trapped and shot hawks.

A more consistent picture was obtained by analysing fat and protein from the hawks that had been killed, which reduced the emphasis on low mass birds from traps. It also permitted inclusion of four males and six females shot during April, May and June, to give samples of eight males and 10 females for the March-June breeding season (Figure 19).

Although there is less information on Goshawk mass and body composition in summer than at other times of the year, the overall pattern agrees with that found in a number of other studies of raptors (Mendelsohn 1986). Mass in raptors tends to be greatest in autumn or winter and to fall to an annual minimum during the breeding season (Newton *et al.* 1983, Village 1983, Hirons *et al.* 1984). This makes it possible to interpret the limited summer data for adult Goshawks with some confidence.

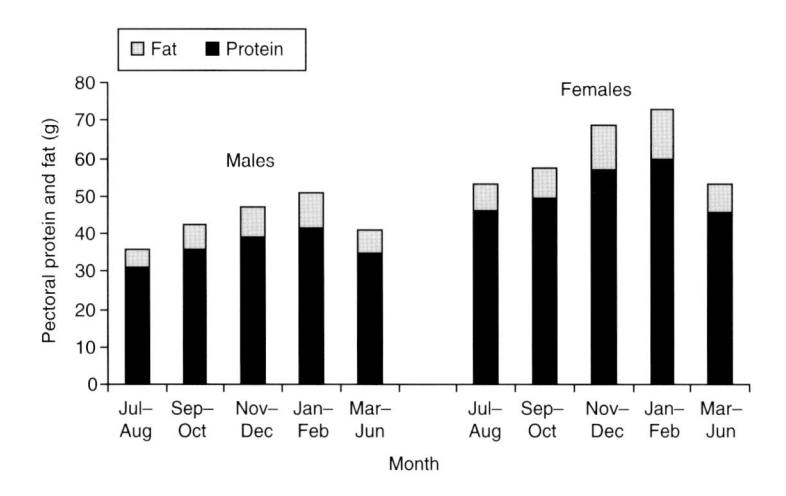

Figure 19. Annual variation in pectoral fat and protein body mass of 161 male and 151 female Goshawks from central and southern Sweden. Data from Marcström & Kenward 1981a.

The average mass of juvenile male hawks dropped from 791g (n=55) in July, when they were trapped near nests while still being fed by adults, to 713g (n=16) when shot or trapped away from nests in August, but had recovered to 782g (n=123) in September. Similarly, juvenile female mass dropped from 1,186g (n=59), when trapped at nests, to 1,105g (n=6) in August, 1,125g (n=91) in September and 1,191g (n=144), above the July mass, in October. Mass then increased to the December–January peak, and then held up until March for juveniles of both sexes. Adult mass showed a similar increase in autumn, to a January peak followed by a decline for males. In contrast, mass of females was still increasing into February. Juvenile mass tended to be less than for adults during winter, although the differences were significant only among females. Fat and protein reserves of both sex and age classes fell again as the hawks began breeding and moulting (Figure 19).

THE BENEFIT OF HIGH MASS IN WINTER

For consistency, the above analyses used mass at first capture, but many of the live-trapped hawks were recorded again some time after being marked and released. Their changes in mass between recaptures mirrored those in the figures. Moreover, records of recaptures could be used to investigate whether body mass influenced ability to survive for at least two months during winter. The greater the mass of juvenile females when marked during October to January, the more likely they were to be recorded again after two months. The recovery rate of 26 females with masses greater than 1,400g was five times greater than for 208 hawks of 1,000–1,400 g and 10 times that of 18 females with masses below 1,000 g (Figure 20). The effect was not significant in males.

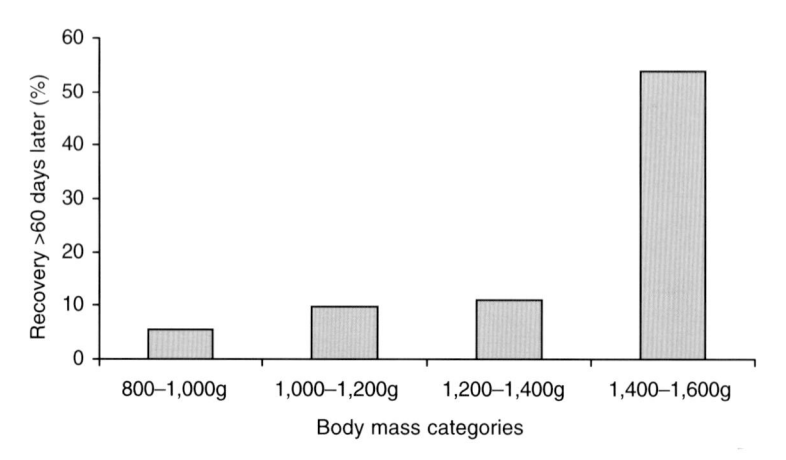

Figure 20. The percentage of female Goshawks in different body mass categories, marked during October–January in central and southern Sweden and recovered after they had survived for two months. Data from Marcström & Kenward 1981a.

The possibility of this relationship between mass and recovery rate resulting from the heaviest hawks being most inclined to re-enter traps or get shot could be ruled out by the data in Figure 18, and further analysis showed that it was not a result of the lightest females being most likely to move to an area where they could not be retaken (Marcström & Kenward 1981a). Quite simply, high mass in juvenile females was associated with good mid-winter survival. There was no significant relationship between body mass and recovery after two months in juvenile males, which suggests that winter mortality was less associated with low mass in males than in females.

The analysis of fat and protein content indicated the value of high mass for survival. Using an equation for calculating energy requirements of non-passerine birds (Kendeigh 1970), Goshawks of average mass in central and southern Sweden require 640 kJ/day for males and 816 kJ/day for females at 0°C, a typical mean daily temperature in mid-winter for the area. These values may be inaccurate (Kennedy & Gessamen 1991) and are in any case for resting birds. However there are also food consumption estimates of wild hawks at comparable winter temperatures (Chapter 7) that convert to energy requirements of 737 and 938 kJ/day respectively (Kenward *et al.* 1981a). At this rate of energy use, the fattest female could have lasted 14 days to starvation level and the best male could have lasted about ten days. The birds would have nearly halved their body mass, losing 60% of their muscle, from which they would obtain about 30% of the energy, and 96% of their fat. At average mass, the hawks could continue hunting for three to four days without food. These survival times would increase in warmer weather, for example to about double the 0°C value at 30°C.

CAUSES AND CONSEQUENCES OF SEASONAL VARIATION IN BODY MASS

With high mass a benefit for survival, why is Goshawk mass relatively low early in the autumn? In juveniles the answer might be low hunting efficiency, but that would not explain why adult mass follows the same trend. A possible explanation is that the adults are moulting. Muscle might be reduced by high demand for particular amino acids required in moult, as happens overnight in moulting Bullfinches (Newton 1968). With reduced muscle power, and flying also impaired by absence of feathers during the moult, low mass could reflect a need to reduce wing-loading. However, mass remains low in Eurasian Sparrowhawks for several weeks after completion of the moult (Newton *et al.* 1983).

In fact, a tendency for mass to increase during autumn and peak in mid-winter is found in many bird species, including ones that finish breeding and moulting long before Goshawks (Newton 1972, Perrins 1979, Pienkowski *et al.* 1979). The best explanation is that high mass increases the energetic cost of flying, which raises the need for food, and also reduces agility for avoiding predators or catching prey (Gosler *et al.* 1995, Macleod *et al.* 2005). However, when weather becomes poorer in winter and daylight foraging time shorter, there is an increasing advantage from having extra reserves. In other words, the most likely reason for the autumn mass increase is that hawks are trading off the cost of increased wing loading against the benefit of having extra reserves.

There would be a particular benefit from low wing-loading in autumn for prolonged flight during juvenile dispersal, as noted by Mueller & Berger (1968). These authors also stressed that although wing-lengths of juveniles tend to be slightly shorter than for adults, the lower body masses more than compensate for this. Moreover they showed that juveniles have longer tails than adults in all North American accipiters. Thus, the average surface-loading of juvenile Sharp-shinned Hawks with tails fully spread was 5.5% lower than for adults. Juvenile Goshawks had lower wing-loading than adults by 3–5% in Swedish Goshawks, with tails longer too from Finnish measurements (Tornberg *et al.* 1999). This should benefit the hunting performance among inexperienced young hawks, as well as their soaring performance during dispersal. Goshawks respond to weather that influences dispersal and migration in a way intermediate between buteos, that travel by soaring, and falcons that tend more to flapping flight (Bildstein 1998).

The highest fat levels among the Swedish hawks were recorded for adult females in late winter, perhaps in preparation for breeding. The situation was similar for Sparrowhawks trapped during breeding by Ian Newton, Mick Marquiss and Andy Village (1983). The mass of female Sparrowhawks increased 15% to an annual peak in the 10–20 days before laying, with the largest clutches and least subsequent desertion among the heaviest females. Half of the mass increase was lost during laying, but the loss accounted for only 11% of the mass of the largest clutches, so females were dependent on good feeding to lay a large clutch. They then tended to maintain body mass during incubation. The mass of breeding male Sparrowhawks fell in the pre-laying period, and again for both sexes during brood rearing.

There are no comparable data from extensive trapping of Goshawks during breeding, but observations of a tame female are noteworthy. At the time when this hawk started to make courtship calls in spring, her appetite increased dramatically for about two weeks (see Chapter 3), but then dropped back to 'normal' some two weeks before laying. She then lost muscle from the breast during oviduct development and during laying of eggs (three the first year and four the next). There is thus a period of fattening before laying in Goshawks too.

It is not clear why the tame hawk reduced her feeding and lost breast muscle shortly before laying, despite a generous food supply. Perhaps it is convenient to draw on this protein reserve for oviduct development and egg albumen, and maybe a breast tending towards concave is better shaped than a plump breast for incubation. The loss of breast condition also coincided with the start of moulting. This hawk lost weight in the two moults before breeding started and even when a single tail feather was lost and re-grew in winter. The links between body mass, moult and incubation, perhaps mediated by the hormone prolactin, deserve further study.

A further observation shows the danger of the laying period for Goshawks, and the importance of female reserves. In 1977, after a winter studying Goshawk predation on wild pheasants near Örebro in southern Sweden, three adult females were radio-tracked into the breeding season. The spring weather was poor that year, and only one female laid eggs. She had been the best hunter that winter, with a small range and no long intervals between kills, but she laid only two eggs where clutches are normally three or four. She had also been killing for herself during the pre-laying period. Her mate was trapped just as incubation started and weighed 720g, well below average for an adult. A few days later he was found dying, too weak to fly. When the female deserted she had received no food for three days, and it was another three before she killed, just as her hunting started to slow up in a way typical of a bird in very poor condition. An incubating hawk that is obliged to start hunting is probably disadvantaged by poor fitness (muscle tone) and unfamiliarity with recent changes in where best to find prey, not to mention the start of moult. Poor reserves at that time threaten not just breeding success but survival itself.

VARIATION IN MASS WITH LATITUDE

The mass of Goshawks trapped and shot in Sweden during the 1970s tended to increase to the north in all age and sex classes (Figure 21). However, after correcting for effects of wing-length, and of sampling in different months and years, the effect remained significant only for juvenile females (Marcström & Kenward 1981a). The increase in winter reserves in the north was relatively small.

Peter Sunde analysed data from 599 Goshawks given to research and veterinary institutes by the Norwegian public during 1988–95. The birds were recorded as

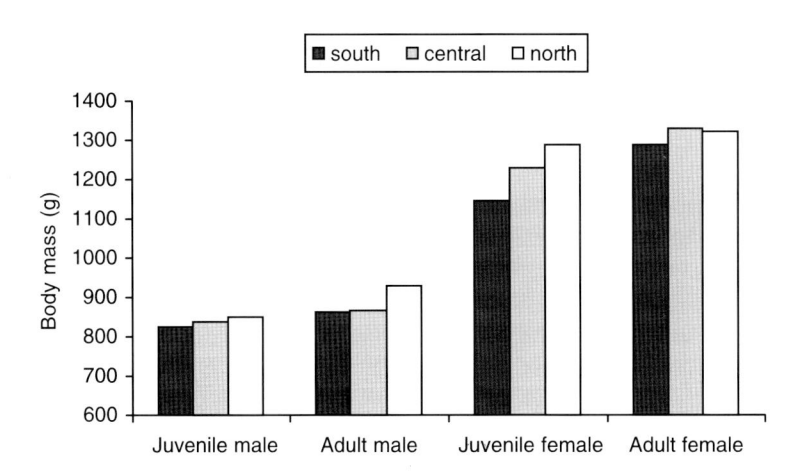

Figure 21. The body mass of Swedish Goshawks live-trapped between August and March in south (Götaland), central (Svealand) and northern (Norrland) regions, corrected for effects of body-size, season and year.

having starved or died in accidents, and were given scores according to the extent of fat and pectoral muscle. He too found that the extent of reserves tended to be least in juveniles and in males and also that the proportion of hawks that had starved in winter tended to increase to the north, especially in males in winter and spring (Sunde 2002). The Norwegian males showed a significant decline in reserves to the north, perhaps because a recent reduction of winter prey for Goshawks in northern Scandinavia (Tornberg et al. 1999) has made conditions for hawks in the north worse recently than in Sweden during the early 1970s. The summer reserves of male and female Norwegian Goshawks were higher in the north.

POSSIBLE ORIGINS OF REVERSED SIZE DIMORPHISM

The sections above show that male body shape confers advantages in agility and load-bearing, whereas females have a greater tendency to accumulate reserves. This sets the scene for considering the reasons behind their reversed sexual size dimorphism (RSD). The big question is why the female tends to be the larger sex among most raptors and owls, but in few other families of birds and mammals. This question had fascinated more than 20 authors by 1980 (Newton 1979, Andersson & Norberg 1981). Many factors could be involved and some may act in concert. Their relative importance is likely to change from species to species, from year to year and as humans modify the environment. The subject is therefore very complex, but worth a quick look at the principal hypotheses.

Some early suggestions were that large size enabled females to protect themselves or their young from the well-armed and supposedly aggressive males, to make males surrender food or otherwise to maintain the pair bond (Hagen 1942, Amadon 1959, Cade 1960, Perdeck 1960, Smith 1982, Mueller 1986). *Dominance hypotheses* lack wide credence as a cause of RSD, because non-cooperation in breeding would be not at all in the male's interest, so that male (and female) behaviour should have evolved to promote harmony. Indeed, captive Kestrel females paired with larger males (from other races) bred just as well as pairs in which the male was smaller (Willoughby & Cade 1964).

Rand (1952), Lack (1954), Storer (1966), Selander (1966) and Opdam (1975) were early in promoting the idea that difference in size would confer each sex with the ability to take different prey, thereby reducing competition during periods of food shortage. Reynolds (1972) and Snyder & Wiley (1976) stressed the value of harvesting different prey populations when food demands are high for the young, and showed that the extent of dimorphism in raptors was related to the proportion of birds in the diet. Newton (1979) documented a general increase in raptor dimorphism as prey agility increases, from carrion eaters, through insect, reptile and mammal eaters to the most dimorphic eaters of fish and birds. He noted that the most agile prey are eaten by few raptor species, so there is little competition from other predators to prevent a species spreading its prey spectrum.

Although *dietary divergence* could be a beneficial consequence of difference in size, it does not explain why the female should be the larger sex. The same applies to hypotheses that females are large for *nest defence* (Storer 1966), or to *compete for males* as a scarce resource (Olsen & Olsen 1987), or that small size increases aerial agility when males *compete for territories* (Schmidt-Bey 1913, Widén 1984a). There are very many bird species, as well as other animals, that defend young and compete for resources without RSD. Similarly, although species that hunt by stealth may minimise *foraging interference* (by alerting prey) if only one pair-member hunts (Andersson & Norberg 1981), this does not explain which sex should hunt.

Storer (1966) introduced the idea that small size would adapt males for foraging because there should be more small prey than large. Others have stressed that small size gives males relatively better hunting agility, load carrying and flight endurance per unit food than females (the *nimble male* hypothesis), and noted in turn that large females are best adapted to store energy for egg production and incubation (the *big mother* hypothesis), because the daily requirements of large birds account for a smaller proportion of their reserves (Selander 1966, Reynolds 1972, Mosher & Matray 1974, Balgooyen 1976, Andersson & Norberg 1981, Mueller & Meyer 1985, Lundberg 1986, Hakkarainen & Korpimäki 1991). Several authors have noted the particular value of reserve storage ability during harsh feeding conditions early in the spring (Lundberg 1986, Korpimäki 1986, Hakkarainen & Korpimäki 1993).

A combination of the *nimble male* and *big mother* hypotheses not only provides reasons for the sexes differing in size, but also for the female being larger: she is the one that lays the eggs. However, they do not explain why raptors should differ in this respect from other birds in which both sexes feed nestlings, so that one partner might be able to provision the other adequately during incubation. Adequate provisioning might be difficult for any species eating small items of food, but why should non-raptors that eat large items, like fish-eaters for example, fail to develop RSD?

The solution may lie in a suggestion by Walter (1979) that raptors would be particularly at risk of injury during strenuous prey capture when eggs are in the oviduct. Eggs and egg reserves also give a female the disadvantage of high wing-loading. Moreover, once she stops hunting to lay eggs, she may also become 'uninformed' about local changes in prey availability, with muscles unconditioned for hunting. It behoves her to continue a nest-minding role as long as the eggs or young need attention and the male can provide food.

On this basis, it is the dependence of raptors on flight for foraging, often coupled with strenuous prey capture, which makes high *female vulnerability* a fundamental for RSD. Priming by vulnerability sets the stage for evolution of big mothers and nimble males. In species where dietary divergence is advantageous, the size dimorphism may then become considerable, and species like the Goshawk adapt in shape too for taking different prey.

SIZE RULES

Later chapters will look more thoroughly at the consequences for diet, breeding and survival of variation in the size and mass of Goshawks. They will show that change in average dimensions of populations can be surprisingly fast (Tornberg *et al.* 1999, Yom-Tov & Yom-Tov in review), which has made the Goshawk a species of particular value for studying evolutionary fitness of individuals.

The changes in average dimensions of populations have implications for geographic variation in size and sexual dimorphism. The size of Eurasian Goshawks increases to the north (Figures 6 & 7). The conventional explanation of this example of 'Bergmann's Rule' is that larger warm-blooded animals have a smaller surface to volume ratio than smaller animals, and therefore lose a lower proportion of their body-heat through their skin in a given time, an advantage in cold northern winters (Bergmann 1847). The rule has been extended also to encompass humidity, on the basis that heat loss should be fastest in damp and cold conditions (James 1970).

In North America, the large size of *apache* Goshawks in the south, and the small size of *laingi* in cool and humid areas, appear to contradict Bergmann's rule. However, Whaley and White (1994) point out that *apache* occurs mainly at altitudes with a much colder climate than at sea-level. They also note that beaks, tails and toes of Goshawks do not support 'Allen's Rule', which predicts relatively short appendages in cold climates (which would again conserve body temperature).

IMPLICATIONS FOR CONSERVATION AND MANAGEMENT

When society permits the killing of animals, in order to provide food or other materials of value for humans or to conserve other species of value, it is appropriate to minimise waste. The use of such animals in research minimises waste of information. Cooperation between Swedish game preservers and wildlife biologists in the 1970s, which Vidar Marcström arranged through the Swedish Hunters' Association, also enabled work to mark and release many hawks to investigate factors affecting survival and movements, and led to the projects on Gotland. Work that started with cooperation concerning dead hawks resulted not only in much new knowledge but also in the development of non-lethal methods of managing raptors (see Chapter 10), funded by game conservation interests.

People who choose to work in the countryside often do so because of a fascination for wildlife, or have interests that can be cultivated for wildlife conservation. Sometimes their practices are unnecessarily destructive. However, is it better to cooperate with such interests and engage them for conservation or to mount campaigns against them?

At a time when so much of the human population lives or works in towns, and thus lacks the direct motivation and opportunity to maintain rural environments of those who live in the countryside, it is important to develop rural human resources that are sympathetic to conservation. Working together with all the rural interests provides the opportunity to change attitudes, develop resources and take advantage of practical skills. The apprentice game warden from one of the estates that contributed Goshawks for analysis in the 1970s, Mats Karlbom, became the keystone for ten years of subsequent research.

That is not an argument in favour of killing hawks, though such killing may sometimes be necessary for conservation purposes. However, where humans kill wild animals for any reason, it seems wise to seek minimal wastage of the biological resource and maximal benefits for conservation from the humans concerned.

CONCLUSIONS

1. Among full-grown Goshawks in Sweden, a combination of wing-length and body mass discriminated between males and females. Around the age of fledging, while wing-feathers were still growing, sexes could be distinguished by combining mass and tarsus-width. Within each sex, variation in wing-length accounted for up to 19% of change in body mass. Wing-length was greater in adult than juvenile hawks, but juvenile tails were longest.
2. At the mean body mass, hawks of each sex were about 9% feather, 7% fat, 24% lean dry matter and 60% water. For individuals, fat ranged from 1% in starved hawks to 17% of body mass in one adult female. Females with maximum body reserves could have survived 14 days without food in mid-winter, compared with ten days for males.
3. Body mass was least in autumn and was up to 32% greater in mid-winter in Sweden. Fat and protein reserves declined into the breeding season. In a tame female, food consumption more than doubled prior to laying, but reserves were lost during oviduct development and egg-laying.
4. Adults had greater mass than juveniles, and trapped hawks weighed less than those that were shot. Juvenile females that were at least 7% above mean body mass when live-trapped survived the next two months five times better than those with less mass.
5. The sexual size dimorphism of Goshawks was greatest west of the Urals, declined to the east of Eurasia and was least in North America. Swedish females were under-powered relative to males, because they had less flight muscle than males relative to the difference in their body-size and greater wing-loading, but had relatively thicker tarsi.
6. Hypotheses to explain reversed sexual size dimorphism in raptors include female dominance, benefits of dietary divergence for feeding young or winter survival, requirements for nest defence or competition for males or territories, foraging

advantages for nimble males and enhanced reserve storage in large females. High vulnerability of females during egg-laying may be the crucial trigger for size increase of that particular sex in raptors.
7. The collection of dead hawks from game wardens, for studies of variation in size and reserve storage, was a successful way to start cooperative ventures that resulted in practical conservation of raptors.

CHAPTER 3

Nesting and laying

Miss Piggy is working her way down the wood that runs from our house towards the River Frome. The strip of woodland is only a stone's throw wide, so although it is hard to see her in the mix of conifers and bare trees, I can hear the bell on one leg whenever she starts and ends a flight. It is March 1992 and her moult starts shortly. The scales onto which she stepped from my glove a half hour ago showed 2lb 12oz, some 1,300g, which is 100g more than the weight at which she would be flown if she was to return to my whistle. So she is not going to come down until the end of the afternoon unless she kills a squirrel, and then only when I throw out a half rabbit from the bag at my side, but she will follow me down the wood. No one in their right mind would have flown a Goshawk like this in the days before radio tagging made it easy to find such independent birds.

She moves to an exposed branch, so I can see her shake out her feathers and stretch one leg before tucking it up into her feathers. A crow finds her and dips a couple of times overhead, making the hoarse rattling call used for predators. I move out into the adjacent field for a better view, and hear her bell. Unusually, she flies out over the canopy, making a 'kak-kak-kak' call. She circles away above the trees, beating her wings stiffly in harrier-like display flight.

Staking a territorial claim on the wood will not endear her to the Buzzards that Sean and I study. I whistle loudly and throw the half rabbit in the air so she can see. She alights in a tree nearby and calls again, quite aggressively, with fluffed out feathers, before dropping to the food, over which she spreads her wings, 'mantling' like a nestling. As I slowly approach to within two metres, she flies not at my fist that proffers a day-old poultry chick, but at my face. I duck, and she lands on the ground, then runs back to the rabbit, with feathers puffed out in excitement. However, her feathers relax when she starts feeding and she can be coaxed onto my glove. There will be no more free flight this spring.

The following spring she is given a nest of about 0.8m diameter, woven from larch and spruce around pegs fixed in a triangular board in the corner of her enclosure. She has food *ad lib* this spring and by 20 March she is eating up to eight poultry chicks a day, calling 'kak-kak-kak' frequently when no-one is in view, occasionally with 'heee-yah, heee-yah' food screams. She comes to the window shelf used for feeding in a very fluffed-out state when anyone approaches, and seizes food aggressively in her feet. She is spending time on the nest and will take twigs from me in her beak, but drops them. By 28 March she has become much more placid, but is still eating voraciously. By 3 April her demand drops to three-day-olds again and she only 'kaks' when she hears me.

On 5 April she flies 'helicopter-like', with tail coverts fluffed out like a powder puff and body held almost vertical, from perch or food shelf to the nest and there pulls at sticks or at food which she now takes from me to the nest. While pulling at these items, she raises her tail almost like a bowing falcon and fluffs her tail-coverts (Plate 12). It seems almost as if she requires my presence in order to eat, taking small pieces from me and eating perhaps a chick at a time (this is similar behaviour to the start of moulting in previous years). I can stroke her, but she remains skittish, in that the handling causes her to fly to another perch, or even to the floor, keeping her coverts fluffed all the time.

By 19 April, when she drops her first secondary, she has become as thin as usual at the start of moult, with her breast tending to concavity. She has become even tamer and can be stroked all over without flying off. I can feel her oviduct starting to distend within her abdomen and her cloaca is visible within the ring of white tail coverts when she flies. The second secondary drops by 21 April and her abdomen contains a noticeably hard 50mm lump. That evening she 'kaks' quietly from the nest where she is waddling about with her abdomen distended, showing grey down as well as white covert feathers. At 08.30 on 22 April she is 'kakking' at the window shelf and there is an egg in the nest. I give her spruce twigs during the day and by the evening she has built these into her nest, almost covering the egg. By 14.30 on 24 April there is another egg in the nest. She is still courting vigorously, with wings held out quivering, like a small bird food-begging, but not eating more than about three of the day-old poultry chicks. For the last week at least she has not liked the chicks with fluff on, but eats them plucked. She does not incubate overnight, but stands by the eggs.

On 25 April the eggs are more obviously in a cup, formed because she has worked new green matter into the nest, along with her first (left) moulted primary. A visitor to see the eggs causes aggressive fluffing out like a ball and fast, shrill 'kakking' on the nest rim. Miss Piggy remains nervous all day. At 09.30 on 27 April there is a third egg in the nest. She has eaten three poultry chicks left overnight and is hungry for another, with 60g of beef. She is still courting, showing a bowing motion like bathing (dipping back and head) more obviously than before. She is incubating when not feeding.

During 28 and 29 April she is courting less and becoming more nervous of me, tending to run about on the shelf after taking one or two of the poultry chicks. She 'kaks' on hearing me and comes off the eggs (on which she is otherwise sitting tight) to the window-ledge. There is no longer much sign of dropping under-tail coverts. Over the next days she becomes even wilder, easily slipping into a run-along-ledge routine that indicates mild angst if I remain while she eats. On 6 May there was a mass of moulted down on the nest. She sits very tight, reluctant to leave even to feed. The eggs are a very beautiful duck-egg blue in the nest, but fade to chalky white within months of being removed.

The following year her voracious eating is earlier, from about 28 February until 12 March. By that time she is showing her tail coverts strongly, has become placid and is taking food mainly with her beak. However, 'kakking' does not start until 9 March, the same week as in previous years. She remains in the placid stage of courtship until 2 April, when I record 'helicopter flight' and the decline in appetite that seems associated with oviduct development. She is ready to be inseminated on 8 April and lays the first egg of four on 17 April, just five days earlier than in 1993. Thus, although her pre-lay fattening is three weeks earlier in 1994, her calling and oviduct development advance by only a week. Sadly, a male that has previously reached the 'chup' calling stage of courtship with another falconer, and might therefore be persuaded to provide semen, is not ready in time for artificial insemination.

TAMED AND WILD

Courtship is the time of year when adult hawks become most conspicuous in the wild, but they are still hard to observe in detail. Observations of Miss Piggy showed a clear sequence of courtship behaviours which involved aggressive demands for food (initiated separately from 'kakking' calls), then nest-building, tail-covert spreading and gentle food-transfer (after presumed fattening), followed by helicopter-flight (associated with oviduct development), copulation invitation and laying (see also Plates 11–13). A Goshawk courting with a human in a three by five by two-metre enclosure is subject to some factors that differ from those in the wild, but her behaviour helps to interpret observations of wild hawks.

Whether a wild Goshawk population persists in an area depends on whether it produces enough young to offset deaths and emigration. The sequence of events that leads up to egg-laying is therefore very important for conservation, as is a consideration of the habitats in which they occur. Habitats and events that lead to laying are described first in this chapter, but many aspects are given further consideration in chapters that follow.

NEST HABITATS

Across their Holarctic distribution, Northern Goshawks breed where trees are large enough to bear large stick nests, open enough for easy access from ground level, yet relatively secluded. Such sites are often within mature forest (Plates 8 & 9). However, Goshawks also nest in strips of woodland along rivers and streams in North America (e.g. White *et al.* 1965, Shuster 1980, Younk & Bechard 1994), reach high density in central European patchworks of small woods, shelter-belts and copses (e.g. Bednarek 1975, Link 1986, Goszczyńnski 2001, Bijlsma 2003a) and even use parkland with more or less isolated trees in some Eurasian cities (e.g. Dietrich 1982, Borodin & Sorokin 1986, Würfels 1999, Rutz 2001, Altenkamp 2002, Dekker *et al.* 2004).

In the far north, Goshawks nest in birch, aspen and larch woods bordering the tundra (e.g. Dementiev 1951, McGowan 1975, Swem & Adams 1992, Doyle 2000, Krechmar & Probst 2003, Plates 10 & 11) or in the pine, spruce and larch woods of the taiga zones. Further south, Goshawk nests are found in the beech and oak woods of Eurasia and in the great variety of North American broadleaved woodlands (reviewed by Bosakowski 1999, Kennedy 2003). In the Mediterranean region, including Morocco, Goshawks breed on the forested sides of hills and mountains, in mixed and pure woodland containing beech, fir, pine and cork-oak trees (e.g. Peus 1954, Smith 1965, Araujo 1974, Perco & Benussi 1981, Mañosa *et al.* 1990, Penteriani 1997). Similar hillside and canyon sites are chosen in the equivalent arid parts of North America, among pines, firs, cedars, oaks and aspens and other poplars. Steep, conifer-covered valley slopes of the Kirghiz Kungei-alatau, the Tien Shan and other Himalayan foothills contain Goshawks at 2,500–5,000m (Schäfer 1938, Korelov & Pfander 1983, Beishebaev 1984), as do the ranges extending south from the American Rocky Mountains.

During the 1990s, conservation groups in North America tried to use this species as a flagship for the conservation of old-growth woodland. This stimulated the assessment of Goshawk breeding habitats in many parts of North America, and also in Europe, leading to an improved understanding of what Goshawks use for breeding. Comparing their use in different areas shows that what Goshawks use in a given place does not necessarily define what they must have. However, comparisons of breeding habitats help discover the limits to what Goshawks will use and thereby to indicate their actual requirements. This chapter starts a process of investigating what Goshawks really need for reproduction and survival.

THE NEST

Goshawk nests are built of twigs and small branches up to about 1.5cm in diameter. The nest width is typically 80–120cm, and new nests may be as little as 25cm deep. Old nests, to which hawks have added material through many years, can measure

160cm across by 130cm deep (Link 1977, Bijlsma 1993) and weigh up to one ton when wet (Holstein 1942). The cup typically averages 20–30cm at its widest and may be 10–18cm deep during incubation. The nest surface becomes flattened and sometimes tilted somewhat to one side by the time the young fledge. The top is covered with leafy twigs prior to laying, and these are added through incubation into the rearing period. A female in the Sierra Nevada of California added twigs until she stopped perching on the nest (Schnell 1958), and leafy twigs are often found on nests with young two to four weeks old (Plates 14–17). Lining with grasses, mosses and bark-chips has been reported (Peck & James 1983, Anonymous 1989) and may be overlooked elsewhere because leafy twigs are more eye-catching.

André Brosset (1981) suggested that the lining material prevents eggs falling into cracks in the nest structure, since this happened when a captive Black Goshawk pair were given no fresh twigs. There is also much less leafy lining in the more finely woven nest of the Eurasian Sparrowhawk. However, Sparrowhawks sometimes use bark-chippings as lining material, possibly to help reduce temperature loss during incubation (Newton 1986), and insulation may be more important for Goshawks, which lay much earlier in the year. Heinz Brüll (1964) reported a 7°C temperature gradient from the top of the nest to the clutch at the bottom of a well-lined nest-cup, leading to a suggestion that hawks manipulate incubation temperature by changing the shape of the cup.

The use of leafy twigs to support and insulate eggs would not explain why Goshawks continue to add them through the nestling period. However, lining material may also prevent food items falling into the nest and providing a breeding ground for pathogens. Nesting females have been observed digging down into the structure between the nestlings (Holstein 1942, Schnell 1958), perhaps to let air in and thus inhibit the growth of anaerobic *Clostridia*. Toxaemia from *Clostridium perfringens* and *C. welchii*, which multiply in meat, has killed captive raptors (Koehler & Baumgart 1972, Kenward 1981a, Wernery *et al.* 2000). The lining material may therefore have several functions, possibly also including the concealment of eggs before the start of incubation.

THE NEST SITE

Where suitable nest trees are scarce, Goshawks sometimes nest on the ground or very close to it. Hawks have nested in birches or dwarf trees at heights of 1–2m at the edge of the tundra in Alaska and Siberia (Englemann 1928, Bent 1937, Dementiev 1951, Swem & Adams 1992, Krechmar & Probst 2003, Plates 10–11). Such low nests are very rare where suitable trees occur, but have also been found in The Netherlands, where one nest was first taken for that of a harrier, and in Norway on the stump of a spruce that was felled one spring after being used as a nest tree (Schweigmann 1941, Haftorn 1971). Exceptional cases also include a nest on a cliff (Grünhagen 1988) and an unsuccessful breeding attempt on an electricity pylon in The Netherlands (Anonymous 2002).

In very poor woodland, nests occur at 4–6m above the ground, but Goshawks usually build at 9–25m. In Schleswig-Holstein, Volkher Looft found that nest height increased from 8m to 24m as tree height increased from 15m to 40m. Rob Bijlsma found a very similar pattern among smaller trees in The Netherlands (Figure 22). Nests are rarely above the lower half of the canopy, especially if it is dense, and typically at the base. This is where branches are of a good size to support the nest and well-spaced for access.

The average nest height in 29 studies in North America ranged from 9m to 21m, with very similar values for 11 studies in Europe (Figure 23). However, the average height of trees with nests ranged only from 14–28m in European studies, compared with 18–43m in North America. In consequence, where heights were measured for nests and nest trees in seven European and 24 American studies, nests were relatively closer to the tops of trees in Europe than in America (Figure 23).

Goshawks nest in trees that are large enough and structured to provide adequate support. Nests occur in a variety of large conifers, including more than 20 species of spruce, fir, larch, pine and hemlock. Among broadleaved trees, nests have been recorded in ash, alder, aspen, beech, birch, elm, hickory, hornbeam, lime, maple (including sycamore), oak, poplar (including aspen), tamarack, wild cherry and willow. In trees with a single stem, such as spruce and larch, the nest is typically supported at one side of the trunk by two or more side branches. Nests quite often take this position in other tree species, but in pines and broadleaved trees a vertical fork in the trunk or a large branch may also be used (Plates 8–9). Nests built away from the trunk on horizontal branches are rare. The two such nests among 419 recorded by Looft & Biesterfeld (1981) in Schleswig-Holstein were possibly built originally by Buzzards, which sometimes adopt this building style.

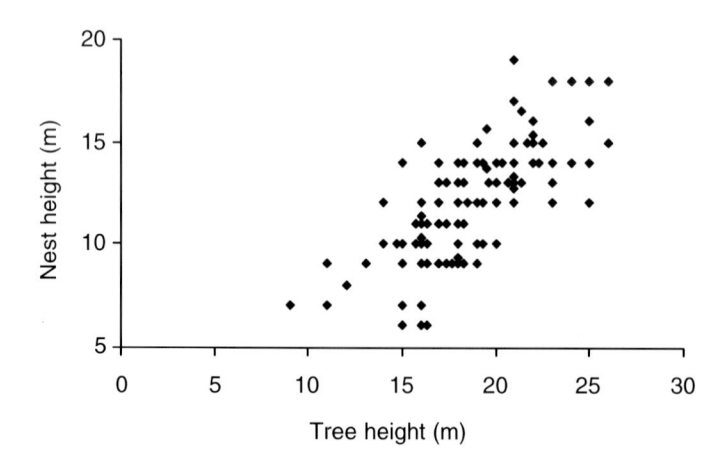

Figure 22. The relationship between height of Goshawk nests and nest-trees in Drenthe province of The Netherlands. Data from Bijlsma 1993.

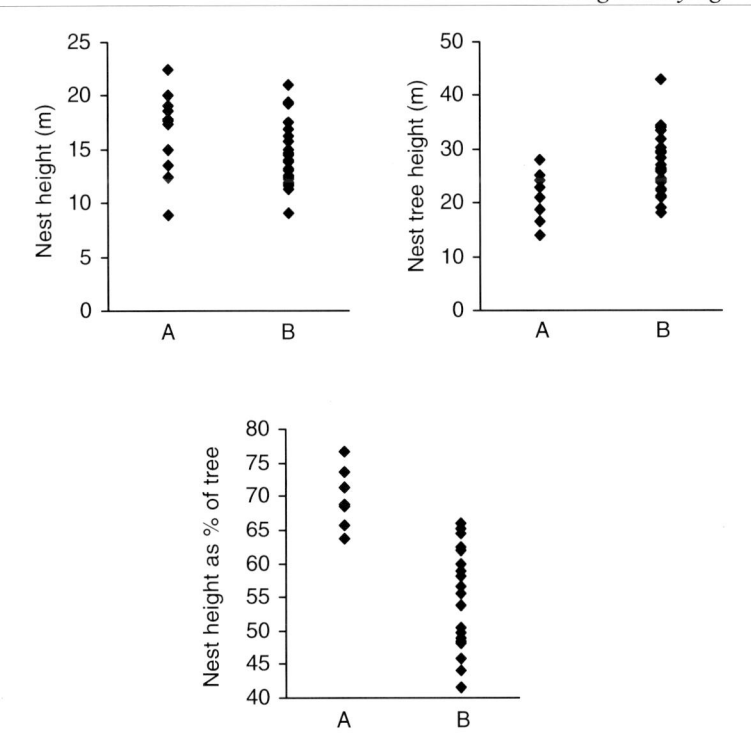

Figure 23. Average height of nests was similar in European (A) and North American (B) studies, but trees were lower in Europe and nests therefore closer to the tree tops ($p<0.001$, sources in Appendix 2).

Goshawks sometimes adopt nests of other species. Buzzards contributed 5% of the nests used by Goshawks in Schleswig-Holstein, including some unusually exposed ones at the edges of woods. Another 2% had been built by Ravens and Carrion Crows, but the majority (93%) were constructed by the hawks themselves (Looft & Biesterfeld 1981). Wild Goshawks will also accept artificial nests (Saurola 1978). They seem most likely to adopt other nests when there are no traditional Goshawk nests. Thus, hawks in Wales took over a deserted heron's nest (Toyne 1997a). While colonising urban environments, Goshawks take over Sparrowhawk territories, often making first breeding attempt(s) in existing Sparrowhawk nests. In Hamburg (and in Cologne), this resulted in Goshawks nesting in some extremely small trees, and in unusual locations. Only in later years (and after the stacked-up sparrowhawk nests, on flimsy support branches, have collapsed) do they build nests in trees that seem 'more appropriate' for Goshawks (C. Rutz, pers. comm.).

Many authors have listed the proportion of nests in different tree species. However, there are relatively few robust tests for preferences, which require data on availability to the hawks of suitably-sized trees in each species. The only generality seems to be preference for different tree species in different areas. Thus, McGowan

(1975) found that 94% of 30 Alaskan nests in stands with more than one suitable species were in birch, probably because these most often had forks large enough to support the nests. Similarly, Speiser & Bosakowski (1987) found 50% of New York and New Jersey nests in birch and beech, significantly more than expected from the availability of these trees. However, at a similar latitude in Schleswig-Holstein in northern Germany, 56% of 506 nests were in conifers although these comprised only 33% of the woodland, with the strongest selection (16%) for larches which were fewer than 3% of the available trees (Looft & Biesterfeld 1981). There was similar disproportionate use of larch for 346 nests in Bavaria (southern Germany) and for 903 nests in Drenthe (The Netherlands); in conifer forests of Oregon, 22% of 82 nest trees were larches, which comprised only 4% of trees sampled at random in the same stands (Link 1986, Bijlsma 1993, McGrath *et al.* 2003). In contrast, beech were chosen for 17 of 23 nests yet provided only 18% of the woodland in the south German Saarland (Demandt 1962). To explain the preference for beech, Fischer (1980) suggested that conditions would be damper in spring at spruce and fir nests than in deciduous species, and this could also explain the preference for larch in other areas. It would be interesting to know the extent to which the tree in which a hawk fledges influences its subsequent choice of nest site.

THE NEST AREA

An area that typically extends 100–200 m from an active nest is used by adults especially intensively for perching and feeding. This is the area to which the male routinely brings food. Its extent may broadly reflect a single flight from the nest and perhaps may also be linked to visibility from the nest. This is also the area in which young tend to confine most activities until their flight feathers are full grown (see Chapter 5). The area may be the whole of an isolated wood, a stand of trees that differs from surrounding woodland due to natural growing conditions or forestry practices, or relatively few trees in a town.

In central Europe Goshawks nest in woods as small as 1–2ha, and nests in woods of less than 10ha are commonplace (Waardenburg 1977a, Looft & Biesterfeld 1981, Link 1986, Bijlsma 1993). At these sites the nest stand is in effect the whole wood, which may well be a uniform plantation. It is rare to have more than one pair in a compact wood of less than 100ha, because Goshawks generally avoid edges and occupied nests are seldom less than 600m apart. An exception is an early record of 400m (Ortlieb 1978). Such close nests may reflect polygamy, as in the case of nests found 200m apart in Hamburg by Christian Rutz (pers. comm.).

The structure of Goshawk nesting areas in woodland has attracted much attention, with early quantitative studies by Paul Bartelt (1977) in the Black Hills of Dakota and Helmut Link (1977) in southern Germany. Interest in conserving nest habitat for Goshawks has motivated some 50 studies in at least ten countries (most listed in Appendix 2). There was early work to compare the nest habitats of the three

1. Adult female Goshawk in level flight, showing strong barring on the underside and rounded tail (Markus Varesvuo).

2. Adult male Goshawk braking and turning, showing area of spread tail (Vincenzo Penteriani).

3. The speckled underside of juvenile female Goshawk, powering into a climb (Tomi Muukkonen).

4. Head of a male Goshawk, showing the pronounced brow-ridge and supercilliary stripe with contrasting dark crown and ear-coverts (Vincenzo Penteriani).

5. Head of the less contrast-rich juvenile female Goshawk (Vincenzo Penteriani).

6. Goshawks are regularly mobbed by smaller birds such as these Hooded Crows (Tomi Muukkonen).

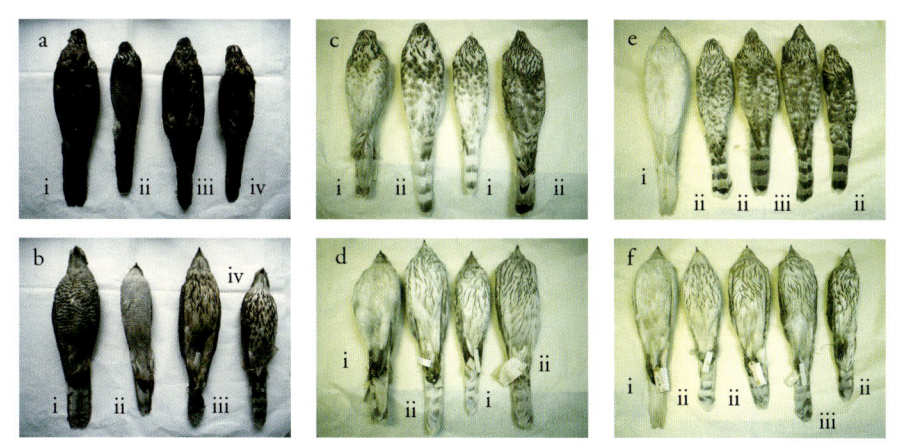

7a. Specimens in **a–f** are as labelled in Moscow Museum. In **a–b**, 2nd-year female (i), older adult male (ii), juvenile female (iii) and juvenile male (iv) of *A. g. gentilis* from back (a) and front (b) show sexual size dimorphism and age differences in plumage. In **c–d**, juvenile male (i) and female (ii) *A. g. buteoides* show speckling on backs (c) and pale lower surfaces (d). In **e–f**, an adult female *A. g. albidus* of the white morph (i) is paler than three male juveniles (ii) and a female (iii).

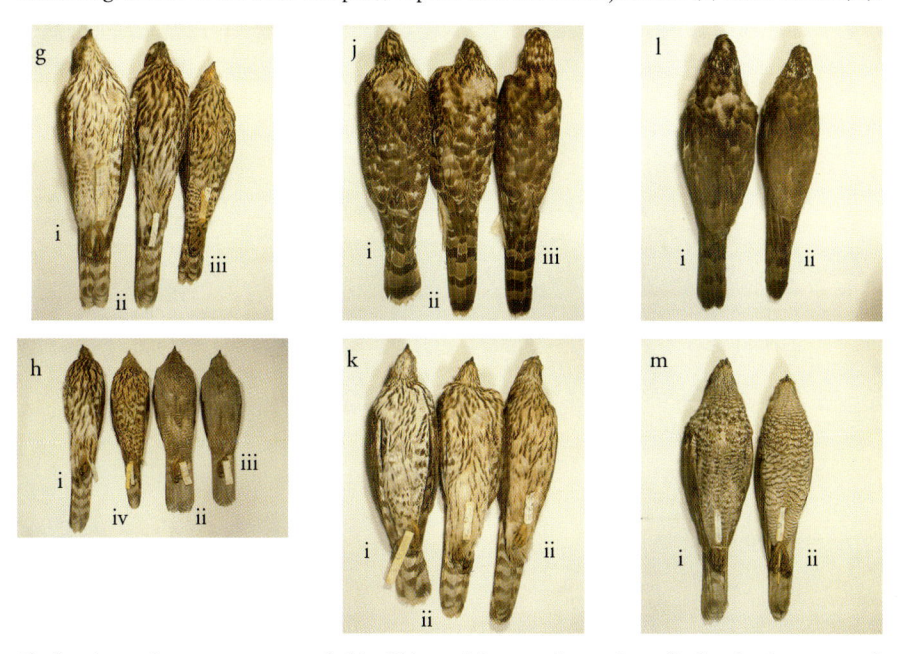

7b. Specimens in **g–m** are as recorded in Chicago Museum. In **g**, a juvenile female *A. g. atricapillus* from Alaska (i) is larger than a male (ii) and female (iii) *A. g. laingi* from British Columbia. In **h**, a juvenile female *A. g. apache* from Mexico (i) is larger than adult female (ii) and male (iii) of the *A. g. laingi* type specimens, but a juvenile male *apache* from Mexico is relatively small. In **j–k**, two Alaskan *A. g. atricapillus* juvenile females (i) are similar in size and colour to a juvenile female *A. g. gentilis* from Germany (ii). In **l–m**, medial streaking of an adult *A. g. atricapillus* (i) distinguishes it from an eastern Eurasian *gentilis* (ii).

8. Goshawk nests: (**a**) in the Kazakh Altai; (**b**) in young Alaskan pines; (**c**) in Pennsylvania; (**d**) in thinned French woodland (Vincenzo Penteriani).

9. An old Goshawk nest in spruce-pine forest on Gotland, reached with the aid of a multi-section ladder.

10. Goshawk nest in low waterside vegetation on the Alaskan tundra (Ted Swem).

11. Closer view of the nest shown in 10, with the female accompanied by a chick (Ted Swem).

12. A third-year female Goshawk in a breeding enclosure invites copulation. Note the white undertail-coverts.

13. An adult female *atricapillus* Goshawk on her nest. The medial streaks show very clearly on her breast feathers, in some cases combining with bars to give (upside-down) "flying hawk" silhouettes (Rory Hill).

14. A fresh clutch of Goshawk eggs in a nest of lichen and pine needles in Scotland (Doug Weir).

15. A Goshawk chick 4 days after hatching.

16. Goshawk chicks after 16 days of development.

17. Juvenile plumage is beginning to appear. This chick is 28 days old and has a leg-mounted radio-tag.

18. Male Goshawk on a pheasant in the snow (Jari Peltomäki).

19. A typical plucking site on Gotland showing gull and pigeon feathers.

20. A pheasant poult killed by a Goshawk at the Frötuna estate.

21. A juvenile female Goshawk from Finland scavenges at a fox carcass in Oxfordshire.

22. Goshawk tracking: (a) Tail-mounted transmitter is attached after fledging; (b) tracking on skis in the Swedish winter; (c) tracking by minibus on Gotland.

23. A radio-tagged juvenile male from Finland that returned to the wild successfully after hard-release in Britain, after being fed near this perch.

Accipiter species in North America (Hennessy 1978, Reynolds 1978). Biologists then started to investigate what Goshawks selected for nesting, by comparisons with the habitat generally available or measurements in random sites (Looft & Biesterfeld 1981, Moore & Henny 1983, Hall 1984, Speiser & Bosakowski 1987). Recent studies have also compared data within and between Europe and North America (Penteriani 2002, Greenwald *et al.* 2005, Rutz *et al.* 2006a).

The average age and density of trees in nest stands tends to be very variable, depending on whether the species are fast- or slow-growing, whether they are deciduous with a crown that spreads with age, like oak and beech, or whether there is an appreciable understorey of young trees. In 24 Holarctic studies (Appendix 2) the tree density varied from 170/ha for mature stands of beech in Italy (Penteriani & Faivre 1997) to 1,345/ha for multi-layer mixed forests of Wyoming (Squires & Ruggiero 1996). Conifer nest trees in lowland Britain averaged 59 years old, with one Douglas fir of only 33 years (Anonymous 1989). Some conifer nest trees in California were more than 250 years old (Hargis *et al.* 1994).

The findings in Vincenzo Penteriani's Italian study (Figure 24) are typical for Goshawks breeding in extensive woodland. Thus, the diameter of the nest tree measured at breast height (DBH) was above average for the nest stand. The nest tree was not necessarily the tallest in the stand, but its large diameter was associated with greater trunk spacing and crown area. The height from the ground to the first branch tended to be greater for trees in the vicinity of the nest than in control sites 150m away, with the result that flight-space (defined as trunk-separation × first-branch-height) was greatest close to nests.

This study, in common with many others, also noted a tendency for nests to be relatively close to paths or other forest openings that may facilitate flying to the nest, including clear-fellings, swamps and heaths, lakes and meadows, roads, railways and swathes cut out along power cables. As a result of the tree-spacing and openings, there

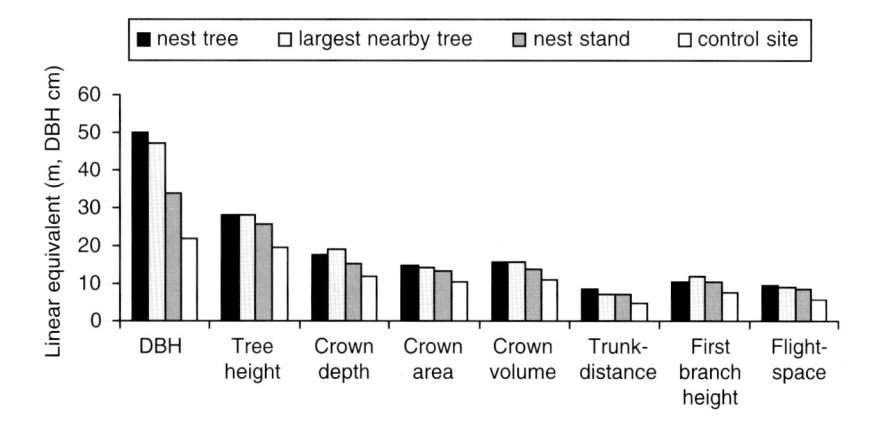

Figure 24. Characteristics of the nest tree and nest area for 30 Goshawk nests in beech-forest of the Italian Appenines. DBH = trunk diameter at breast height. Data from Penteriani & Faivre 1997.

is reasonable flight access from the nest to ground level, where prominent boulders, stones or roots of fallen trees or even low branches may be used as plucking points. Feathers, fur and bones can often be found on one or two of these prominences (Plate 19). They are usually those within 200m of the nest, with reasonable all-round visibility and especially between the nest and any nearby open area which may be used as an unrestricted line of approach by a hawk laden with prey (Figure 25). A hawk carrying heavy prey has severely reduced manoeuvring ability, partly because it must fly fast to obtain adequate lift and partly because of added momentum resulting from the extra weight. A trained Goshawk flew into a tree and killed itself while carrying a pheasant it had just taken in the air (R. Roberts, pers. comm.).

In fragmented woods, Goshawks need not fly long distances through trees to their nests. In Schleswig-Holstein, where woodland is scattered and amounts to only 8% of the land, Looft & Biesterfeld (1981) reported that 82% of 471 nests were less than 200m from woodland edge, 27% being within 50m of the open fields. Similarly, in Bavarian woods that contained only one pair of Goshawks (typically <300 ha), 70% were within 200m of edges. However, in woods with more than one Goshawk pair (and >200 ha), only 9% of nests were that close to woodland edges (Figure 26). In all cases, 1–2% of nests were in edge trees. All the 15 'edge nests' were in conifers,

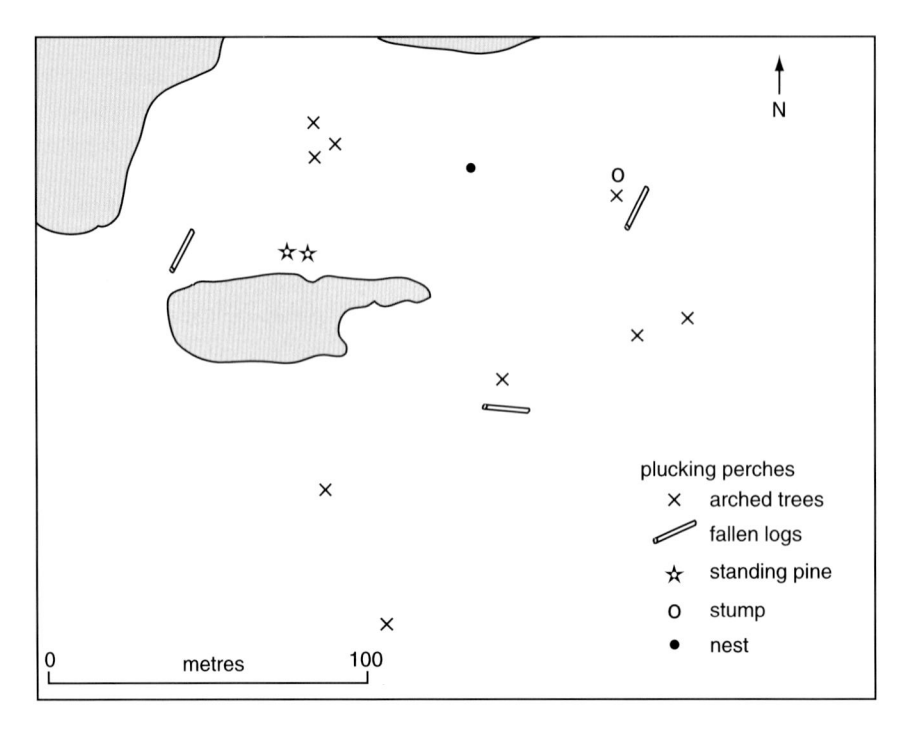

Figure 25. A Goshawk nest area in California, showing the nest tree, nearby meadow and a glade as potential approach routes (in grey) and prominences where prey had been plucked within a short flight of the nest. Redrawn from Schnell 1958.

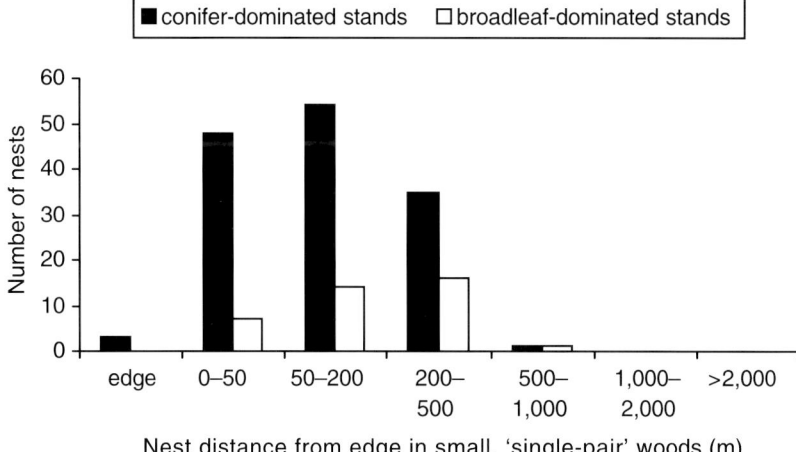

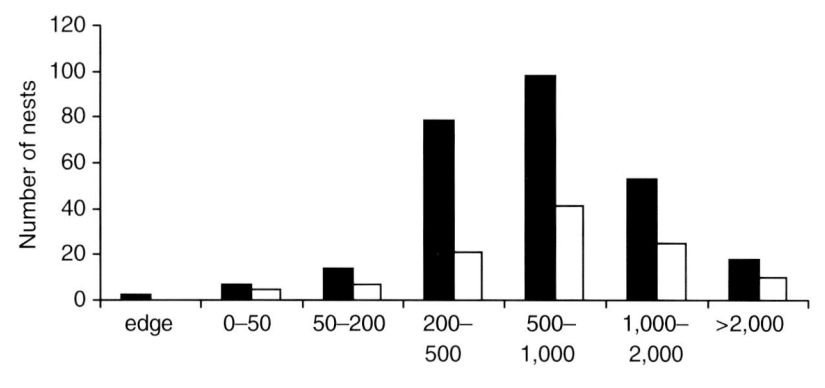

Figure 26. Conifer-dominated stands were used for most of 559 Goshawk nests in Bavaria, with broadleaf-dominated stands used increasingly away from edges, especially in small woods that contained only one pair. Data from Link 1986.

which (except for larch) would have offered most seclusion before leaf-out. The proportion of nests in broadleaf stands increased into the woods. In general, the hawks chose seclusion if given the opportunity, but did not require it.

In general, Goshawk nest areas spread across landscapes, neither favouring flat areas nor slopes (Penteriani 2002). However, nest areas that are on slopes tend to favour the middle and lower parts (McGowan 1975, Link 1986, McGrath *et al.* 2003), perhaps because these require least effort when delivering prey from the valleys. There is also a strong tendency for sites on slopes to face north and east, except at high latitudes and altitudes. Thus 65% in Alaska (McGowan 1975) and 54% in Norway faced south (Selås 1997a) and, in Arizona, there was a tendency for those at high altitude to be the most southerly (Crocker-Bedford & Chaney 1988). The relatively many studies at high altitude or latitude in America probably explains why more sites with southern exposure were recorded there than in Europe (Penteriani 2002).

Richard Reynolds, whose study of the three accipiters in Oregon was one of the first to note a general north-to-east aspect of nest sites, suggested that this direction gave cool, damp conditions that favour tree growth (Reynolds 1978). A more detailed analysis of thermal conditions by Melissa Siders (1995), who worked with Pat Kennedy (Siders & Kennedy 1994, 1996) much further south in the Jimez Mountains of New Mexico, supported this suggestion. Helmut Link (1977, 1986) stressed the importance of avoiding heat-stress for the young from excessive exposure to the sun, which is more likely on steep southern slopes. He described an exposed south-facing nest originally built by Buzzards, in which all three young of a juvenile female Goshawk died in hot weather within a week of hatching.

The canopy of the nest tree and its neighbours can provide shelter from the sun. Many authors have noted a relatively strong closure of the canopy at Goshawk nest sites, typically of 60–96%, with only two of 30 Holarctic studies recording lower averages (Appendix 2). Those exceptions, of 50% canopy closure from Alaska and 52% from British Columbia (Iverson et al. 1996, Bosakowski & Rithaler 1997) were also the most northerly records, and hence where protection from overhead sun would be least important. Canopy closure may give protection against nest detection by avian predators as well as from the sun. Use of dead trees has been recorded for up to 15% of Goshawk nests in North American studies (Porter & Wilcox 1941, Dick & Plumpton 1998, McGrath et al. 2003), but is rare in Europe.

In summary, the stand of trees within 100–200 m of the nest often differs noticeably from the surrounding forest. It tends to contain larger and more widely spaced trees, but with high canopy closure. The trees have few large branches below the canopy and a sparse understorey of smaller trees in parts of the stand usually thins out within 50m of the nest. There are interesting questions about the cues used by the hawks when selecting sites. Perhaps they often select trees that are free of leaves during nest-building because these provide good visibility and perching opportunities. Perhaps variation in selection between tree species and for canopy cover can be related to thermal conditions for nest sites at different latitudes, altitudes and slopes.

NEST AND AREA OCCUPANCY

In the area around the active nest, and in nearby similar stands, there may be several alternative nests. These are recognised as alternatives because they are appreciably closer together than the distance between adjacent occupied nests and not more than one in a group is occupied each year. During an 18-year study in Bavaria, 44 of 57 Goshawks nest areas had more than one nest; 27 had two, 10 had three, five had four, one had five and one pair had as many as 11 alternatives, which were identified by finding Goshawk feathers and prey remains under the nests or embedded in them (Link 1977). This is an average of 2.6 nests for the 77% of areas with more than one nest.

The positioning of alternatives can be seen in an earlier study by Zygmunt Pielowski (1968), which recorded an average of 2.4 alternatives for five multi-nest areas during ten years in a Polish forest (Figure 27). Similarly, there were averages of 2.6 nests at 28 areas studied during nine years in California (Detrich & Woodbridge 1994) and 2.3 nests at 59 areas studied for six years on the Kaibab plateau in Arizona (Reynolds & Joy 2006). The same nest was used the following year 54% of the time in Poland and one nest was used continuously for eight years. Studies of 18–28 years in Dutch Goshawk populations also had 50–52% repeat occupancy, with up to six repeats at one site (van Haaff 2001, Bijlsma 2003b). A nest in Denmark was used continuously for 17 years (Schiøler 1931).

The alternative nests in Poland were 70–800 m apart, with 54% repeat occupancy and average nearest-neighbour distances between sites only 1.4km. Studies with 50–52% repeat occupancy in The Netherlands had similarly dense Goshawk populations. However, nearest-neighbour distances averaged 3.25 km in the Californian study, where nests up to two kilometers apart were considered to be alternatives (average spacing 273m) and repeat occupancy was 44% (Detrich & Woodbridge 1994). On the Kaibab Plateau, mean nearest-neighbour distance was

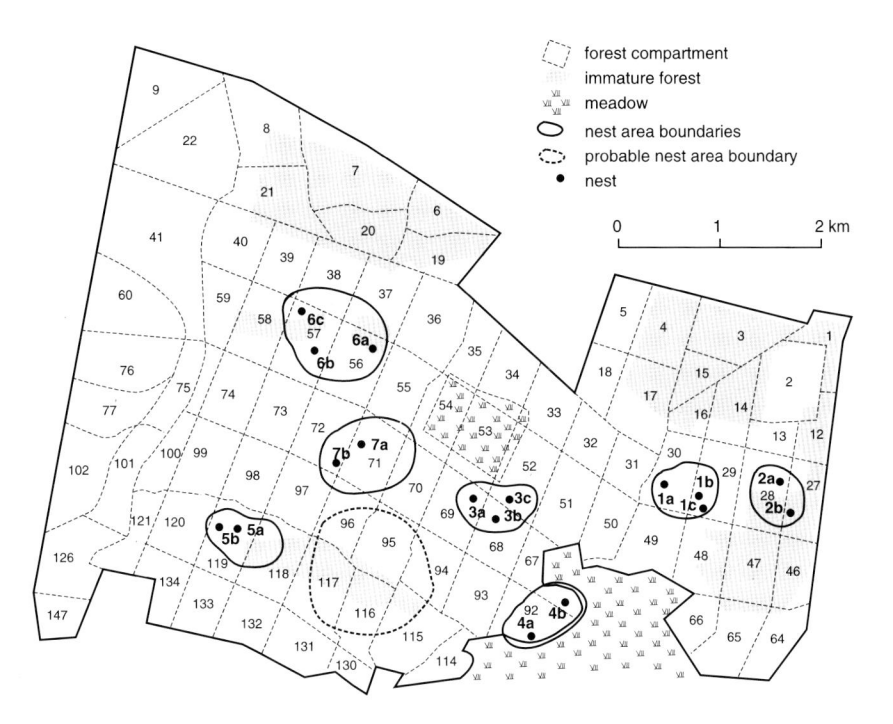

Figure 27. Alternative nests used by pairs of Goshawks in the Campinoski forest near Warsaw in Poland during 1956–65. Redrawn from Pielowski 1968.

3.9km, nests up to 3.4km apart were treated as alternatives (median spacing 285m) and only 33% of nests were re-occupied the following year (Reynolds & Joy 2006). This raises a question of whether hawks become less constrained in nest choice as nest density decreases. Alternatively, low density may be associated with increased turnover of adults and the new recruits tend to choose new nests.

It is not clear why Goshawks frequently alternate between different nests. Identifying hawks by feather patterns or marking shows that change is not simply a result of deaths or failed breeding. Fleas and mites are frequently found in old nests, and Brüll (1964) suggested that intermittent use of nests reduces the build-up of parasites. This does not explain why the same nest is sometimes used year after year, unless subtle differences in nest structure and exposure can sometimes cause especially low survival of parasites. Perhaps some nests vary in attractiveness from year to year because of differences in water-logging or snow-cover. There is scope for further analysis of data from marked birds to indicate the relative contributions of nest characteristics, breeding success, change of adults and other factors to the use of alternative nests within nest areas.

There is also variation in occupancy of individual nest areas. Variation from year to year is important for understanding how food and weather may influence Goshawk breeding and is considered further in Chapter 8. Long-term change in occupancy can indicate the importance of forest structure.

In the Olympic Mountains of Washington State, Sean Finn, Dan Varland and John Marzluff (2002) noted that continued occupancy of 30 historical Goshawk sites was predicted by relatively high canopy closure and crown depth (see also Figure 24). However, the strongest predictor of site success was shrub cover, which was close to 40% at unoccupied sites, compared with 19% at 12 occupied sites and 11% at the eight that fledged young. Shrub cover, which was also only 11% at Kaibab sites (La Sorte *et al.* 2004), has seldom been quantified comparatively, although Tim Kimmel and Richard Yahner (1994) found the extent of ground cover reaching one metre to be much lower at Goshawk nests (25%) than at control sites (47%). In Oregon, continued occupancy of 15 among 51 historic Goshawk sites was associated with abundance of older forest stages with canopy closure of more than 50% (Desimone 1997, DeStefano *et al.* 2006), which fits the general tendency for Goshawks to select areas of large trees with high canopy closure (Greenwald *et al.* 2005).

COURTSHIP BEHAVIOUR

The onset of spring courtship is in January in the south of the Goshawk distribution, but a month or two later in the north. On Gotland, the peak of first registration of radio-tagged hawks at nests where they bred was in March (Figure 28), before laying in April.

It is likely that paired Goshawks often visit their nest areas outside the breeding season, but they would not have been recorded in the weekly checks on Gotland

Nesting and laying 71

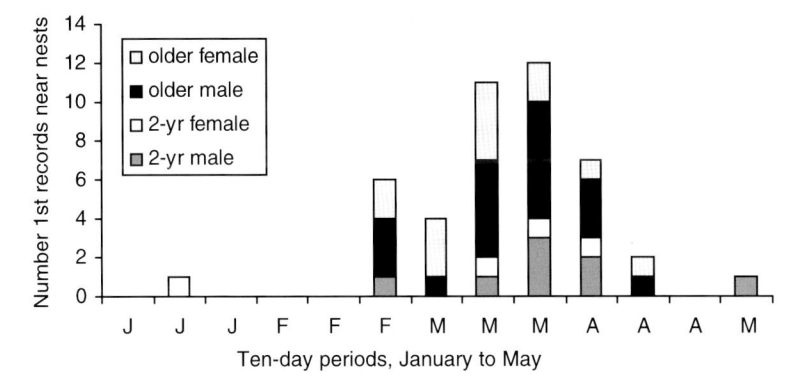

Figure 28. First records of 44 radio-tagged adults within 500m of nests where they bred. Observational bias may have delayed the pattern by one of the ten-day intervals.

until their presence at nests was quite persistent. Nevertheless, these data indicate that nest areas were strongly tenanted at least a month before egg-laying. There is some indication that second-year hawks, which were breeding for the first time (no first-year Goshawks bred on Gotland), may have settled slightly later than older hawks. One such male was definitely not at the nest before the end of April, and perhaps replaced another male there.

Courtship flights and calls, and even nest-building, also occur in the autumn. Pertti & Seppo Sulkava (1981) recorded that a 10–20cm layer of fresh twigs was added to two Finnish nests in September and October, and a complete new nest was built at one site. In the same area, there was a weakly significant tendency for the first records of building at four or five nests to be earlier in years with warm March temperatures (Sulkava *et al.* 1994).

FLIGHT BEHAVIOURS

Three types of flight display have been described for the Northern Goshawk. Demandt (1927, 1933) reported displays above the canopy on sunny, relatively windless days, with the long, main tail feathers held together and the snow-white under-tail coverts spread so wide that the hawks appeared to have a short, broad tail with a long dark strip extending from the centre. A slow, deliberate wing-beat was used in one type of display flight, with the wings held unusually stiff and straight, so that the hawks, with so much white at the tail-base, could easily be taken at a distance for male Hen Harriers. On one occasion this rowing 'harrier-like' flight was performed by a female for a couple of hours. She flew in all directions over part of the forest, settling occasionally but then soaring up again, calling frequently. This flight has a warning component, since Kollinger (1964) describes its use by a female approaching an intruding female: the intruder fled, and the resident female seemed to emphasise her slow-deliberate wing-beats as she returned.

Demandt also noted an undulating flight, like a Woodpigeon but with even sharper descents and ascents 'as if rebounding from an invisible elastic surface' (Joubert & Margerit 1986) with wings almost closed. This undulating flight, sometimes called 'sky-diving' may end with a stoop into the canopy, sometimes from more than 200m above. Both members of a pair may perform this display together, a faint rushing noise being heard from nearby as the heavier female dives (Fischer 1980), and a male has been described attempting it in an enclosure (Wiggins 1971).

Schnurre (1963) reported zigzag chasing flights between the trees. Similar chases have been described for hawks in avairies (Brehm 1969) but have not been widely observed in the wild and may therefore represent misplaced aggression rather than a normal display. Nevertheless, some observers believe that courtship flights within the woods, with or without chasing, are much more common than those which take place above the canopy and are therefore most easily seen by humans (Siewert 1933, Holstein 1942).

Link (1986) provided an elegant depiction of a display flight (Figure 29). He noted during his observations that undulations were performed 70 times by males but only 22 times by females, which also flew them less intensively, but that rowing flight was twice as frequent among females. He also describes a mildly undulating

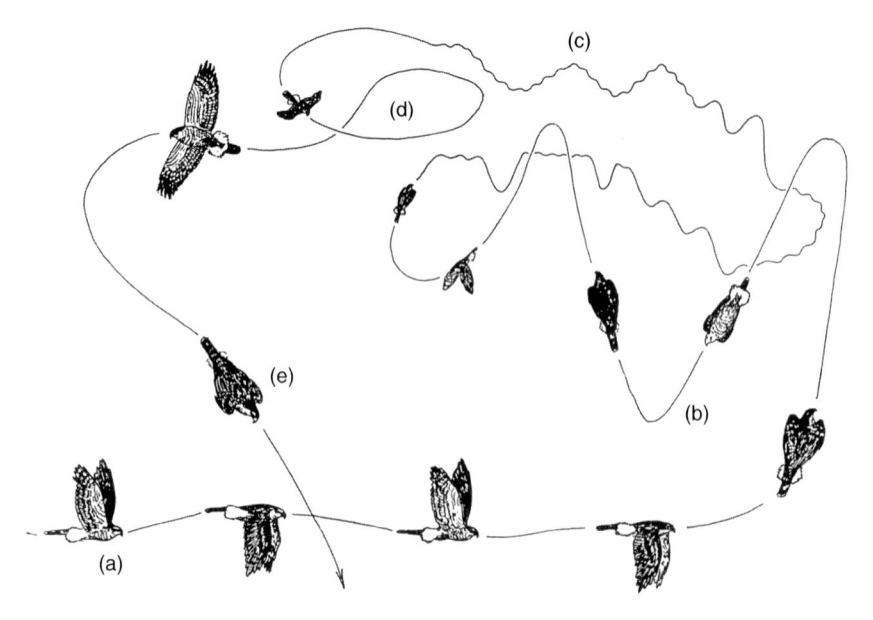

Figure 29. In display, harrier-like rowing flight with stiff wings (a) often introduces undulating 'sky-diving' flights with more folded wings (b) or in variations that may include intervals of straight flight and even stepped 'pot-hook' flights (c) and soaring in the nest area (d), ending with a dive (e) to the nest. Redrawn from Link 1986.

'wave-formed' rowing flight, and the rowing flight with neck held in a heron-like 'S' to elevate the head and maximally expose the pale breast as a territorial threat display. This behaviour was accompanied by wailing calls.

CALLING

To the human ear, and from sonograms as given by Jeserich in Glutz von Blotzheim *et al.* (1971), there are three main categories of call, all of which can be heard during courtship and breeding. The most frequent calls, which are used mainly by adults, are given in 'arousal' situations and are all based on a repeating short 'kak', sometimes described as a chattering 'kek' (Gromme 1935, Penteriani 2001). These include the alarm call, 'kak-kak-kak-kak-', in which the call has its highest pitch and most rapid rate (of 4–5 syllables per second), the courtship 'contact' call in which the repeat is a slower 'kak... kak... kak... kak...', and probably also the male call associated with food delivery, which is a croaking 'guck' (Schnell 1958) or 'chup'. This call may be associated with head-bowing when a male is perched and is sometimes repeated after several seconds or fast enough to be disyllabic.

The second type of call is given in situations of 'need'. Its earliest form is as a faint, high pitched, almost whistling 'whee' that is repeated by young when feeding is anticipated (Schnell 1958). Young close to fledging give the much louder and longer 'heee-yah' wail, which becomes especially prevalent in the post-fledging period (see Chapter 4). This call normally becomes rare as hawks disperse, but is occasionally given by juveniles to adults in winter and trained hawks easily become conditioned to give this call (*ad nauseam!*) when they expect food (Chapter 9). The adult female uses a very similar but deeper call during courtship, and a softer version when she is close to the male, in which context it has been considered a 'dismissal call' after provision of food (Schnell 1958).

Figure 30. A pair of Goshawks in display flight.

The third category of call is given when hawks are in situations of 'oppression'. Thus young in the nest may give a soft 'twit' call when approached by a dominant nest-mate, and a trained bird may repeat the call several times when startled. A hawk subject to physical aggression in the wild, or constraint through being grasped when trained, may give a trilling scream which is in effect a high intensity repeat of the 'twit'. These calls imply recognition of the oppressor, because they are not usually given by trapped hawks until the trap and release experience has been repeated at least once.

The daily and seasonal pattern of calling by adult Goshawks has been studied in detail by Vincenzo Penteriani (1999, 2001) at eight nests in France. In agreement with Holstein (1942), he found that calling of adults at nests started up to 45 minutes before sunrise. Periods of duet calling, which started either by wailing of the female or 'kakking' of either sex, were twice as frequent in the hour before sunrise and the three subsequent hours as at other times of day. Penteriani recorded all calls at those times in each month of the year (Figure 31). The peak season for calling was the three months before egg-laying in late March, with reduced calling during incubation, a further decline to the low levels of winter in the post-fledging period and no calling at nests between mid-November and mid-December.

This courtship calling makes the nest area of a shy forest species remarkably obvious, which suggests that calling serves an important signalling function. Perhaps the male shows by a duet with his mate that his efficiency allows time to spare from provisioning. Data from Penteriani *et al.* (2003) suggest that pairs that fail may call less during January and February than those that rear young, and that successful pairs may start calling earlier in the day during March. Among a less sedentary population of hawks in Utah, there was little calling until about six weeks before laying (Dewey *et al.* 2003). Daily duration of the intermittent calling was little more than an hour, centred on sunrise.

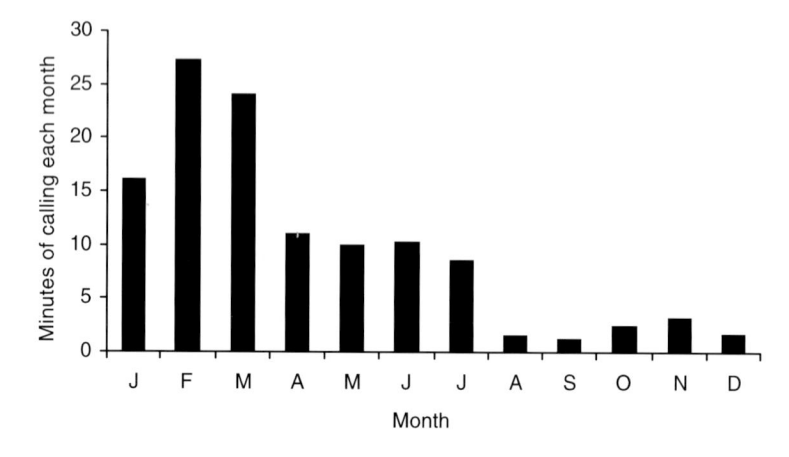

Figure 31. The total duration of Goshawk calls during the hour before sunrise and three following hours for eight nests in Burgundy, France. Data from Penteriani 1999.

TERRITORIALITY

Once a pair has formed, a female has reason to prevent other females approaching her nest. Even if they do not usurp her, they may obtain food which she needs from the male. Thus female Goshawks drive other females from the nest site, although they may court visiting males (Kollinger 1964, Ziesemer 1983, Rutz 2005a). These authors also report that males tend to drive away males, but court with birds of the opposite sex. Dominic Kollinger described a 'classical' alternating boundary dispute between males, in which an intruding male was chased back towards its nest and then turned to chase away the defender-become-aggressor.

As in other aspects of Goshawk behaviour, there is likely to be considerable variation in the aggressiveness of individual hawks and at different stages in the breeding season. Kollinger (1964) found that a female which twice drove away an (experimentally released) female before laying would tolerate the intruder while incubating. Goshawk territorial encounters seldom seem to result in physical contact. One hawk usually retreats when the other approaches in harrier-like warning flight, or flashes its white underside at the intruder. However, actual fights have been observed in the courtship period, with hawks falling grappling to the ground as they attempt to strike each other with their talons (Schnurre 1956, Bednarek 1975).

Interactions between hawks are also seen away from breeding sites and outside the breeding season. These may involve one hawk stooping at another hawk while soaring, or flying at another perched hawk which then retreats. Such interactions have now been observed frequently enough between radio-tagged hawks to show that they do not drive birds away from an area. About 30 interactions between hawks in winter never resulted in birds leaving their ranges, nor was there a tendency for females to displace males or adults to displace juveniles even temporarily from one spot (Kenward 1977, Ziesemer 1983).

NEST-BUILDING

Holstein (1942) believed that the male Goshawk was responsible for building new nests, with the female sometimes helping at old ones. However, this opinion probably gives too little credit to the female. Zirrer (1947) and Kollinger (1964) considered that the female was mainly responsible for the engineering of the nests they observed, while Ortlieb (1981) reported female Goshawks that brought much material themselves, even re-inserting twigs fetched by the male. Both the adults fetched nest material but the female alternated between bringing larch twigs, which were most flexible for weaving into the nest, and stiffer sticks. The hawks settled on thin branches, and pulled twigs off in their beaks, flapping their wings for balance. Sometimes small branches broke off under the hawks' weight. Twigs were pushed into the nest structure with a wriggling motion of the head, a behaviour that was already developed in nestlings (Brezinski *et al.* 1978, Schnell 1958). The removal of twigs from the crown of neighbouring trees sometimes left conspicuous broken ends near the nest.

Trained female hawks kept with or without males have started making nests by themselves (Mavrogordato 1937, Pritchard 1970), and one such hawk produced five nests in cooperation with a human handler (Muschiol 1964). Females are therefore probably at least as important for nest building as their mates, although there may be considerable individual variation, as in other aspects of Goshawk behaviour. In Sparrowhawks, both sexes construct the nest, which is typically built afresh each year and may take 100 hours of work (Newton 1986). Goshawks may re-use nests because there a fewer trees suitable for their large structures. However, Goshawks require relatively more energy to construct their large nests and build earlier in the year than Sparrowhawks; it would be interesting to see if new nests are built more often in years with abundant food.

Serious nest-building may start no more than ten days before laying, and a new nest can be built in as little as eight days. However, building typically commences more than a month before laying, sometimes being interrupted by spells of snow or heavy rain, and nests are occasionally built in the autumn (Sulkava & Sulkava 1981). Building may continue during laying and incubation, especially when the original nest has been abandoned due to disturbance just before laying (Fischer 1980). Brüll (1984) describes a female which finally decided on an old crow nest. When she laid, her head and tail could be clearly seen projecting over the nest rim, but the nest was built out to normal Goshawk size within a couple of weeks. In Finland, where there is often 30–70cm snow cover in March, building may start 1–2 weeks earlier at the least snow-covered nests than at others, and the first sticks may be laid on top of the snow (Sulkava 1964).

PROVISIONING

Although the disproportionate contribution of male hawks to provisioning has long been recognised (e.g. Brüll 1937), its importance during courtship has not been studied in detail. Perhaps pairs vary in whether the male frequently brings food to the nest area for the female before the start of calling or display flights or nest-building. Pairs in enclosures usually do some nest building, often in cooperation, before food transfers start and one captive pair even fed simultaneously on the same carcass (Pritchard 1971).

Food is sometimes transferred between hawks in flight, but usually at perch sites or on the ground, and can occur several times a day if prey is small (Brüll 1937, Fischer 1980). The male utters the contact call when he reaches the nest area, sometimes as no more than a soft 'guck' or 'chup', and may be answered by the female. He has usually eaten the head of the kill before bringing it, perhaps primarily because that ends all prey movement but maybe also because brain is rich in phospholipids for sperm production.

Courtship provisioning is a dangerous process for the male, who must allow the female to approach him with food on her mind. A male often seems wary of his mate when he brings prey, and may fly off with a distress trill when she approaches. Male hawks are sometimes killed during courtship in enclosures and in the wild (Kollinger 1964 and Chapter 9).

On Gotland, an adult female was trapped on the remains of a radio-tagged male that she had killed some two months before egg-laying. His body was found by Mats Karlbom late on 11 February 1983, after apparently being killed that morning. The extensively plucked and well-eaten body lay near a bloodstain that penetrated five cm into the snow between marks of a single wing-beat and some compressions which suggested a landing point after the male had been seized. There was no evidence of other players in this drama. The male had been ringed in 1981, and was in excellent condition (975g) when tagged on 22 October 1982, as was the trapped female (1,575 g). His crop was full of rabbit, but there was no sign of a rabbit carcass. As he could not have carried the two-kilogram carcass of a freshly killed adult rabbit (there would have been no young rabbits on Gotland in February), perhaps his mistake was to approach a hungry female when in an obviously sated state.

The aggression shown by this female recalls that of Miss Piggy until she had received much spring food from me. The wild female had not bred previously with this male, so he was unproven. The female that nearly died at the start of incubation after losing her mate on the Swedish mainland (Chapter 2) shows the risk of having to become dependent on a male. At a time well before laying, a female might increase her likelihood of breeding by promoting replacement of a seemingly incompetent adult male in her home-range. Closer to laying, and therefore with other competent males perhaps already committed to sites and mates, this would be a more risky tactic. Better prospects for a female with an inadequate mate might then be to continue taking what he can offer but also to look elsewhere, with the fall-back option for a long-lived species of not breeding that year.

The asymmetric provisioning puts much pressure on the male. If he can feed the female with time to spare for nest-building and other courtship activities, he is likely to be a genetically sound partner. In raptors with sexual asymmetry in provisioning, the quality of display flights may give initial evidence of male fitness, but his provisioning is probably a more useful cue. In Northern Goshawks, the 'proof of the pudding is in the eating', and initial female aggression can be seen as a mechanism for not breeding with a mate until he has provided sufficient food.

COPULATION

Goshawks sometimes mate on the ground or on the nest, but the female seems usually to use a perch near the nest when inviting the male. She holds her head low, and arches her body, with wings slightly dropped and white undertail coverts well fluffed out (Brüll 1964), sometimes also uttering a soft two-tone 'kuk-kuk' call. He runs or flies to her and lands on her back, with his legs bent and feet clenched, flapping his wings for balance (Holstein 1942, Brüll 1964). His tail twists sideways and down while hers twists sideways and up for their two cloacas to meet. The male and female usually call during the ten or so seconds of mating (Wortelaers 1951). From his work on urban Goshawks in Hamburg, Christian Rutz (pers. comm.) describes pairs wailing in concert, with an obvious climax towards the end, 'yeeeep-yeeeep-yeep-yeep-yep-yep'.

Vagn Holstein (1942) estimated that Goshawks copulate at least ten times a day from 1–2 months before laying, some 500–600 times a season. Anders Møller re-assessed the original notebooks and proposed that sperm competition explained this copulation frequency (Møller 1987, Birkhead & Møller 1992). It is high even by the standard of colonial species, such as seabirds, which have a high risk of extra-pair copulation while males are absent. Frequent copulation ensures that most sperm in the female's cloaca are from her mate, especially during the fertile period shortly before egg-laying. As Goshawk males left females for periods that averaged 140 minutes in Holstein's notes, and quite often visited adjacent nest areas around egg-laying on Gotland (Figure 32), frequent copulation is an important sperm competition strategy.

In Hamburg, Rutz (2005a) observed nine intrusions by males at two pairs, which in four cases did not lead to eviction. One intrusion resulted in undisturbed copulation despite being in view of the resident male, which flew to perch alongside but did not interfere. However, this occurred some 30 days before egg-laying, when the female would not have been fertile but copulation frequency is building to a peak (Figure 33). Females too make excursions to other nests at that time (Figure 32), as also noted for Sparrowhawks radio-tagged in Edinburgh by Mike McGrady (1991).

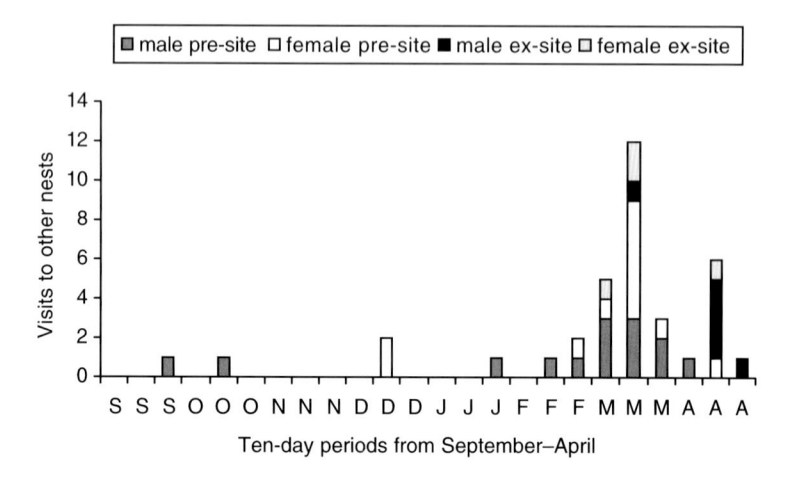

Figure 32. Records of 22 radio-tagged adults on Gotland within 500m of nest sites where they did *not* breed. Both sexes made 'pre-site' visits to other sites prior to first records at their breeding site, but subsequent 'ex-site' visits to adjacent sites were mostly by males.

Although the peak period for copulation started 40 days before egg-laying in Vagn Holstein's notes (Møller 1987), a peak from 30 days to five days before laying (as observed by Rutz) may be more typical and many hawks on Gotland were not consistently present that long. Therefore 100–300 copulations per season may be more typical for most Goshawks. A decline in copulation frequency occurs just before laying (Figure 33), with eggs fertilised only by insemination several days before laying (Schulz 1981). Males were also more attendant in the days before laying than during incubation (Rutz 2005a). Occasional copulations occur later in the breeding season (Holstein 1942 and Figure 33).

The Northern Goshawk appears to be at an evolutionary extreme of mate-testing behaviour. Females are aggressive until they have been well provisioned. They call in duet with males in the dawn twilight that is good for hunting (Chapter 6). Promiscuity also puts pressure on the male, but is a little puzzling in terms of mate choice unless females also operate a 'copulation for food' exchange. However, it seems that frequent mating, combined with enhanced male presence just before laying (which was when Miss Piggy needed less food), works well against extra-pair fertilisation. DNA-testing of adults and young at 39 Goshawk nests in Arizona showed that only one in 77 young had not been sired by the resident male (Gavin *et al.* 1998). Perhaps the frequency would increase in areas with less than the 3.9km nearest-neighbour spacing that Reynolds & Joy (2006) reported for the Kaibab Plateau. Mate-choice is discussed further in Chapter 5.

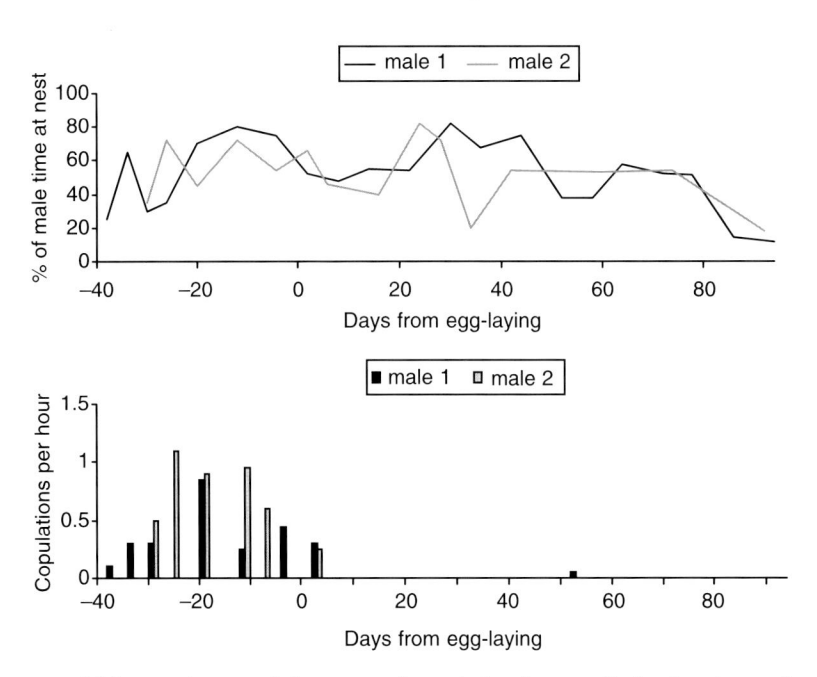

Figure 33. Male attendance and frequency of copulation for two Goshawk pairs nesting in Hamburg. Data from Rutz 2005a.

LAYING

Goshawks usually lay clutches of three to four eggs, less commonly two or five eggs and very rarely as few as one or as many as six. Measurements of 163 eggs from Belgium and Germany varied between 52 and 65mm long by 41 to 48mm wide, with an average of 57 × 44 mm (Glutz von Blotzheim *et al.* 1971). Fresh weights were generally 50–60g at laying, which represents 3–5% of female bodyweight, with extremes of 35g and 75g. Bijlsma (2003b) recorded an egg from an elderly hawk in the Netherlands that measured only 32 × 23mm. Such eggs are very rare. This one was by far the smallest among 1,011, from 326 nests.

Fresh Goshawk eggs have a greenish or bluish tinge and very occasionally a slight flecking of lime-yellow to grey-violet. Erythristic mottling with red has also been recorded (Ivanovsky 1998). Fischer (1980) suggested that their fading to white makes Goshawk eggs less attractive to collectors than are the beautifully mottled eggs of falcons.

Santi Mañosa (1991) found that measurements from 67 clutches of eggs at La Segarra in Spain related significantly to the age of the female, the success of the egg and the date and order of egg-laying. The volume of 17 eggs laid by females in their first spring was on average 13% less than 183 eggs from older hawks, and 11 failed eggs had 12% less volume than those that hatched. Eggs laid at the very beginning and end of the season were relatively small, and the last eggs laid in 45 clutches were about 4% smaller than the previous egg. In clutches of four or five, the first eggs too tended to be small.

The hawk may be on the nest for as little as ten minutes while laying the first egg in clutches, with subsequent eggs laid at two- to three-day intervals. There are occasional gaps of four days, or even five days, perhaps due to food shortage and cold weather. Two five-day intervals were recorded when temperatures dropped to between –5°C and –12°C during laying in Finland (Huhtala & Sulkava 1981). The sequence of laying combines with incubation behaviour to be important for chick development and survival (Mañosa 1991, Bijlsma 1993, 1996), in ways that are considered more fully in Chapters 4 and 8.

LAYING DATE

The date of egg-laying varies with the age of the female, latitude and temperature. In Schleswig-Holstein, the 10% of breeding females that were in their first spring laid on average eight days after the females in adult plumage (Looft & Biesterfeld 1981). In Drenthe (The Netherlands), 46 pairs with first-year females laid eggs ten days later than 139 pairs with older females (Bijlsma 1993). Jan Drachman & Jan Tøttrup Nielsen (2002) found that first-year females laid only six days late in Denmark, but reported a tendency for retarded laying by females more than nine years old. Several recent studies that have recorded senescent effects (e.g. Risch *et al.* 2004, Kruger 2005) are discussed in Chapter 8.

Data from 24 European studies show a strong trend for laying to be delayed in the north (Figure 34). In rural areas at low altitude, laying is delayed by about four days for every 3° northward. In North America, laying in late April and early May seems quite common in the north and in southerly populations at high altitude (McGowan 1975, Reynolds *et al.* 1994, Younk & Bechard 1994, Doyle 2000). The laying date also varies with altitude in Europe (Dobler 1991). Hauri (1963) found that Goshawks high in the Swiss Alps started to lay in early May, whereas those in the lowlands began at the start of April, a month later. Laying in mid to late May also occurs in Siberia (Johansen 1957, Fischer 1980).

Relationships between egg-laying, weather and food have been investigated in detail in two areas of lowland conifer forest in Ostrobothnia (Finland) by Seppo Sulkava and Kaukko Huhtala (Sulkava 1964, Huhtala & Sulkava 1976, 1981) and Risto Tornberg (Sulkava *et al.* 1994). Laying dates were recorded during 21 years in an area of central Ostrobothnia. The average start of egg-laying each year varied between 11–27 April. There were weak tendencies for laying to be delayed when temperatures had been low in January and February (as well as for nest-building to be delayed by low temperatures in February and March). In Drenthe, Rob Bijlsma (1993) found that laying dates that varied between 31 March and 8 April during eight years were quite strongly delayed by cold March weather. Similar strong effects on laying dates of cold weather prior to laying have been noted by Looft & Biesterfeld (1981) in Schleswig-Holstein and by Drachman & Nielsen (2002) in Denmark.

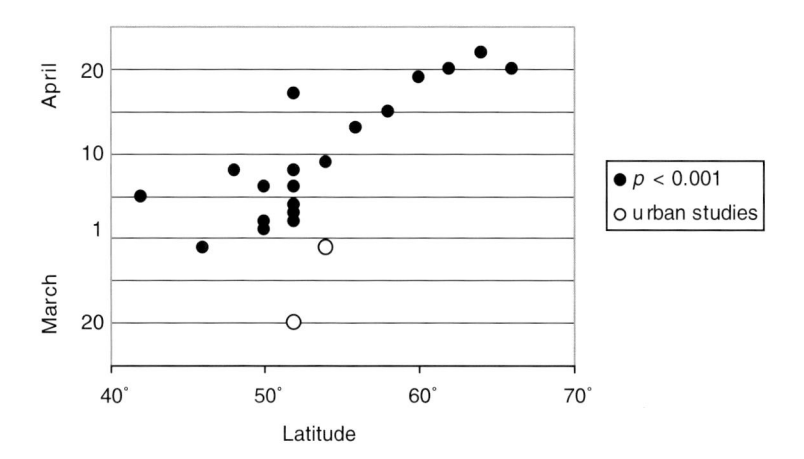

Figure 34. The date of laying of the first egg in 24 European studies. Two studies where hawks laid relatively early were in urban areas. Sources in Appendix 2.

CLUTCH SIZE

In Ostrobothnia, there was also a tendency to lay smaller clutches when temperatures had been cold in January and February (Sulkava *et al.* 1994). However, a much stronger effect was for Goshawks in southern Ostrobothnia to lay large clutches in years when large numbers of woodland grouse were counted the following August in this area (Figure 35). The grouse were counted well after Goshawks had laid eggs, and included adult grouse available in spring as well as their young. The laying of large clutches by hawks may therefore have stemmed partly from spring conditions that favoured production both of Goshawk eggs and young grouse, as well as from the availability of adult grouse as food during courtship and laying.

The factor that linked most strongly with clutch size in Ostrobothnia was the laying date. Across clutch sizes that varied from 2.7 to 4.2 between years in central Ostrobothnia, a 12-day delay in laying was associated with the laying of one egg less (Figure 36).

The remarkable strength of the relationship between clutch size and laying date raises the question of whether food supply acts directly to produce large clutches, or through laying date. This question has been examined experimentally in European Kestrels (Dijkstra *et al.* 1982, Meijer 1988), but is difficult to address in the Finnish material because laying date and food supply were estimated in different areas. However, there is quite strong synchrony between nearby areas of Finland in prey numbers and their effect on Goshawks (Ranta *et al.* 2003). If prey data for southern Ostrobothnia are used together with laying date in central Ostrobothnia to predict clutch size, there is an even better fit than for laying date alone (explanation for variance increases significantly, from 85% to 89%, without a significant interaction between the two factors).

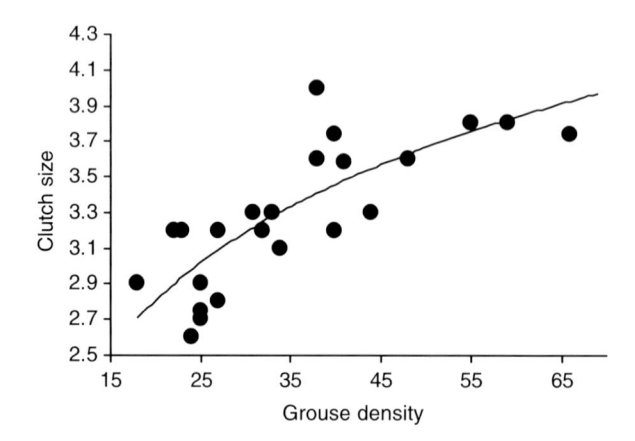

Figure 35. The clutch size of Goshawks in southern Ostrobothnia increased in years when many grouse were counted in August; $p < 0.001$. Data from Sulkava *et al.* 1994.

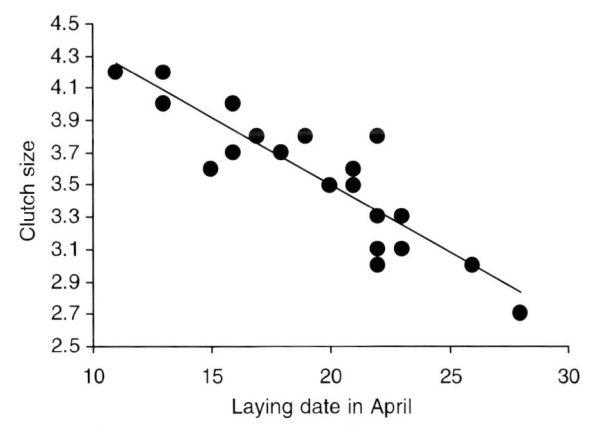

Figure 36. The average size of Goshawk clutches in central Ostrobothnia declined when laying was delayed. Trend line $p < 0.001$. Data from Sulkava *et al.* 1994.

Thus, Goshawk clutch size related independently to laying date and food supply in Finland. There is also evidence for direct effects of food on laying date and clutch size in other studies. A review by Ian Newton (1998) found that the laying date advanced in 20 of 21 experiments with extra food given to breeding birds, while clutch size also increased in nine of them. As clutch size is also affected by weather that delays laying, and clutch size is also a predictor of brood size, a number of studies have shown that spring weather can affect Goshawk productivity (see Chapter 8).

Young females tended not only to lay late, but also to lay small clutches. As clutch size in Ostrobothnia was reduced by one egg for each 12-day delay in laying, the late-laying of 8–10 days for first-year females in Schleswig-Holstein and Drenthe could be equivalent to 0.7–0.8 fewer eggs. In fact, 28 clutches laid by first-spring hawks in Schleswig-Holstein averaged 3.3 eggs, just 0.4 eggs less than the 3.7 eggs in 398 clutches from older hawks. In Drenthe the 52 clutches from young females averaged 3.1 eggs compared with 3.5 in 143 clutches from older hawks. Again, the age-linked reduction of 0.4 eggs was about half that predicted by data from Finland. However, it took a delay of one month to reduce clutch size by one egg in Drenthe (Bijlsma 1993), so relationships between date and size of clutches may reduce with latitude.

Since clutch size can be so greatly influenced by food and weather, and shows such great annual variation, it is perhaps unsurprising that there is no latitudinal trend in the number of eggs (Rutz *et al.* 2006a). In 30 European studies, the averages varied between 2.8 and 4.0 in 29 cases, very similar to the annual variation in Ostrobothnia (Figures 35 and 36). An exceptional low value of 2.1 (Schönbrodt & Tauchnitz 1991) may have reflected an unusual recording method, the use of mirrors on poles. The grand mean for these studies is 3.3, similar to the 3.2 for five studies in North America that ranged from 2.8 to 3.8 eggs (Tufts 1961, McGowan 1975, Reynolds & Wight 1978, Root & Root 1978, Lee 1981a).

It is interesting to note that the average size of clutches fathered by six males from Drenthe in their first year was fractionally larger than for adult males (Bijlsma 1993). Perhaps a young male must be an exceptional hunter for a female to accept breeding with him.

REPEAT CLUTCHES

Raptor breeders recognise that if a clutch of eggs is removed within a certain time after completion, the birds will go through the later stages of courtship and lay again. If the eggs are taken beyond the critical time, the birds do not usually re-lay, presumably because an irreversible change has occurred in the controlling hormone system. The 'point of no return' seems to be 10–20 days after clutch completion in large falcons, for which removal of the first clutch is used to raise productivity for captive and wild stocks (Cade & Temple 1976, Monneret 1978), the birds re-laying after about two weeks. Goshawks may not re-lay as readily as large falcons, since a captive pair whose eggs were taken 10–12 days after completion only completed a second clutch in one of three years (Haddon 1981). However, another female re-laid regularly for six years when her eggs were taken immediately after completion. The captive hawks re-laid an egg after 19 days, and minimal intervals of 15 and 17 days have been observed after loss of clutches in the wild, although 25 days until the first egg of a second clutch has been recorded (Zirrer 1947, Fischer 1980, Looft & Biesterfeld 1981). The normal re-laying interval is probably 15–20 days.

IMPLICATIONS FOR CONSERVATION AND MANAGEMENT

This chapter considers an important aspect of Goshawk conservation, namely the habitat they use for breeding. There has been considerable concern in North America about the extent to which Northern Goshawks require old-growth forest. The concern about forest habitat reflects a desire to conserve areas of pristine forest and mature semi-natural woodland from timber harvest and has involved several law suits (Kennedy 2003, Squires & Kennedy 2006). In areas like the Olympic peninsula in Washington, trees in damp mild conditions on good soils can grow to awesome heights forgotten in Europe. The trees become giants, with a distant puff of canopy 40–50m above the ground. The United States Endangered Species Act was seen as a way to preserve extensive mature forest, as habitat needed for Goshawk nests and across their extensive hunting areas.

Preserving such areas is a human heritage issue, and may be vital for species other than Goshawks. However, a comparison of Goshawk habitats in Europe and North America raises questions about the needs of Goshawks for extensive forest. In well-wooded areas, the types of nest tree and nest stand used by Goshawks do not differ substantially between Europe and North America. Nevertheless, nesting habitat that Goshawks *use* in well-wooded areas is not necessarily what they *need*.

In Europe, Goshawks are at home in small woods fragmented by farmland. It is usual to record the proportion of woodland in European study areas and 25 of 81 studies had less than 20%. Most American studies describe extensive woodland in their areas without estimating a percentage, but among nine that do (Appendix 2) only two record less than 50% (Figure 37). Moreover, Goshawks have established breeding populations in at least seven European cities, including 5–6 pairs each in Amsterdam and Groningen (Netherlands), 14 or more pairs in each of Berlin, Cologne, Hamburg and Saarbrücken (Germany), and some 35 pairs in and around Moscow (Rutz *et al.* 2006a). In Hamburg, Goshawks bred successfully in a solitary tree in a residential area, with another nest just ten metres from a five-storey building (Rutz 2001).

Goshawks can breed in relatively small trees and fragmented woods in North America, so they probably do not require the structure of extensive woodland for breeding there any more than in Europe. However, the stringers of Utah and Wyoming are naturally fragmented woods, not the anthropogenic mix of woodlots and farmland in central Europe. These habitats, and cities, are used by Cooper's Hawks in North America (Craighead & Craighead 1956, Rosenfield *et al.* 1995, DeCandido 2005) but not by Goshawks. Goshawk population status in changing habitats will be addressed more fully in Chapter 10, after considering the ways in which they may be affected by humans and other species, including their predators, competitors and prey.

An important aspect of this chapter for management is the cues that can be used for finding Goshawk nests, which is necessary both to study breeding biology and

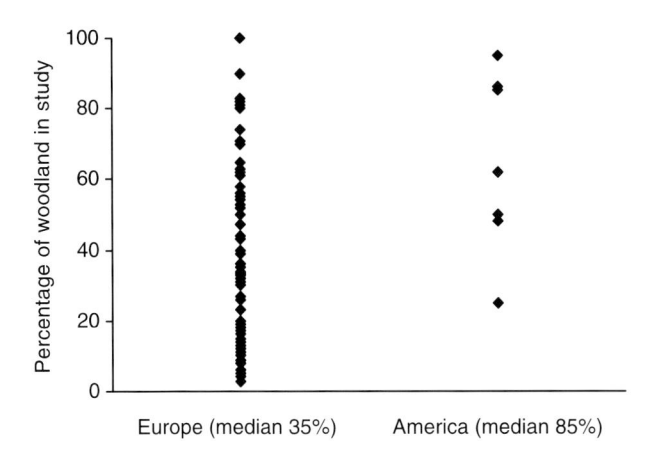

Figure 37. The percentage of woodland in areas used to study goshawk breeding is less in Europe ($p = 0.005$) than in North America. Sources, from Rutz *et al.* 2006a, are listed in Appendix 2.

to mark young for year-round studies or for management purposes. At the start of a study, valuable tips about past nests are often available from hunters, foresters or previous researchers. Nest sites are also indicated by opportunistic observations of display flights or the calling of adults and young, or revealed by radio-tracking (see Chapter 5). However, the two main approaches for systematic recording are to search on foot or, more recently, to detect hawks that respond to grid-based broadcasts of adult alarm calls and juvenile wailing (Hennessy 1978, Fuller & Mosher 1981, Kimmel & Yahner 1990).

For broadcast surveys, Pat Kennedy and Dale Stahlecker (1993) recommended stations at 260m spacing with 130m offset on adjacent transects spaced 300m apart, using alarm calls during the nestling period and wails during the post-fledging period. Broadcasting for ten seconds in three directions at 120°, and then listening for 30 seconds for a response, then repeating each direction separately with listening, took three minutes at each station and gave a 75% detection rate for single visits in Arizona and New Mexico. This was similar to an 80% detection rate during the nestling period in Pennsylvania (Kimmel & Yahner 1994). Detection reduced to 40% for single surveys in denser forest in the state of Washington (Watson *et al.* 1999), albeit rising to 80% with 3–4 visits. McClaren *et al.* (2003) found food delivery calls of males to be no more effective than alarm calls during the nestling period and juvenile wailing during the fledgling period on Vancouver Island, which gave up to 75% single-visit detection, but recommended a change of transect spacing to 200m before fledging and 400m later, listening for nine minutes/station. These tests in areas with known nests indicated that the technique can be 50–75% effective for detecting females at failed nests during (but not later than) the nestling period.

A technique that can be used to a set protocol by relatively inexperienced operators is helpful for cost-effective standard surveys across wide areas (Joy *et al.* 1994). The provision of taped calls shows inexperienced observers what sounds to expect in response. Care may be needed to avoid confusion in areas with avian mimics, such as jays (Kennedy & Stahlecker 1993), although 'kakking' calls by jays in unsurveyed areas can be another sign that hawks are present (C. Rutz, pers. comm.).

For anything more than an index of abundance, broadcast surveys must be followed up by finding nests on foot. During ongoing studies and to monitor possible population declines, there is also a need to record whether each site is active in years following discovery. The morning peak in courtship calling provides a convenient indication of occupancy before laying, and has been used for decades for this purpose in northern Europe, where calls can be heard up to a kilometre away in relatively open areas on cold, still mornings (Seppo Sulkava, pers. comm.). Penteriani (1999) reported detecting calls at 100% of 27 occupied nests with one listening visit for 35 minutes from 30 minutes before sunrise. Dewey *et al.* (2003) obtained 90% detection rates at 20 sites for two years in New Mexico, and recommended listening twice (with a two-week interval) at a location within 200m of alternative nests for 135 minutes from 45 minutes before sunrise during the 45 days before laying.

With experienced observers, the effectiveness of the traditional search methods should not be underestimated. Kennedy & Stahlecker (1993) noted that merely listening at sites for three minutes gave 46% detection of wailing juveniles. On Gotland, on approaching a stand where nesting was possible, we listened and also made calls without electronic aids. A guttural imitation of the adult 'kak' calls sometimes produced a response. Blowing through a blade of tough grass held between knuckles and base of the two thumbs to imitate juveniles also produced responses.

Surveys always started with known sites, looking first in the nest used the previous year and then at known alternatives. The fresh foliage of leafy twigs used to line nests can often be seen from the ground when checking for site occupancy. However, the presence of down from the moulting female is a better indicator of incubation, because alternative nests may be lined with fresh material, and greenery from autumn courtship may be preserved fresh in hard winters (Sulkava & Sulkava 1981).

If this did not reveal a nest in the courtship or incubation period, our approach was to sweep rapidly through the area looking for other signs, minimising disturbance by taking no more than 15 minutes. We did not look for new nests, but primarily for (i) plucking points (ii) feathers or down shed by the moulting female or growing young and (iii) the white excreta or 'mutes' which may lie as streaks 20–50cm long squirted a metre or two from plucking points, or spattered over a wide area from favourite perches near the nest. These signs are not a completely reliable guide, however, because mutes get washed away and there are sometimes no prey remains to be seen away from the nest until the post-fledging period, when the young hawks are plucking food throughout the stand.

Similarly, searches of other suitable areas were not initially directed at finding nests, which are not easy to spot in conifers and mixed woodland, but at detecting hawk sign or provoking calls. In pure deciduous woodland it is relatively easier to spot nests before leaf-out. If old nests were not occupied, or new ones spotted by chance at visits early in the season, stands known from sign to be occupied were searched more thoroughly after young would have hatched. During the nestling period food is often taken directly to the nest, and some hawks apparently pluck mainly away from the nest area, perhaps to reduce drag while carrying the carcass or to avoid drawing the attention of predators to the nest stand. There may therefore be less sign on plucking posts than during the courtship and post-fledging periods, but mutes and prey remains accumulate under the nest.

Whether the initial detection of nest sites is by systematic broadcast surveys or by routine searches on foot, it is first necessary to decide which trees or stands are too small to be worth searching. This decision is often planned with forestry maps or more recently with Geographic Information Systems (Kimmel & Yahner 1990, Johansson *et al.* 1994). Some nests may be missed in both types of survey. Hawks may not respond to broadcasts, and in routine searches there is a very human tendency not to search as thoroughly (if at all) below a minimal spacing threshold between occupied sites. At least three studies have addressed the question of whether broadcast surveys may be less likely than routine searching to detect nests in some habitats, but without finding significant differences (Daw *et al.* 1998, Rosenfield *et al.* 1998, Boyce *et al.* 2005). However, the latter study estimated that four broadcasting visits were necessary to establish lack of occupancy with 95% reliability, compared with three visits for traditional search methods.

Christian Rutz (pers. comm.) notes that in urban habitats, every stand of more than ten possible nest trees has to be checked. It is insufficient to focus exclusively on obvious places, such as parks, cemeteries and hospital grounds. Aerial photographs can identify backyards with nice stands. All that seems marginally acceptable for Sparrowhawks must also be checked for Goshawks, as must the horribly inaccessible little woodland patches between the rails of transport systems or behind fences of water supply stations.

CONCLUSIONS

1. Goshawks build large stick nests next to the main stem or in forks of conifers and broadleaved trees. Nest cups are usually lined with fresh leafy twigs which provide support, insulation, concealment of the pale eggs and perhaps sanitation after hatching.
2. Nests in European studies had the same height range (usually at 9–21 m) as in North America, but trees were smaller in Europe and nests closer to their tops. There was a preference for nesting in larch and other deciduous trees in some areas. Less than 10% of Goshawks nests are built by other species, mainly buteos and corvids.
3. Nests are typically in tall trees in stands of large trees, with ample sub-canopy flight space and low shrub cover. Nests are seldom close to woodland boundaries, but often with small openings nearby. Nest sites on slopes favour the lower parts, with cooler northerly and easterly aspects favoured in the south.
4. Goshawks nest in extensive and naturally fragmented woodland. In Europe they also use woods in farmland, and in parks and even solitary trees in at least seven cities. Sites tend to have at least two alternative nests, on average 200–300 m apart. Change of use is every other year on average, but continuous use of single nests for up to 17 years has been recorded.
5. Courtship typically starts 4–6 weeks before egg-laying, with first-time breeders slightly later at nest sites than older hawks. Courtship calling and nest-building sometimes occur in autumn. In courtship displays, males give undulant, sky-diving flights more than females, which favour harrier-like display flight with stiff wing beats.
6. Call types are 'kakking' in alarm and courtship arousal, wailing in feeding situations and twittering when oppressed. Courtship calling at nests occurs mainly at dawn and reduces after 2–3 hours, peaking in the 2–3 months before egg-laying.

7. The male feeds the female during courtship. She may be aggressive to potential mates and may visit other nests before settling at a breeding site. Copulation can occur ten times daily and 100–500 times pre-laying, which decreases risk of fertilisation by visiting males.
8. Clutches contain 1–6 eggs, mostly 3–4. Eggs laid last in clutches and by first-year females tend to be small. Eggs are laid between late March and early May in Europe, later by about four days for every 3° northward but also delayed at high altitude. Laying, which can be even later in Siberia and North America, is delayed by cold weather and for first-year females.
9. Clutch size is increased by early laying, prey abundance and age of females, but not by change in latitude. New eggs may be laid 15–25 days later if a clutch is lost at the start of incubation.
10. Nests are found for research and management by systematic searching for signs of site occupancy (kills, excreta, feathers), which can be preceded by detection of calls and response to broadcast calls on survey transects.

CHAPTER 4

Incubation and rearing

The pine woodland of eastern Gotland is pleasantly cool on a hot June afternoon in 1987. With the Land Rover parked on a nearby track, I'm checking if there are young at site 4. The three alternative nests have been used for all seven years of the study, and an earlier view from the ground had noted greenery on last year's nest. It is one of the highest nests, some 15m above ground in the fork of a large pine that lacks branches for 12m. This section of forest has soil a bit deeper than areas where the limestone breaks through nearby. Large columnar trunks are a little cathedral-like, with a scattering of well-grown understorey spruce that create visual walls, while sunny patches on the bracken and bilberry leaves are like light from 'windows' high above.

As I approach the nest tree, the female flies past, 'kakking'. There is one wisp of down to be seen on a nest twig, but no 'whitewash' on the ground. Perhaps she failed this year, but I must check. It takes several tries to flick the safety strap round the trunk so that I can climb a tree that is otherwise impossible to hold to, and too big for the chains of our two-metre aluminium ladder sections. I hate the flakes of bark that come loose to stop my climbing irons gripping, and rest for several minutes to catch breath on reaching the lowest branch. The nest contains only the tail of a recently eaten squirrel.

These are days before health and safety requirements wisely oblige two-person climbing teams, so it takes 40 minutes of lone searching to find the new nest. It is some 300m away, behind spruce trees that make it more exposed towards the track than the old nest area. Copious 'whitewash' is immediately noticeable on the bracken.

Fortunately, this is an easy climb, in a smaller pine with branches that are dead but firm to grasp from the ground. There are three young, with feathers starting to hide the down on their backs and heads, so they are nearly four weeks old. At this age they spend time standing and can move about the nest, but are easy for one person at the nest to hold down gently while measuring wing-lengths. The lengths enable estimation of hatch dates, and hence when the young will be 30–35 days old. That is our target age for returning to take measurements that will define their sex, when there will be little further increase in tarsus width, and to fit leg-mount radio tags.

INCUBATION

Incubation is done mainly by the female, who develops a brood patch of highly-vascular skin which may be as much as 15cm long and 5cm wide (Holstein 1942). The breast feathers are not shed, but move aside to enfold the clutch as the hawk settles with a rocking motion on the eggs. Rising to turn the eggs regularly has been recorded at 4–5 hour intervals during daylight hours (Nick Robinson, pers. comm.).

The female may sleep while incubating, with her head resting forward on the nest rim, but spends much time looking alertly around. If humans approach she seems to press down onto the nest, drawing in her feathers in the fashion of a frightened hawk, and may not leave the nest until the tree base is struck. Other females leave unobtrusively before human visitors get close, but some will continue incubating while a person climbs into a hide in a nearby tree (Schnell 1958). Females are more reluctant to leave the nest late in incubation, perhaps partly because of greater commitment to their investment in breeding and partly as a result of habituation to previous disturbance.

The male has little or no brood patch, and takes over only when the female leaves to eat, as described by Siewert (1933:56): 'About midday the female leaves quite soundlessly, although her 'ka-ka-ka' can then be heard from a little way off. At once there comes the brief 'chuck-chuck' of the male, from a slightly different direction; the female answers, and then they both call from the same spot. Ten minutes after the female left the nest the male suddenly appears, calls twice 'chuck-chuck', and settles quickly onto the clutch. But he stands up again at once, pushes the eggs carefully beneath him with his beak, and rocks himself down onto them again. He turns the eggs again after two minutes and then sleeps, sinking his head onto the twigs so that only his glistering blue-grey back can be seen'.

Horst Siewert then describes the male waking to watch nearby small birds, but: 'Suddenly the female is there, quite unannounced. The male rises once, calls a quiet 'chuck-chuck', and glides away in an elegant sweep. The female lets herself onto the eggs, shuffling onto the nest cup on her heels, with her feet clenched. And so incubation continues'.

Transfers also occur through the male calling and leaving prey at a point in view of the female and then immediately departing or flying to the nest. The female may be reluctant to leave the nest when the male brings food, especially at the end of incubation or when young are small and the weather is bad. Some males then bring the prey and lay it on the nest rim, as they often do when the young are old enough to feed themselves. At the other extreme is the female which was seen to leave the nest and fly 'kakking' at the male which had perched nearby without bringing food. He had a well-filled crop, and was settling to digest this with one foot drawn up into his breast feathers, when the female importuned him to leave. A food transfer took place 15 minutes afterwards, perhaps of the male's original prey which he had fetched from elsewhere (Fischer 1980).

Males seem normally to incubate for less than an hour, while the female feeds or occasionally bathes. Nick Robinson (pers. comm.) reported a female incubating for 96% of 575 hours of daylight observation, with the male sitting for just 39 minutes. When using a characteristic 'horizontal posture' signal from radio tags (Chapter 5) to find nests of two males on Gotland, I had to run to get close within their 5–10 minutes of incubation. A breeding female that was also radio-tagged had a 'feeding' signal (see Chapter 5) that was appreciably more hurried during this time than in winter.

In contrast, some males seem particularly keen on incubation, as shown in a fine series of photographs taken by Klaus Stülcken in 1952 (Brüll 1984: 249), in which a female at first returns with a green twig and finally pushes her mate off the eggs. Stülcken (1943) describes another female that deserted her mate on the eggs for nearly five hours. The male stood up and called three times during this period, the female returning after three hours to call back from a short distance away but then departing again. The eggs were chipping at the time, which may have disconcerted the first-year female. Females do not normally leave their nest unattended for more than a few minutes, for example to defecate, although Ortlieb (pers. comm. in Fischer 1980) reported a nest where the female was often missing for hours. On the tenth day of incubation, in cold weather with snow-showers, the eggs were left for eight hours. Despite this chilling, and the considerable risk of predation of the eggs by corvids, they eventually hatched.

Food is brought to the nest one or more times a day during incubation, probably depending on the size of the prey. However, there can also be gaps of as much as three days with no food brought (Brüll 1964), and at these times the female's own reserves are important. Holstein (1942) described a female continuing incubation for five days after losing her mate. She was then fed artificially for ten days by placing food on the nest rim every other day, until she started to receive food from a first-year male.

HATCHING

Since incubation does not become intensive until most of the eggs have been laid, all the young usually hatch within two or three days of each other. The duration of incubation is best measured from the laying of the last egg to hatching of the last young. This period represents the time for an egg to hatch under full incubation, and is typically 38 days although 35–43 days have been reported (Welander 1924, Holstein 1942, Bijlsma 1993).

A most thorough investigation of incubation was by Santi Mañosa (1991), who visited 40 nests in Spain repeatedly during laying to mark eggs. At 38 days after the first egg was laid (an incubation period also found in eggs in incubators), he again started repeated visits to record the asynchrony of hatching. He found that the last two eggs hatched an average 1.7–2.0 days apart across different clutch sizes, which was similar to the average laying interval of 1.9 days. However, the interval declined sharply for subsequent eggs (Figure 38), confirming a tendency for females to start incubating after laying two eggs.

As a consequence, the total incubation time of a clutch tended to increase with clutch size, from 36.5 days for two-egg clutches to about 40 days for clutches of three to four eggs and higher with five eggs (Mañosa 1991). This was not the whole story, however, because the time taken for all chicks to hatch varied from zero to five days for the 20 clutches of three eggs and from one to six days for 13 clutches with four eggs. When hatchlings showed little variation in age, females had probably incubated at an earlier stage in laying.

These results were confirmed by Rob Bijlsma (1996), who monitored laying and hatching of every chick in 16 Dutch nests. Incubation times increased from

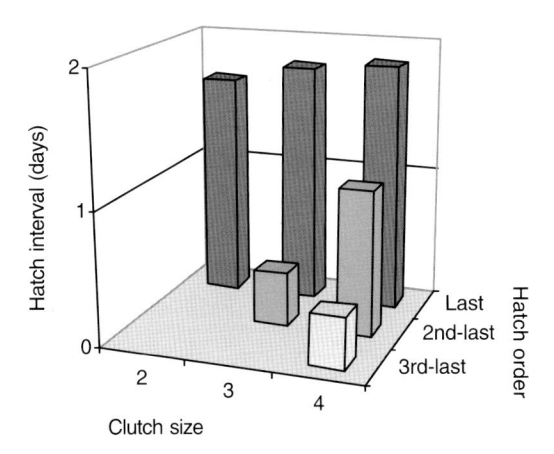

Figure 38. The hatch interval was long for last-hatched eggs (approaching the laying interval of 1.9 days) and larger clutches, indicating that incubation started later for large clutches and always after laying of the second egg. Data from Mañosa 1991.

38.0 days for fourth eggs, through 38.7 days for third eggs to 40.1 and 41.3 days for second and first eggs respectively. The more intensive incubation of larger clutches results in shorter incubation times. Reynolds (1975) proposed that incubation from last egg to first young took 30–32 days in his Californian study. This was probably an underestimate, perhaps as a result of being based on relatively few nest visits (four or five visits each year). More data are needed to confirm whether or not American Goshawks have shorter incubation periods than those in Europe.

Goshawk chicks call from within the shell as much as 38 hours before hatching, a faint 'chep, chep, chack, peeep, peep, peep' (Fischer 1980). Some hours later the shell is pipped, a small outward dent being made through which air can probably circulate to the chick more efficiently than by diffusion through the shell into the air cell. The chick may then rest overnight, and only complete the first crack into a circle the next day.

During this time the adults are clearly aware that something is happening beneath them. Siewert (1933 : 60) wrote: 'After ten minutes she half rises, with distinct signs of unease, turns a bit and settles again. At 12.43 she rises, preens her tail briefly and pushes at one of the eggs. I can hear no peeping this time. After brooding for another 20 minutes, she pushes again at the eggs beneath her. This restless behaviour continues for some hours. Despite the wind, she seems to feel the heat of the sun and opens her beak to pant from time to time. At 14.15 I can see clearly that something is happening beneath the female, because now and then her body rucks upwards a little, and thereby provides more room for the young emerging from the shells beneath her.' This female also showed her disturbance by facing in an unusual direction as she tried to settle on the clutch, and by her reluctance to leave the eggs when the male brought food. Other females, in contrast, seem reluctant to cover the eggs at this time (Stülcken 1943). The young appear bedraggled as they emerge, but their snow-white down soon fluffs out as the female broods them, with her body raised somewhat above the level used in incubation.

Failure to hatch is related to clutch size. Only four of 102 clutches monitored by Santi Mañosa (1991) were of single eggs, and all were abandoned. Clutches with just one egg may represent a severe problem affecting the female, such that no incubation occurs. Some 19–20% of 53 clutches with two or three eggs were abandoned, but just 7% of 46 clutches with four or five eggs. However, 31% of the larger clutches failed to hatch at least one egg, compared with only 9–10% partial hatch failure in clutches of two or three eggs. Just 7% of eggs were considered infertile, without significant difference based on clutch size. Such eggs often become buried in nest material after the others have hatched. In the detailed study of North American Goshawks on the Kaibab Plateau by Richard Reynolds and Suzanne Joy (2006), buried eggs were infertile at 12 of the 15 nests where they occurred.

Brüll (1964) noted that the egg shells of hatched raptors rapidly disappear without the females being seen to carry them away, and wondered if they were eaten. Brosset (1981) saw that Black Goshawks crushed them up and dropped them into the nest lining. This may well be the fate of Goshawk egg-shells too, thereby removing chick-smelling objects that might otherwise draw attention of tree-climbing mammals to a nearby nest.

FEEDING THE YOUNG

The first feed may occur within a few hours of hatching, but only when the young provide the correct stimuli. The female watched by Horst Siewert (1933) brought prey to the nest shortly after hatching, something she had never done before, but took it away again when the young failed to call or raise their heads. When she again brought prey, a few hours later, the young called, 'an almost whistling 'weee, weee, weee' as their raised heads lolled weakly from side to side'. Brüll (1984) stressed the importance of the contrasting white down with dark beak and eyes as feeding stimuli, and noted how a Sparrowhawk he observed once laid food over the eyes of young which were too satiated to take food in their beaks.

The female gives the young tiny morsels of flesh, turning her head on its side so that food is more easily taken from her beak. Intestines are not normally eaten and are often removed before the prey is brought to the nest, as are feathers. Jay Schnell (1958) describes the feeding of slightly older nestlings: 'The female held the prey tightly with the inner toes of both feet. The outer toes were used for support and their contact with the nest platform afforded the balance needed while she was tearing the prey. The hooked beak was closed into the flesh and portions of meat were torn away with an upward pull and a simultaneous twist of the head. The food was then held out to the begging nestlings, who reached forcefully for the female's beak, removing the portions of meat'. What the young do not eat, the female usually swallows herself or removes from the nest. Schnell reported that the female he observed appeared to portion out the food to each member of the brood, but Siewert stresses that the strongest young were fed first.

There may be six to ten feeds a day at first, since the young have small crops and can store little food. The feeds occur when the male brings food, or when the female fetches food from a cache. Caching has been ascertained by several authors, who have observed females (i) leaving the nest with no sign or sound of the male, and often in no great hurry, and (ii) returning rapidly with food which can sometimes be recognised as an unfinished carcass which the female had earlier removed (Siewert 1933, Schnell 1958). Schnell recorded ten cases of caching, probably in nearby tree crutches or alternative nests, the food usually being fetched again 1–9 hours later but twice remaining cached overnight. Caching only occurred in the period before the young started to feed themselves.

Working with Klaus Stülcken at a nest in Schleswig-Holstein, Heinz Brüll (1964, 1984) set up a hide to record prey deliveries and photograph hawks at one study site in 1950, at a nest where the birds were used to such disturbance. Recording started during the laying of five eggs, and daily recording continued for four months, to the fifth week after young had fledged (Figure 39a).

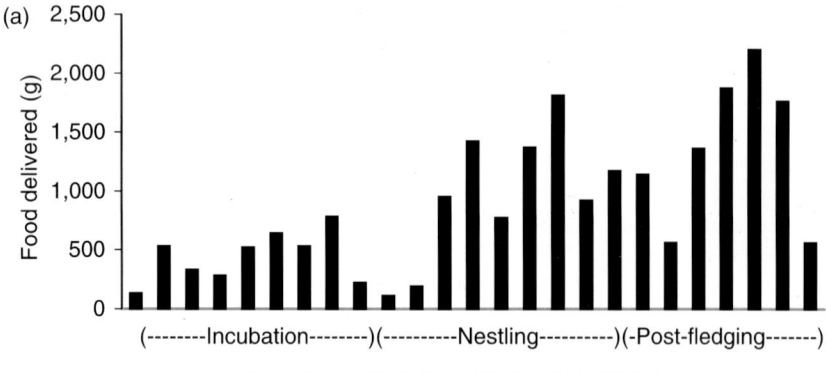

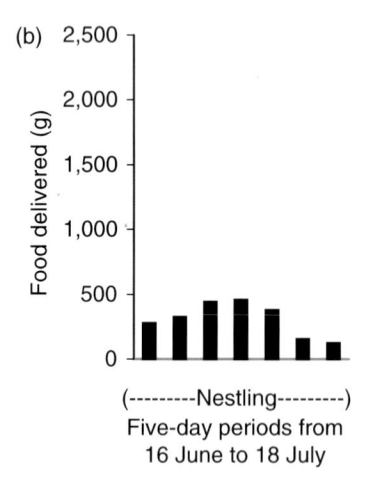

96   The Goshawk

Figure 39. The average weight of food delivered per day in five-day periods, at a nest in Germany (a) that reared four of five young and a nest in California (b) that reared two of three young. Data from Brüll 1964 (a) and Schnell 1958 (b).

This was a site at which the delivery of adult rabbits provided abundant food, except for a short period around hatching. The shortfall at that time may have contributed to the rapid disappearance of one chick. Broods of five are very seldom all reared to fledging. However, the rate of prey delivery increased again during the rearing period to average 940g daily. In comparison, the male observed by Jay Schnell (1958) reached a peak five-day delivery rate of only 460g, which then declined to below 200g (Figure 39b), at which point one of the three young hawks died. Data in Rutz (2003) indicate that deliveries need to average about 250g per young hawk each day during rearing.

Whereas Brüll (1964, 1984) was recording about one prey delivery per day during incubation, the rate increased appreciably during rearing. At that time Schnell (1958) was noting three to six deliveries on most days. The rate may increase or decrease with time, and thus seems to depend as much on availability of prey as on age of the young (and hence their food requirements). In Finnish forests, Sulkava (1964) found that the mass of food brought per day tended to increase through the rearing period, although the prey delivery rate did not, because the most frequent prey (tetranonid chicks) were themselves increasing in mass. If the male's rate of prey delivery falls, contributions from the female may become important, as found in Cooper's Hawks and Eurasian Sparrowhawks (Snyder & Wiley 1976, Geer 1979). In Schnell's study, the fall in the rate of prey-delivery by the male coincided with a lack of American Robins and Steller's Jay nestlings, which provided 66% of the total male prey.

More recent studies have sought to automate the recording of prey deliveries with still cameras that the arriving hawk triggers (Tommerås 1986), and most recently with continuous video recording (Grønnesby & Nygård 2000, Lewis 2001, Smithers 2003). Rogers *et al.* (2006) collected 12 hours of video data a day (staggered records during 15-hour days) throughout the nestling period at ten nests in the Apache-Sitgreaves National Forest in Arizona during 1999–2000. Prey deliveries peaked in the morning at 0.44 per hour before 11.00, compared with 0.3 prey delivered per hour later in the day. Although delivery rates decreased during the rearing period, the mass of individual prey items increased by more than enough to compensate, so that the delivery rate increased from 300 to 900g/day, averaging 640g/day for broods that hatched 2.3 and fledged 1.9 young.

The mass of the prey brought in Finland averaged 750–950g/day in different years, a total of 34–43kg fed to each brood over a 45-day rearing period. Brood sizes were 2.5–3.5 young across years, and the maximum of 950g is close to the 940 g/day received by Brüll's brood of four. Smithers *et al.* (2005) found that chicks in single broods got 510g per day in Minnesota, but relatively much less each in larger broods. The daily biomass for each chick was 320g per day overall, and about 300g per chick in Arizona. The biomass needed by broods of 2–3.5 from hatching to independence (about 45 days post-fledging) is 60–100 kg. This may include part of the 300g/day required by the parents (Rutz 2003).

PARENTING

Although the male often transfers prey to the female away from the nest at the start of rearing, later he usually delivers prey to the nest and may even feed the young himself. This is beautifully described by Klaus Stülcken (1943): 'When the male next announced his presence with prey, something quite extraordinary happened. Unusually, his mate failed to fetch the prey, but remained resting on her perch and didn't even call when the male flew past her to the nest. His feet held a rather superficially plucked jay, which he set down carefully. His next behaviour was unprecedented. With a powerful bite he pulled off the prey's head, removed a reddish morsel in his beak and held it haphazardly above the downy mound of young. The male had decided he wanted to feed them. I say 'decided', because the young had all buried their heads in the well-lined nest cup and were showing neither contrast markings nor hunger cries to stimulate him. The recently fed chicks were deeply asleep. After holding up the morsel for a moment, the male tapped the nearest nestling with the bridge of his beak. The baby hawk sat up and hurriedly grabbed and swallowed the food. The male immediately tore a fresh morsel, and the nestling turned towards him, the two others also stirring.' . . . 'Suddenly the male crouched with lowered wings over his food, in the typical 'mantling' pose used to guard against a potential robber. The female, his own mate, clattered onto the nest, and in an instant the male was gone.'

Although Uttendörfer (1939) reports a male that reared four young from their second week when his mate disappeared, it seems unusual for males to do much feeding of the young (Boal 1994b). If the female is lost in the first two to three weeks after hatching, prey usually accumulate as a stinking, fly-blown pile at the edge of the nest and the young starve 'in the midst of plenty' (Brüll 1964). This situation can also be recognised by the lack of fresh remains at nearby plucking sites, and the lack of fresh greenery on the nest. Looft & Biesterfeld (1981) report five cases of rearing that continued when females disappeared at 12–20 days, with young fledging successfully from four of the nests. These young were approaching the age at which they could feed themselves, and may have received enough from the male to carry them through. However, just as females sometimes attract new males if they lose their mates during incubation, so bereaved males sometimes attract new females, and this was known to occur at one of the nests.

Three cases of female replacement during incubation were recorded by Looft & Biesterfeld (1981). A juvenile female was incubating a clutch of four eggs on 24 May after the original adult female had disappeared on 20 May, but the only nestling that hatched was eventually reared by a third female, another adult. In both other cases, where moulted feathers of another female were found on the nest after the original occupant had probably been shot, the eggs failed to hatch. Rob Bijlsma (1991) recorded a change of female between March and July in 12 of 97 breeding attempts in the Dutch province of Drenthe. Adult females were replaced ten times by an immature, while adults replaced immatures twice. In one case an adult with five eggs was replaced within days of laying, after which the adult male too was replaced by an immature and all five young fledged. The gain for the replacing hawk is presumably the opportunity to breed the following year at the site, with perhaps also a benefit from gaining additional experience through fostering a brood.

A female normally broods the young or perches on the nest rim for much of the first week after hatching (Plates 13 & 15), watching her offspring and removing blood-sucking and other flies that settle on them (Schnell 1958). She starts reducing her presence at the nest in the second week, although she may still remain in the nest tree for much of the time (Figure 40). Young are brooded at night for at least two

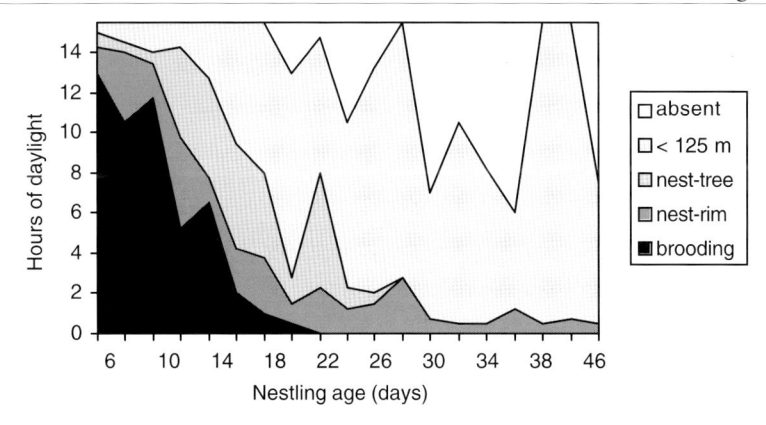

Figure 40. As young Goshawks develop, the female reduces her presence at the nest. Data from Schnell 1958, with three days adjustment to reconcile development stages.

weeks, and the female roosts in the nest tree for longer, also covering them as best she can to protect them from rain or strong sun. Stülcken (1943) describes the female spreading the wings to shelter 17–19 day old young during a storm.

During the third week, the female is spending more time out of the nest tree. Schnell (1958) describes a peak in provision of green sprigs at this time, with as many as six bouts per day involving 1–7 sprigs at a time. The female is important for nest defence too, and may fight with intruding Buzzards and other raptors. After the female had been killed, young at a Swedish nest were killed by another female Goshawk.

When the female calls in alarm, the young flatten themselves down onto the nest. They probably learn to recognise dangerous beings in this way. Breeders of raptors have noticed that young which could see humans and hear their parent's alarm calls were subsequently much more afraid of humans than those reared in visually isolated enclosures (Koehler 1970). Defence of the nest against humans is relatively common in North America, where females whose nest trees are climbed may cause 'deep gashes in arms and legs' and strike blows 'with such force as to rip heavy overalls and to cause flesh wounds which bled freely' (Dixon & Dixon 1938:4). In Europe, the female often remains silent and out of sight, sometimes giving the alarm call from tree tops or in flight above the canopy, and very rarely approaching within shotgun range. No team member was attacked in seven years of visiting nests on Gotland. However, attacks do sometimes occur in Europe, in one case involving both parents when chicks were young (Schiermann 1925). More recently, attacks have very occasionally been noted in the Netherlands, Norway and Britain (Rutz *et al.* 2006a, A. Folkestad, pers. comm.).

It is hard to be sure how much the female contributes to provisioning, because food she brings to the nest may have been transferred at a distance by the male.

Schnell (1958) estimated that the female provided 15% of the prey, and females provided 9–12% of food during the nestling period at two nests in Alaska (Zachel 1985). In Wyoming, where intensive radio-tracking was combined with nest observations to assign deliveries to a particular bird, females apparently caught 29% of the prey items (Good *et al.* 2001). In contrast, some broods may depend entirely on the male (Younk & Bechard 1994).

Ward and Kennedy (1996) suggested that food abundance during the nestling and fledgling-dependency periods affected young Goshawk survival not only by limiting starvation, but also by allowing the adult female Goshawk to spend more time in the nest stand as a protection from predators. Indeed, the nest attentiveness of experimentally fed females was greater than those without food supplements (Dewey & Kennedy 2001).

DEVELOPMENT OF THE YOUNG

Excellent descriptions of development of the young are given, with photographs that can be used for estimating ages in the absence of measurements, by Horst Siewert (1933) and Clint Boal (1994a). The reduction in female brooding during the second week after hatching (Figure 40) coincides with development of a new, greyer coat of down, and the nestlings beginning to move about in the nest hollow, supported on wing stubs and tarsi. The young suffer in heat, panting and lying on their sides away from each other, whereas they huddle together in the cold. They gather in a semi-circle around the female to be fed, turning away and lying down again with heads outward when they are satisfied. From the first few days they are able to turn and raise their rumps to squirt their mutes over the edge of the nest, and by the end of the first week they orientate effectively to the edge of the nest to do so. As a result spots of 'whitewash' can be seen under the nest for the first time.

At the start of their third week (Plate 16) the young are beginning to try to stand, albeit weakly, and can hold onto the nest with their feet if handled. Their main flight feathers and some scapulars are emerging, and they have already reached about half their full weight (Figure 41). The females especially are growing at their fastest in the third week, and are becoming distinguishable from males in size and tarsus-width, although it may still be difficult to estimate the sex unless both are available in a nest for comparison. They are starting to make preening motions, and peck at things that catch their attention in the nest, especially red objects, although they do not yet feed themselves.

The third week is also a time at which there is an increase in risk of death, both from predation and starvation (Figure 42). Risk of death from starvation is presumably increased by growing demand for nutrition, and this may also increase the risk of predation as a result of increased calling or other activities.

The risk of starvation increases with the order of hatching, and can exceed 50% for the last arrival in broods of four or five chicks (Figure 43). The risk of death from

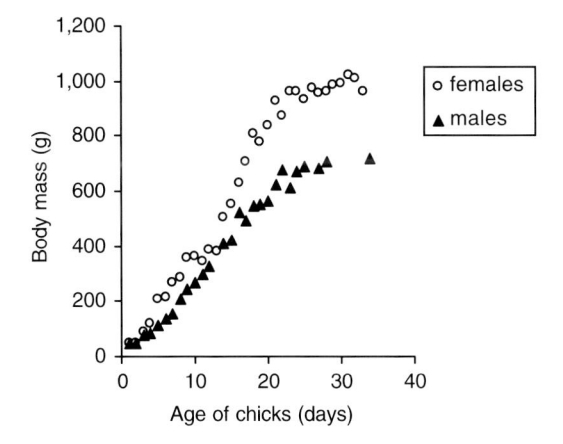

Figure 41. The gain in mass with age of nestling Goshawks, showing how males and females diverge strongly during their third week. Data from Bijlsma 1993.

natural predation does not alter with order of hatching, not least because a visiting predator will often kill the whole brood.

As a result of a spread in size due to asynchronous hatching, one nestling tends to fall behind its nest-mates in growth if food is short. This is a process known as 'runting'. An undersized chick is then liable to be pushed aside by its larger sibs at feeding time so that it only gets fed once they are satiated (Siewert 1933). Once a runt becomes too weak to feed, it may be killed or simply be trampled into the nest structure.

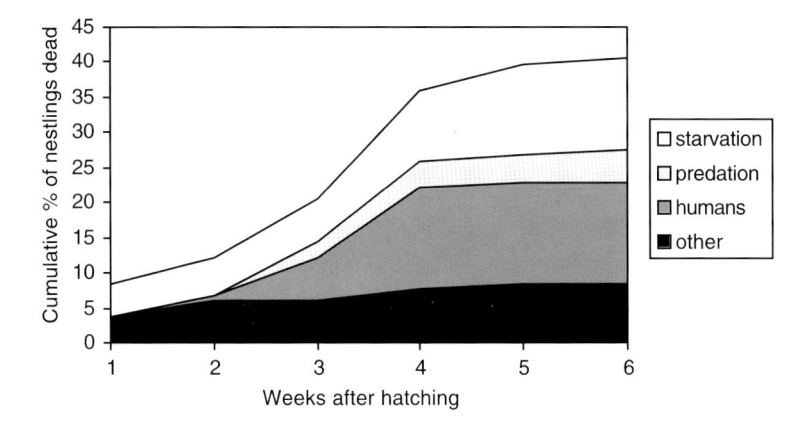

Figure 42. Among 131 nestling goshawks recorded by Santi Mañosa (1991) in Spain, deaths occurred especially in their third and fourth weeks, due mainly to humans, other predators and starvation. Few nestlings died in their fifth and sixth weeks.

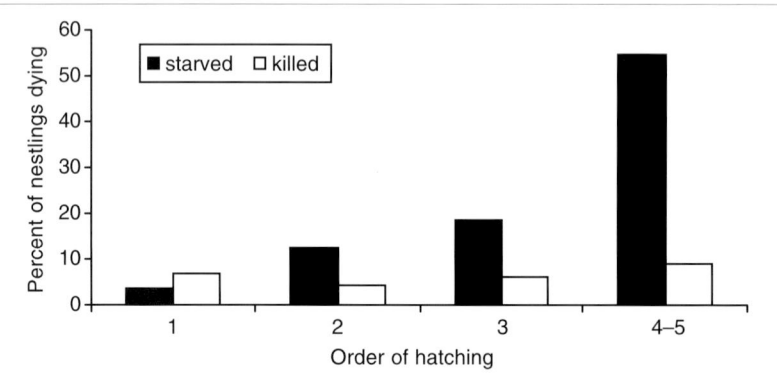

Figure 43. Among 17 deaths from starvation, six occurred among the last-hatched in 11 broods of four to five hawks. Data from Mañosa 1991.

Seen in the light of kin selection theory (Trivers 1974), the runt dies rapidly because it shares half its genes with each of its nest-mates. During food shortage the runt's 'selfish genes' are more likely to survive if its siblings remain healthy than if the runt continues to compete for food and the whole brood suffers malnutrition. It is actually the female that predisposes the brood to runting, by continuing to lay after she has started full incubation and thereby producing at least one chick that starts smaller than its siblings.

However, the small chicks are tenacious of life, which keeps the runting system flexible in responding to any improvement in food supply. Lee (1981a) described one of a brood of four in Utah which, despite having been half the size of its nest-mates at one point, eventually fledged 14 days after the first. Looft & Biesterfeld (1981) found that all of four runts survived when fed and experimentally moved to nests with small broods of similar age. They also recorded a ringed runt that was shot as a breeding bird next year. The variation in hatching intervals observed by Mañosa (1991) suggests that females might even control the process actively, perhaps by delaying incubation, and thereby reducing asynchrony, during repeated breeding with a highly competent male.

Nestlings at the start of their fourth week are starting to show the largest primaries, and contour feathers are appearing over their backs and wings. They can move about the nest well, and during this week they become quite strong on their legs. They start to pull at food for themselves, although they still receive most from the female. They may now make the whistling scream 'weeay, weeay' when she goes to fetch food from the male.

The young are starting to exercise their wings, stretching them together above the backs, and flapping them for several seconds at a time, holding on to the nest with their feet. Visual and muscular co-ordination are now advancing rapidly. The young watch what is going on around the nest and spot their returning parents at a distance.

They can shoot out a foot to grab food, and deliver a hard peck with the beak or remove flies from the heads of other nestlings. Blood-sucking *Simulium* flies have been associated with deaths of raptors with similar ecology in parts of North America (Smith *et al.* 1998) but do not cause problems for young Goshawks (Doyle 2000, Krechmar & Probst 2003).

With the development of feeding behaviour the chicks may also show aggression towards each other, as described by Schnell (1958) 'At 2.50 pm, the male left a nestling robin in whole condition at the nest. Nestling 2 at first pecked inquisitively at the yellow areas at the game of the beak. Nestling 3 joined in. Most unexpectedly a change occurred in Nestling 2. It aggressively attacked Nestling 3, uttering a sound resembling the female's cackle in rhythm and tone.' Use of the alarm call by 20-day-old young may be unusual, as it is rarely used against humans visiting the nest at this stage (although hawks close to fledging often call in alarm when the nest tree is climbed). The other young responded by moving to the nest rim and facing outwards, with their heads held low and rumps towards the aggressor. Schnell suggested that this 'defence stance' protected the head from attack.

By the start of the fifth week (Plate 17) the chicks have developed most of the typical Goshawk behaviour patterns. They may 'mantle' over food, spreading their wings and tail over it while facing away from their nest-mates. If acute aggression occurs, one youngster may give the trilling distress scream. At calmer times a youngster may bring its foot up to scratch the side of its head. It may also stand at the side of the nest and stretch one wing and the leg on the same side stiffly downward and backward, balancing on the other leg. This stretch is called 'warbling' by falconers, and in fully feathered hawks the tail is also fanned to the side and down, covered by the primaries of the stretched-out wing to form a smooth surface of feathers for several seconds before the hawk relaxes and refolds its wing. The tails of the young hawks have as yet little feather to spread, but they are gaining muscular control of their tails, which may be waggled two or three times from side to side before the nestlings lie down again after a bout of wing-flapping. Adults often wag their closed tails too, as they settle on a perch after a flight, or at the end of a bout of preening, and the comically goose-like wagging gives every impression of contentment.

Now that the young are eating more roughage, they are bringing up mucous-bound pellets of fur and feathers from the prey. The pellet travels up from the stomach with the aid of sideways swings of the head and neck, and is finally expelled with a more vertical retching movement, either dropping in the nest or flicking sideways over the edge.

At 30 days there is still some down on the crown and behind the eyes, but the upper surfaces are otherwise well-feathered. The lower surface, on the other hand, is still down-covered apart from well-developed contour feather tracts on either side of the breast. The young can now flap their wings for half a minute at a time, doing little jumps and running about the nest. They are also beginning to make the 'heee-yah, heee-yah' food scream, although this becomes really strident only after fledging, when it can be heard a kilometre away across open ground. Parents bringing prey to the nest at this time do not loiter. Often a squabble for ownership develops, with

104   *The Goshawk*

another nestling taking the food from the hawk that claimed it, which may threaten with the trilling scream before letting go. Wings may be spread and head held back in the threat posture, which in the fully fledged bird involves the feathers at the back of the head also being raised like a crest. Risk of death from starvation is presumably increased by growing demand for nutrition, and this may also increase the risk of predation as a result of increased calling or other activities.

In the sixth week the young become 'branchers'. The down is gone from their upperparts, except for a few wisps on the head, although there may still be a fair amount on belly and legs. Their wing exercises may last for 10–15 minutes, lifting them 20cm off the nest. Although they still spend much time lying down, at other times they stand on the edge of the nest and start to move onto nearby branches in the nest tree. Their primaries are two-thirds grown although their main tail feathers are less than half exposed, with the blue-sheathed blood-and-lymph-filled bases of the main feathers now hidden by the wing and tail coverts. The divergence of wing-lengths is relatively small (Figure 44). Males are therefore further towards completion of feather growth than females and leave the nest 1–3 days ahead of their sisters.

SEX RATIOS

Among the 813 young sexed by Looft & Biesterfeld (1981) in Schleswig-Holstein, 56% were male (a highly significant excess). The predominance of males tended to increase in years when nest success was high (Figure 45).

Similar observations of excess males have been linked to years with large brood sizes elsewhere in Germany (Bezzel *et al.* 1997a) and to large broods and early laying

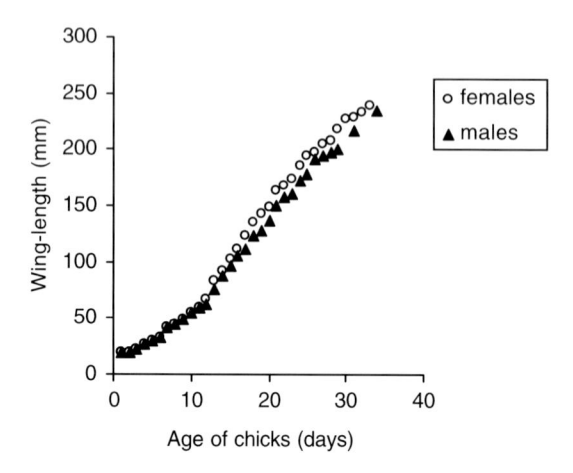

Figure 44. Wing-length grows linearly from about day 12 to fledging (from day 38), when it is closer to full size for males than females. Data from Bijlsma 1993.

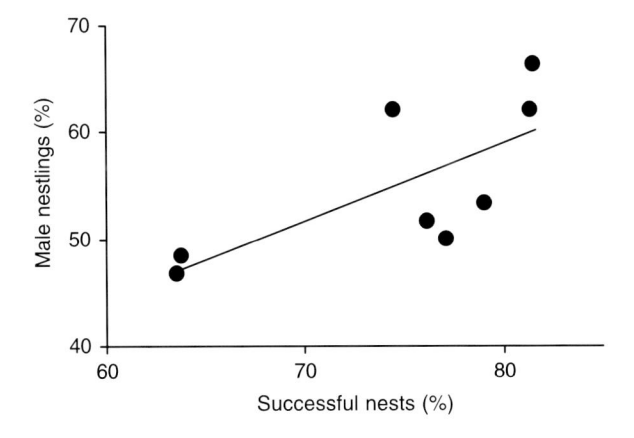

Figure 45. The tendency for the sex ratio to favour males in years when nest success was high in Schleswig Holstein. Trend line $p = 0.05$. Data from Looft & Biesterfeldt 1981.

in Finland (Wikman 1976, Byholm *et al.* 2002a), but to pairs that laid late in the Netherlands (Bijlsma 1993) and to years with low nest occupancy in Arizona (Ingraldi 2005). The relationships between breeding success and predominance of males raised the question of whether good conditions result in more male Goshawks being hatched or more males surviving after hatching. The sex ratio at hatching, known as the primary sex ratio, can vary quite strongly in birds in ways that have been hard to explain (Komdeur *et al.* 1997, Komdeur & Pen 2002).

In Finland there has been very extensive ringing of Goshawks and also monitoring of game populations throughout the country for many years. Patrik Byholm brought these data together to study relationships between 5,455 Goshawk broods and counts of woodland grouse in 50-km squares during 1989–99. The sex ratio in Goshawk broods linked extremely significantly to grouse numbers, with no residual effect of brood size once the variation in this important food was taken into account (Byholm *et al.* 2002b). Goshawks hatched an excess of males when spring food supplies were good.

The overriding role of food supplies in determining primary sex ratios can account for regional differences in factors that relate to brood sex ratios, such as the linkage to early laying in Finland but late laying in the Netherlands. Korpimäki *et al.* (2000) showed that sex-ratio related strongly to parental condition in Kestrels, such that links between laying date and sex-ratio that changed between years of food abundance and scarcity were lost after parental condition was taken into account. Thus, late laying would produce many male Goshawks if food such as young rabbits was increasing late in the laying period.

Christian Rutz (2005b) was able to demonstrate a direct link between good food and abundant males at the family level among Goshawks in Hamburg. Broods raised on a diet rich in pigeons not only contained more males than other broods, but also had better recruitment of those males, and were heavier. Sex-ratio in broods was not related to laying date, brood size or age of breeders.

Although an effect of food supply on primary sex ratio may usually explain the skewing of goshawk sex ratios towards males, the mechanism remains unknown. Moreover, it is not the whole story. When offspring are sexually dimorphic, a difference in secondary sex ratio, at the end of a rearing period, may also reflect a reduced ability of parents to rear one sex (Clutton-Brock *et al.* 1985). Mats Karlbom climbed to record clutch sizes for 67 Goshawk broods that were monitored on Gotland, so we were able to assess sex-ratios in relation to brood sizes and brood losses (Kenward *et al.* 1993a). There was a slight excess of males in 22 broods that lost no eggs or chicks, but males were dramatically reduced to only 25% of the 16 nestlings in ten broods that had declined by two or three young (Figure 46).

Some broods on Gotland were quite short of food. The size of successful broods on Gotland averaged only 2.2 chicks, substantially less than the 2.9 young in successful nests throughout Finland during the 1990s (Byholm *et al.* 2002b). The killing of Goshawk young by their nest-mates has been described in many studies (e.g. Ruthke 1929, Schnell 1958, Estes *et al.* 1999), and there were several instances of this 'cainism' (siblicide) on Gotland. Chicks were found dead or moribund at 10–15 days after hatching, with head injuries consistent with pecking by sibs (Figure 47). A young male was found alive on the ground under a third nest just before fledging, and replaced in the nest with its male sibling; on the next visit the latter was dead under the nest, with a probable claw wound through the cranium.

When females have grown larger than males, they are likely to dominate such encounters. Thus, a primary sex ratio biased towards males when spring food is abundant can be complemented by a bias against rearing of males as a result of siblicide when food is poor in summer. However, any tendency of brood reduction to favour females is relatively weak compared to the effects of food supply on primary sex ratio (Byholm 2003).

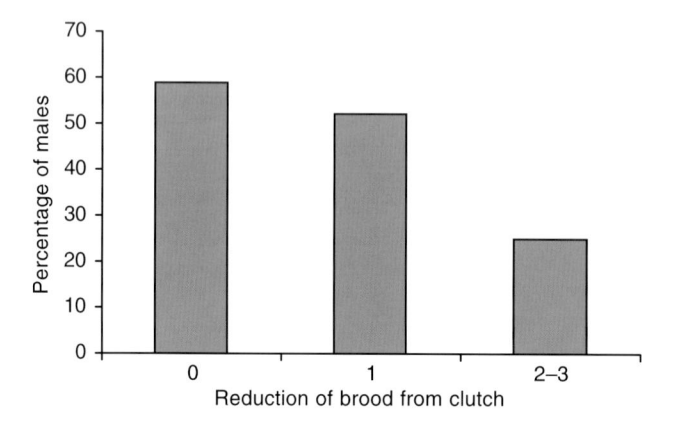

Figure 46. Among 67 broods on Gotland for which clutch-size was known, the proportion of males was least ($p < 0.05$) in those that were most reduced at fledging. Data from Kenward *et al.* 1993a.

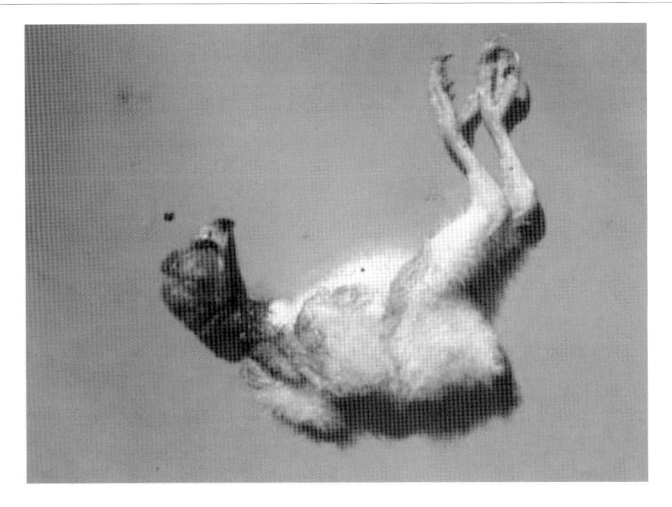

Figure 47. A 10-day-old Goshawk killed by a sibling on Gotland (Mats Karlbom).

In North America, where sexual size dimorphism is less than in Europe (Chapter 2), there is evidence of a contrasting effect, in which poor breeding conditions favour males. Ingraldi (2005) found that among 76 broods studied for six years in Arizona, the ratio of males was highest in years when few hawks bred and in broods that fledged only one chick.

FEATHER GROWTH AND MOULTING

The increase in wing-length of nestling Goshawks, during its prolonged linear phase (Figure 44), is mainly due to growth of the longest primary feather. Measurements on 12 male and 19 female nestlings on Gotland, for which the hatch dates were known, showed a constant rate of about 8mm per day in male Swedish Goshawks and 9mm in females.

In adult Goshawks, the progress of moult is strongly associated with breeding. Although some down is shed in advance, the main feather moult of females starts during laying, which suggests a strong hormonal link between the two processes. Information on moult has been presented by Heinz Brüll (1964, 1984) from seven moults of two trained hawks and hide observations with systematic collection of feathers during eight and 12 years at two nests in Germany, and by Chris Reading (1990) from seven moults of a trained German hawk, with confirmation from Henny *et al.* (1985) of similar moult patterns in North America.

The earliest main feathers to moult are the innermost primaries, which are called the first primaries though they are at the back of the wing. For breeding females, these are generally shed at about the start of laying (Brüll 1984). The primaries then moult and re-grow in sequence to the front of the wing. In the wild German females, the first four primaries were typically shed within three weeks, with the first three in particular at intervals of 2–9 days (Figure 49). Since these four primaries are 200–300mm long, growth at 9mm per day takes about a month to complete and a large gap appeared in the wings of these females. It is an advantage for the incubating female to get as much moulting as possible done early, provided she receives plenty of food from her mate. The remaining primaries of the nesting females were shed less rapidly, and typically after some delay, with only one in growth during much of the period in which nestlings were raised.

The feathers of the males were not shed so rapidly at the start of the moult, although several smaller ones at the front of the wing could be growing at once towards the end (Figure 49). Whereas female feathers could be collected in the nest stand, and often at the nest, it was rare to be able to find feathers of males, which spent much time hunting elsewhere. However, their moult sequence could be pieced together by finding a feather from either wing, because recoveries from the females and the captive birds showed that the left and right primary feathers were seldom shed more than three days apart.

Moult start dates (when primary 1 was shed) could be compared for six wild and three trained hawks over a number of years (Figure 50). This shows that individuals tended to start moulting in the same two- to three-week period each year. The start period for each hawk was between late March and late May and differed considerably between individuals. There may have been a slight tendency for the males to start later than the females.

The duration of moulting, from shedding the first primary to the tenth, averaged 154, 156 and 174 days for three wild hawks, compared with 103, 104 and 130 days for three trained ones. This indicates that the trained hawks completed their moult faster than the wild birds. The moult started as fast for the trained male and two females as for wild females, and tended to decelerate slightly, but lacked the gap present for wild hawks (Figure 51), perhaps because non-breeders generally lack a gap or because the trained birds were well fed.

Figure 48. A healthy brood of Goshawk chicks.

Incubation and rearing 109

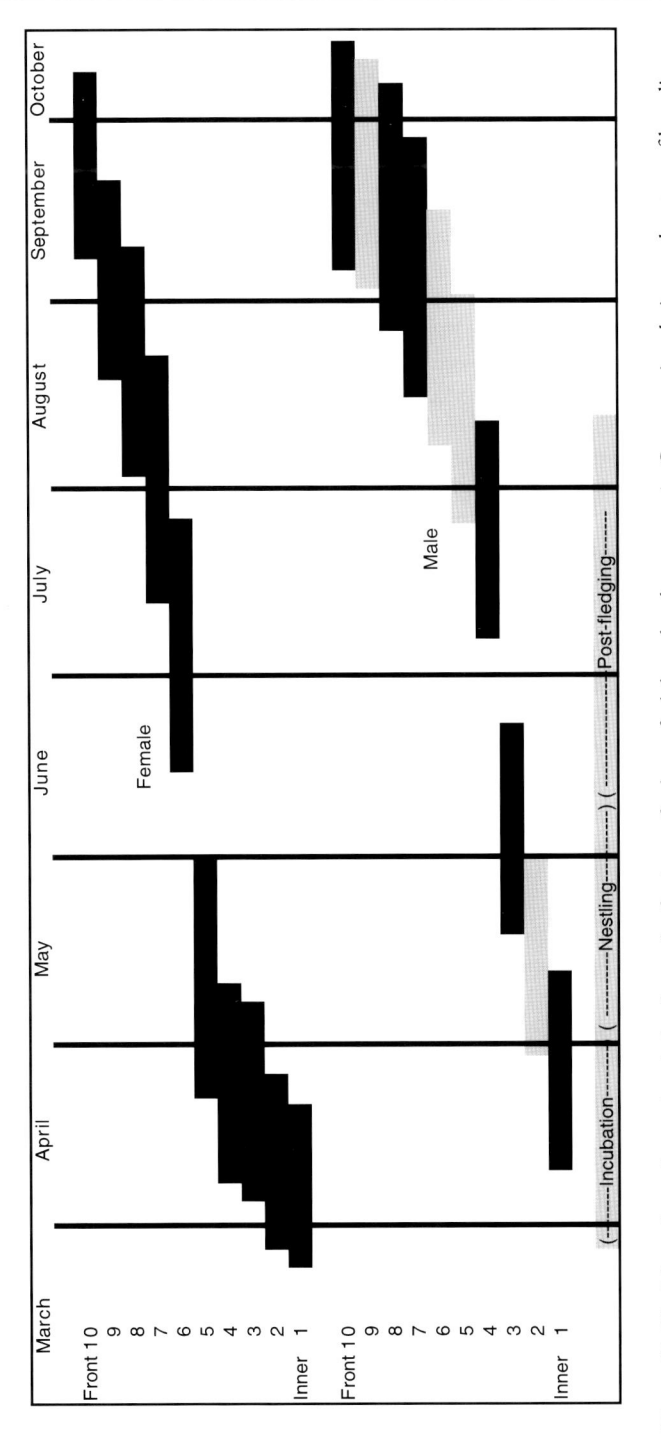

Figure 49. Black bars show observed periods of growth of primary feathers of adult goshawks at a nest in Germany, in relation to the stage of breeding, with grey bars for feathers with unknown moult dates. Redrawn from Brüll 1984.

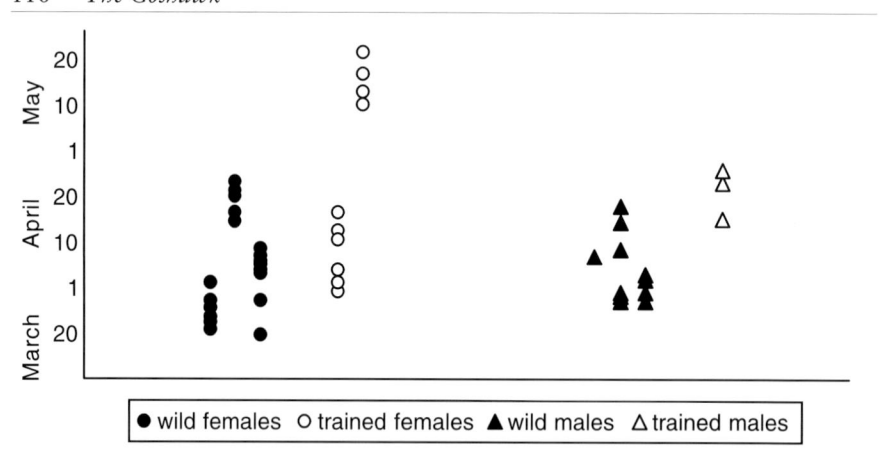

Figure 50. The date of shedding a first primary feather for wild and trained Goshawks monitored for 1–9 years. Data from Brüll 1984 and Reading 1990.

On Gotland too, wild Goshawks often stopped moulting completely while rearing young, such that hawks trapped then had no part-grown feathers. The moult presumably started again later where it left off. However, some adults were trapped in winter with incomplete moult of their primaries. When trained hawks do not complete the sequence of primaries, the moult next year starts 'before their turn' for the two-year-old feathers, such that the primaries contain two moult centres and complete their moult early.

The moult creates two types of risk for breeders. One is of impaired flying ability if the female has to start feeding herself through loss of a mate while many primaries

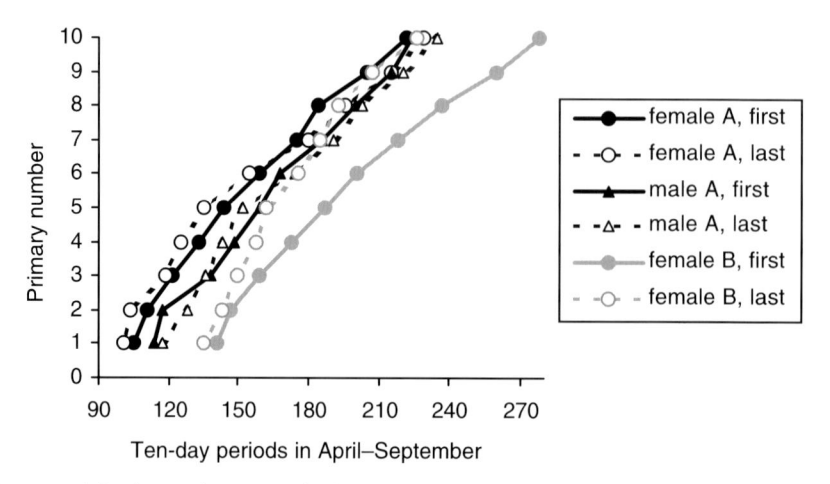

Figure 51. The date each primary feather was shed in the first and last moults of three trained Goshawks. Data from Brüll 1984 and Reading 1990.

are growing during incubation (Figure 49). Perhaps rapid moulting during incubation only occurs if the female is very well fed, as was true for the trained hawks. The other risk is of having to make do for two years with the same feathers. This may help to explain why the feather shafts of older hawks become thicker (Reading 1990) and hence presumably stronger, and why females moult as fast as possible during incubation. The role of food supply and reproductive hormones in control of moult would make an interesting study.

The tail moult usually starts when three or four pairs of primaries have been shed, and the secondary moult shortly thereafter. Secondary feathers and rectrices (tail feathers) do not moult in such a regular sequence as the primaries. The innermost pair of rectrices seem almost always to moult first in first-year hawks, and often thereafter, while the second innermost feathers often moult late, but the order for the other pairs is variable and not always symmetrical (Figure 52). The moult of secondary wing feathers seems to start at 2–4 foci along the wing and moves inwards, but superimposed on this pattern is a tendency for feathers not shed in one year to be shed early in the next. The limited data from the wild hawks suggested an even less predictable tail and secondary moult sequence than in the trained birds. Trapped wild Swedish hawks quite often did not complete even the primary moult by the time moulting ceased in late September or early October.

During the last two-thirds of the 4–6 months of the main feather moult, the contour feathers are shed too. The primary and secondary coverts are shed in sequence with the large feathers, while the upperwing coverts are moulted from posterior to anterior wing surface. The body feathers moult first in the middle of the breast and back, the moult sequence then moving forward and backward so that the forehead feathers and tail coverts are the last to be replaced. The contour feather moult is usually completed on the chest and back, although occasional streaked juvenile feathers remain very conspicuous on the barred breasts of second-year hawks.

Sex	Year	Left 6	5	4	3	2	Centre 1	1	2	3	4	5	Right 6
Male	1	B	E	D	C	F	A	A	C	D	E	F	B
	2	C			B	D	A	A		C	D		B
	3	D	A	B	E		C	A		E	B	A	C
Female	1	B=		D	B=	E	A	A	C	D	E		B
	2	D	A	E	C	F	B	A		B	E	C	D
	3	C	D	E		B	A	D	F	A	C	E	B
	4	B	F	E	A	D	C	B=	B=		C		A
	5	A		D		C	B	D		A	B	C	
	6	B	C	F	A	E	D	D	B	F	C	E	A
	7	C			B		A	A			B		C

Figure 52. Letters indicate the sequence of moult in the left and right tail feathers of two trained German Goshawks, with retained feathers in grey. Data from Brüll 1984.

A few old brownish contour feathers usually remain on the wing or tail coverts of second-year birds, and the presence of worn feathers among the plumage suggests this incomplete moulting is common among adults.

If the hawks are stressed by food shortage while feathers are growing, a conspicuous pale line of weakness may form across the feathers. This fault bar can be seen by holding main flight feathers up to the light. Fault bars can also be caused by psychological stress, at least in trained hawks, and are therefore often called 'fret marks' by falconers, or 'hunger traces'. In first-year hawks the bars run at the same level across all the feathers, and must be a threat to survival because feathers tend to break at these points. Once one feather has broken, there is even less support for others with fault bars at the same level, and so these have an increased chance of failing too. It is worth recording any fault bars when marking young hawks, as an indication that they have experienced food shortage. If shortage is especially severe or other damage occurs, feathers may even pinch-off completely, and perhaps regrow later if the hawk survives.

IMPLICATIONS FOR CONSERVATION AND MANAGEMENT

The risks that breeding creates for the female are great. She is particularly vulnerable during incubation, through reduced body reserves and flight impairment from the moult. This may explain why disturbance during incubation so readily causes desertion. It is a time when researchers and foresters need to be especially careful with prolonged activities that could cause brood failure. Toyne (1997b) recorded desertion at four of five nests disturbed by forestry work before young were ten days old. However, the passage of vehicles and people on nearby tracks is not harmful, and may even help habituate the adults to unavoidable disturbance. Goshawks breeding in towns can become very resistant to disturbance, even from humans that stop near their nests, but may retreat when binoculars are raised to watch them (Rutz *et al.* 2006a).

In Britain, the Forestry Commission recommends no forestry operations within 400m of nests between February and the time when down from moulting females appears on nests, some ten days into incubation, and then not within 300m until appearance of appreciable 'whitewash' under nests indicates that young are about ten days old, and then again not within 200m until young have left the nest area (Petty 1996a). The risks from disturbance by humans decline when young are old enough no longer to need brooding. Prey can be delivered during intervals between periods of human presence, although distraction of adults from hunting and other brood care activities is clearly undesirable.

If young are to be marked for monitoring after they fledge, the fifth week is a good age. They are strong enough to be handled with little risk of injury, but not yet likely to try leaving the nest as the tree is climbed. Few deaths are likely after this week (Figure 42), so the numbers represent productivity at fledging. Mass is at a peak (Figure 41), as an index of condition of the young. Body measurements will not change greatly, despite incomplete calcification of some bones and continued growth of feathers. Tarsus growth is virtually complete by day 25 (Mañosa 1991) as a key for sexing, and a harness for radio-tags can be applied without risk of becoming too tight (Chapter 5).

The time for a marking visit can be estimated by assessing the extent of 'whitewash' under the nest, if the young are small, or by observation from the ground. However, in order to time the marking accurately, an earlier climb to the nest is advisable. Although young can be aged into three-day categories by using photographs (e.g. Boal 1994a), a more rigorous method for research is to measure the length of the wing from the carpal joint to the tip, while held flat against a wing-rule. The primary feathers of healthy young grew at 8–9mm/day during 12–30 days after hatching (Figure 44). Age was estimated by adding (from equations that regressed winglength on date) a start-up measure of 3.6 days for males and 4.9 days for females (Kenward *et al.* 1993b):

$$age_{male} = 0.125 \text{ (wing-length in mm)} + 3.60$$
$$age_{female} = 0.111 \text{ (wing-length in mm)} + 4.94$$

These relationships estimated age with two-day accuracy for 74% of the nestlings and with 3.8-day accuracy for all of them. Growth rates need checking in smaller Goshawk races, omitting from calibration any birds that die as runts or from disease before fledging. An interesting question is whether growth rates are slower where Goshawks are smaller, or whether Goshawk feathers generally grow 8–9mm/day and such hawks fledge younger.

With hatch dates estimated, a schedule can be prepared to visit each nest again when the young are 28–35 days old. Moreover, laying dates can be estimated by adding 38 days to ages estimated by wing-lengths. In 16 Dutch broods, hatch order always reflected laying order, even when young hatched on the same day (Bijlsma 1996).

CONCLUSIONS

1. Incubation is almost entirely by the female, although some males take over for an hour or so while she feeds and can even become reluctant to leave the eggs.
2. Full incubation starts after laying the second or third egg and typically lasts 38 days from last egg to last hatching, which can be up to six days from the first to hatch.
3. The male's provisioning rate increases after hatching. Average mass of food delivered per day varies with brood size from less than 400g to more than 900g. The male is rarely seen to feed the chicks from his beak. Males typically transfer food to the female away from the nest early in rearing and later deliver direct to the nest.

4. Females brood extensively during the first week after hatching, mainly remain in the nest tree in the second week and are frequently out of the nest stand only after three weeks. In America, they often attack humans that visit nests, but rarely do so in Europe.
5. 'Whitewash' becomes noticeable under nests in the second week after hatching, when young are growing grey down. Most deaths occur in these two weeks, especially among late-hatched young from large broods. Females diverge markedly in size from males in the third week. Young become well feathered in their fifth week and fledge at about six weeks old.
6. It seems that primary sex ratios tend to become male-biased in Europe when food is abundant, and that male chicks are especially vulnerable to siblicide if food is scarce during rearing. However, an American study found that poor conditions favoured males.
7. Primary feathers of young hawks develop at an average of 8–9 mm per day, and complete growth after fledging. Adult females may shed half their primaries during incubation, but both sexes often stop moulting during brood rearing and continue in the autumn. Moult of primary feathers follows a regular sequence outward, unlike that of secondary and tail feathers.
8. Prolonged disturbance within 300–400 m of nests in rural areas should be avoided until young are in their second week, and within 200m until they disperse. Young can be aged by wing-length to estimate dates of hatching and to plan visits in the fifth week for sexing, recording productivity, assessing condition when full grown, and marking.

CHAPTER 5

Markers and movements

There are several ants crawling under my shirt, and midges too are starting to find exposed skin under the pile of brushwood. Fifty metres away, a box-trap is fastened under the nest where a young male Goshawk has been fostered. The bird had travelled by air from Finland to Scotland, along with several other downy young, for an experimental adoption by Sparrowhawks. It is now more likely to become a breeder if moved to join other hawks released by more traditional hacking in Argyll. If it can be caught.

The Goshawk is flying round the clearing, with much screaming. During the last four hours, its calls have told me of its approach several times, and then it has gone away downhill again. It is hungry, because to feed even a single Goshawk has been hard work for the Sparrowhawks. There is meat in the bottom of the trap, and to accommodate such a large bird in a device designed for small hawks, the trigger bar has had to go. I must close the trap by pulling on a cord.

The young hawk finally settles on the nest, and peers acutely down at the food below. After hesitating for several minutes, it drops to a cut branch placed to slope down to the trap. After two weeks on the wing, the hawk can move confidently.

There is none of the perceptible re-balancing after landing from early flights, and it loses little time in running down the branch towards the trap. However, it can hardly fit into such a small space.

At last, the hawk can no longer resist the bait. It drops in and I pull down the door with its latching wires, tying the cord from the trapdoor tight for double safety. The scratches from the brushwood don't matter as I rush for the trap. It is a fine July evening in 1973 and Herman Ostroznick owes me a pint of beer.

MARKING AND TRAPPING

To study movements and predation adequately outside the breeding season, hawks have to be identifiable as individuals. Traditionally, they have been marked with rings and recorded through recoveries after their journeys. To that approach can now be added the use of more conspicuous visual tags and identification from feathers shed around nests, or the use of DNA techniques and micro-transponders. However, the tracking of chosen individuals during their travels, and systematically rather than by chance sighting or recovery, requires radio tags. Application of these markers for Northern Goshawks will be considered at the start of this chapter, followed by the findings they have provided.

Rings, radio-tags and other markers can be attached to hawks in the nest. However, much of the marking for studying the movements of individuals in detail has involved trapping. I have thought long and hard about whether or not to describe techniques for trapping Goshawks. The argument for secrecy is that unscrupulous people may use traps illegally.

I have decided on transparency, for three main reasons. Firstly, when hawks are caught legally, for instance under licence for research or to protect poultry or game, this will be done most efficiently and with least risk to the hawks if the best techniques are widely known. Secondly, if people decide to remove hawks illegally, then it is better if they catch them alive, with the opportunity to release them elsewhere, than if they use guns or poison and thereby also put non-target species at risk. Finally, in most countries in which raptors may be kept for aviculture, falconry or in zoos, the birds must now be officially marked to confirm that they were obtained legally. There are forensic DNA techniques to test whether there has been tampering with markers (Chapter 9). Large penalties for illegal procurement provide further strong discouragement for unlicensed hawk trapping.

TRAPS

Goshawks are usually caught for research or management with box traps, spring-nets, hung-nets and nooses (see Figure 127 on page 275). You will need to visit the research literature for diagrams and details of these traps, because there is space here only for general principles.

Box traps are usually baited with live pigeons(Figure 127, back row), which are protected from the hawks in a compartment with food, water and shelter from the weather (Meng 1971, Karlbom 1981, Kenward *et al.* 1983). The most effective types are those into which hawks can walk from the side. Unfortunately these side-entry traps are the most expensive to build and cumbersome to move about, and must use heavy mesh to prevent access to hawks or bait by powerful mammals. The easiest box trap to carry to nest sites is a Swedish falling lid design (Figure 127, left). The entry of a hawk into the capture compartment pulls a wire attached to the roof of the bait compartment, which lifts a pivoting hook and lets the lid fall under gravity. This trap requires half the building time and materials of the side-entry (falling end) type (Figure 127 right)and another sprung-roof design from Sweden (Meng 1971).

When used in winter, away from nests, these top-entry traps could be raised above the ground to hinder access by mammals. They were positioned in conspicuous locations, such as corners of 'hard' edges where woodland met open spaces, ideally just next to a tree or snag with good perch sites. In three study years at two sites they caught 43 hawks with 24–34 trap-days per capture, compared with 29 hawks at 7–16 trap-days per capture for the side-entry traps (Kenward *et al.* 1983). Each capture of 32 fledged young at nests took 9–10 days for top-entry designs in the first year (Karlbom 1981), but was much faster in later years. With one trap under the nest and two or three at plucking posts, it was often possible to catch all the brood in two or three days with two or three trap-visits per day (live traps should always be visited this often). The adult female could often be caught near the nest when young were half-grown, and the male in traps 100–200m from nests along his likely access route(s).

The easiest of all traps to carry was a Swedish spring net design that folds flat to an 80cm square. This trap was set with a fresh kill as bait and more often than not caught the hawk when it returned to feed again (Kenward *et al.* 1983) The design was ideal for catching specific hawks such as radio-tagged birds or hawks that were killing poultry or game (Figure 127, front right).

There is also a spring trap design with semicircular metal bows which are set as a flat netted circle over a small bait compartment (Figure 127, front centre). This 'butterfly trap' is not safe unless the radius is at least 90 cm (C. Rutz, pers. comm.), because otherwise a hawk's wing can be broken between the metal bows when they spring together vertically, even if they are padded. The other traps have never killed or seriously injured a hawk in several hundred research trappings, except for an adult male that was eaten by his fledged brood when caught in a spring net near the nest. Thus, spring traps are not safe near nests.

Nooses are frequently used to catch raptors. In the 'bal-chatri', a compartment holding bait is covered with nylon monofilament nooses (Berger & Hamerstrom 1962). There should be two layers to the compartment, so that a live mouse or other bait cannot tangle in the nooses. Although such traps are highly effective when dropped from cars near perched buzzard and kestrel species, Goshawks are less easily approached (except perhaps in towns). Nooses for Goshawks need a breaking-strain of at least 15kg, and the trap should be attached to a fixture or weight with elastic so that it cannot be carried away. When a bal-chatri was used to catch adult hawks on a nest, after temporary removal of young, it inadvertently caught both male and female together and the male subsequently deserted (Karlbom 1981). Both adults at five nests were caught quite rapidly with a single, elastic-loaded noose set on dummy eggs during incubation on Gotland, but again the males deserted. Thus, nooses on Goshawk nests have been less satisfactory than with Sparrowhawks and Peregrine Falcons (Newton 1986, Mearns and Newton 1984).

A more successful method of catching hawks at nests has been to suspend nets vertically in the path taken by hawks to attack a large tame owl, or dummy owl, placed nearby. Nets with relatively coarse mesh (3–6cm) are used, 1.5–2m high and held to poles with clips or limp wires that pull away readily to enfold the attacking hawk. Nets are placed in a box round the owl, or in a 'V' that protects the owl from possible approach paths, and the operator hides nearby. The technique was used initially for catching Northern Harriers with Great Horned Owls (Hamerstrom 1963). There are further details for positioning and tensioning nets in Bloom (1987).

Success in catching female Goshawks exceeded 78% in five studies, compared with 24–74% success for their mates (Bloom *et al.* 1992, Detrich & Woodbridge 1994, Erdman *et al.* 1998, McCloskey & Dewey 1999, Reynolds & Joy 2006). A stuffed owl moved by an operator hidden beneath (and giving owl hoots) proved to be as effective as live owls (McCloskey & Dewey 1999), and model owls animated by radio-control have been used for other raptors (Jacobs 1996). However, this hung-net method has not been used with great success in Europe, where Goshawks are less aggressive. Dummy owls did not provoke attacks in Hamburg (C. Rutz, pers. comm.), but Krüger (2002a) found that females always attacked a live or dummy Goshawk.

Before discovering the convenience of spring-nets, I used soporifics to catch three radio-tagged Goshawks, by placing barbiturate sedatives in kills to which the hawks returned to feed and then waiting two hours for the drugs to take effect (Kenward 1976a). On the first occasion, the hawk could not fly and was caught when it ran into water. On the second it went to roost quite low and was captured with a waxed noose at the end of a fishing rod. Both these hawks recovered quickly from the drug but their capture was risky. The third hawk responded more strongly to the same dose of drug, was asleep by the kill and took 24 hours to recover. It died about two weeks after release, which raises suspicion of organ damage from the drug. This method is best avoided.

Although box traps are relatively straightforward and safe to use, the other methods should not be tried without consulting those with prior experience and ideally being trained by them. Training is advisable even for using box traps, because no manual can convey all the finessing needed for success, including where to wedge the grass to stop triggering by wind, marking to deter mammals and how to scatter feathers to suggest prey injury. In Europe, North America and many other countries, licences must be obtained before trapping any raptor species. The welfare of the hawks and any bait animals should always be a primary concern.

IDENTIFICATION BY NATURAL CHARACTERS: FEATHERS AND DNA

The rapid moulting of the innermost 4–5 primary feathers by breeding females (Chapter 4), combined with variation in colouration and size between individuals, enables these feathers to be used to monitor continuity of female occupancy at nests. This approach, pioneered by Paul Opdam and Gerard Müskens (1976), was validated with other markings (Ziesemer 1983) and has been applied in many European studies. A marked change in pattern or length of the feathers indicates a change of female reliably for birds at least three years old. For younger birds, considerable change in pattern occurs between years, with some size increase from juvenile to older plumage.

Kühnapfel & Brune (1995) indicated that changes in colour and pattern of feathers moulted in the first two years are a reliable guide for age. Juveniles tended not only to have a yellow-buff tinge to the pale parts of their feathers and totally distinctive breast feathers (Chapter 1), but the 2–5 transverse bands on main flight feathers tended to be easily discerned forward of the vane and to have sharp boundaries at least on the edges away from the base. Bands on juvenile tail feathers tended to have a white fringe that was lacking in adults. In the second year, pale regions on primaries remained slightly off-white and bands remained just discernible forward of the vane, but edges behind the vane were becoming diffuse especially towards the base.

In older hawks, any residual bands had very diffuse edges and were not discernible forward of the shaft, and pale areas behind the shaft were tending to disappear. Feather patterning and size remains sufficiently constant in older hawks for any marked change of the same feather in a series to be a reliable indicator of change. However, detecting change of female or lack of change within years 1–3 is less certain. Moreover, it is not clear how well these characteristics would apply outside Europe especially where pale Goshawks are more abundant (Chapter 1). Rust & Kechele (1996) noted that in feather series for 2–12 years from 178 females in southern Germany, 3% were of a type that remained paler and with greater bar contrast into adulthood.

A more rigorous approach would be to use DNA micro-satellite techniques for individual 'fingerprinting' (Parkin 1987, Gavin *et al.* 1998, de Volo *et al.* 2005). Although feathers are mostly keratin, the polymerase chain reaction (PCR) can be used to obtain adequate DNA from the remains of blood vessels or other adherent tissue in moulted feathers (Taberlet & Luikhart 1999, Rudnick *et al.* in press). As access to PCR techniques improves, it could be used to check the efficacy of feather-based identification at nests, and to register recruitment of hawks from which an initial sample was obtained in the nest.

ARTIFICIAL MARKERS FOR VISUAL DETECTION: RINGS (BANDS) AND DYES

Rings are the most widely used artificial markers for birds. Although some Goshawks are ringed at trap stations, most are ringed at the nest. Hawks less than five weeks old are not very adept at slashing with their feet. They can be pushed down on their stomachs by placing one hand on their back, then sliding the other hand to grasp their tarsi. Some people prefer to ring nestlings up the tree while others lower them in a bag to the ground to be weighed, measured and ringed by an assistant. An advantage in ringing at nest level in Scandinavia is that the mosquitoes are less troublesome than on the ground. In damp summers, mosquitoes gather in a following cloud on the way through woodland. Although it must be a tiny proportion that ever get a blood meal and concomitant chance to breed, none seem prepared to waste their opportunity.

Rings of different size are used for male and female tarsi. Most countries use split rings for wild Goshawks, which occasionally results in ring loss. Two Goshawks that had other markers in Germany were later net-trapped without their rings (Ziesemer 1981a), and at least one Swedish hawk with a similar ring was found after several months with the ring edges several millimetres apart, despite having been carefully abutted originally. The alloy of the rings was some 2mm thick, and it seems inconceivable that hawks could open such rings, bearing in mind the force required to do so by hand. Nevertheless, more than 100 hawks carrying radio-tags were retrapped on Gotland with rings intact, so the problem is rare and may only arise with some makes or batches of rings.

These metal rings are typically about 1cm deep, with impressed writing that can only be read in the hand. Raptors such as Goshawks can also be marked with colour rings that are 16–30mm deep and carry large digits for reading at a distance. Reynolds & Joy (2006) used two-digit colour bands, readable at 80m with a 20–40× telescope, for recording the return of adults to nests on the Kaibab Plateau in Arizona.

Fridjof Ziesemer (1982) developed an artificial complement to natural feather markings. A 3–4 digit stamp, with 5–8mm numbers that rotate on a rubber loop, can be used to mark codes with protein-fast dye on the undersurface of all primary, secondary and main tail feathers of nestlings and captured adults. Use of the correct dye (STK-BA 4710 from Hagedorn & Dänicke of Hamburg) is crucial. Secondary and tail feathers are not all moulted each year, so a bird can be identified in this way from shed feathers for up to two years.

To avoid risk of duplication and to obtain maximum help from networks of volunteers, those wishing to use any type of visual marker should first consult their national ringing authority. These organisations are always a valuable source of knowledge and may provide practical training, which is often a legal requirement.

MARKERS FOR ELECTRONIC DETECTION: MICRO-TRANSPONDERS AND RADIO-TAGS

Radio tags are the most widely used electronic markers for Goshawks. However, it is worth mentioning another class of Radio Frequency Identification Device (RFID) that has been used in research on wild falcons (Kenward *et al.* 2001a). Programmable Integrated Transponders (PITs), which are available as rounded cylinders measuring just 2×12mm, can be inserted under the skin at the base of the sternum to register individuals by returning a unique code when scanned by a high frequency activation signal. The returned signal is a product of resonance, rather than a power cell, and is therefore too weak to be detected beyond a few centimetres. Detectors can be set at nests to detect continued occupancy of breeders and recruitment of raptors marked as nestlings, but feather-based techniques are probably preferable for Goshawks. Eventually, micro-transponders may be used mainly as electronic passports for trained hawks (Chapter 10).

The details of attachment, tracking and data analysis for radio-tags are comprehensively covered in other publications (White & Garrott 1990, Kenward 1987a, 2001, Millspaugh & Marzluff 2001, Fuller *et al.* 2005). Valuable advice can be obtained from equipment manufacturers that specialise in research on raptors, and from experienced researchers. To avoid adverse effects on hawks, and the repeating of mistakes in data collection that reduce efficient use of scarce research funding, training is indispensable.

Initial studies involved tracking VHF (Very High Frequency) tags from vehicles, including aircraft, and on foot. However, Goshawks have now also been tracked with UHF (Ultra High Frequency) tags that transmit to satellites. Tags of 10–15g, representing 1–2% of hawk weight, can be mounted on legs and tail feathers. However, the smallest UHF tags are 15–20g including harness materials, and are too bulky to be mounted other than as back-packs, which keeps their mass close to the centre of lift. Harnesses of Teflon ribbon are ideal, with neck loops and body loops that can be adjusted in length while keeping a fixed ratio to each other, and separated by a strap along the keel that prevents pressure on breast muscles (a suitable design is in Kenward *et al.* 2001a). The correct fitting of harnesses is a welfare issue, so training is essential.

Training is not required to fit VHF tags to legs of nestlings, provided the design is satisfactory. A leather strap attached round the leg by rivets has proved satisfactory (Plate 17). The problem with these tags is to maintain an antenna length that neither breaks off nor impedes the bird and yet gives an adequate signal. A short antenna is adequate for tracking until dispersal, and can be used to monitor foraging after some practice in an area with excellent access, but signals are inadequate for tracking dispersal.

Tail-mounted VHF tags are satisfactory for tracking juvenile dispersal. A tag and a light antenna is sewn and glued firmly to one main tail feather, with attachment threads looped and tied round the base of the other so that the weight is shared but the feathers can moult independently (Kenward 1978a). This works so well that on one occasion, when the second feather moulted without the feather to which the transmitter was glued, the new feather grew down through the thread loop (F. Ziesemer pers. comm.). The transmitter is usually mounted on the two central rectrices (Plate 22a), but may be attached to the second innermost, with loops round the third, since the second innermost normally is late to moult (Chapter 4). Tags at 1.5% of bird body mass are ideal, because feathers tend to moult prematurely (typically after a few days) if there is more than 1% loading on each.

The type of radio-tag required depends on study objectives. Leg-tags fitted to nestlings are fine for recording locations prior to independence. Improved long-life designs may permit studies of recruitment and turnover at nests with these tags. Tail-mounts can be fitted only after feathers are full grown, which requires trapping after fledging, and last only until moulted, but give strong signals and are ideal for posture-sensors. These can be mounted to give signals that distinguish perching, pause-hunting flights, soaring flights, feeding, incubation and death (Kenward *et al.* 1982). Such signals are also available from VHF back-pack tags which can be fitted (just) before fledging and also send strong signals. At 3% of bird body mass, a female Goshawk in northern Europe can carry a 35g back-pack with four-year battery life.

VHF back-pack tags and tail-mounts on perched raptors of Goshawk size can be tracked routinely at 3–10km from mast-mounted vehicles, and are detectable at 40km in line of sight from hill-tops and aircraft. If locations are triangulated from within 1km with a three-element handheld antenna (Plate 22b) or within 2km using a vehicle-mounted five-element antenna (Plate 22c), location resolution is 100m which is adequate for most habitat studies. Finer resolution, for studies of sociality or predation, requires a closer approach and more skill to avoid disturbance. A solution for tracking Goshawks through extensive dispersal and inaccessible terrain is the use of UHF tags tracked by satellites in the ARGOS system (Sonsthagen 2002, Sonsthagen *et al.* 2006a, b, Underwood *et al.* 2006). The resolution of data from such tags is not better than 1–2km, and tags cost $4,000–$5,000 compared with $200–$250 for VHF tags.

Back-pack tags require very careful mounting, because they have caused abrasion, weight loss, nest desertion, reduced flying and increased mortality in other bird species. Of 36 UHF tags on adult female Goshawks in Utah, 21 produced stationary, 'cold' transmitter readings before the following April (Sonsthagen *et al.* 2006a). Such signals indicate either transmitter detachment or death, and there was no successful breeding among 11 survivors tracked the following summer. However, back-pack tags have lacked adverse impacts in other studies of goshawks, albeit mainly using VHF tags that are smaller than the UHF tags tracked by satellite.

A comparison of re-trap rates for 113 juveniles with tail-mounted radio-tags and 238 marked only with rings on Gotland gave a high probability (0.93) that the radios had no adverse effect on survival (Kenward *et al.* 1999). However, only two colour bands on 14 adult males fitted with tail-mounts in Arizona were recorded at nests in following years, compared with two on seven males with back-pack radios (Reynolds *et al.* 2004).

It is intriguing that Goshawks are the subject both of an unusually thorough 'no-impact' finding and what may be the only published claim of reduced survival caused by tail-mounted radios on land birds. However, Reynolds *et al.* (2004) noted that all their males continued breeding successfully after tagging and, as the difference in disappearance between years was marginally significant and depended on multivariate analysis with small samples, they recommended further tests rather than ceasing to use tail-mounts. The following parts of this chapter, and those that follow, show the benefits of using tail-mounted radios for studying behaviour and demography of the Goshawks on Gotland.

MOVEMENTS IN THE POST-FLEDGING DEPENDENCE PERIOD

On Gotland, young Goshawks were recorded out of the nest tree from their 39th day onward. The average age on which they had definitely fledged was 44 days for 34 males and 46 days for 33 females (Kenward *et al.* 1993a). However, as checks were at intervals of two to three days and hawks may have fledged and returned to the nest between checks, typical fledging dates are probably 40–46 days. Earlier fledging might be expected where there are nearby trees with good perches.

The first flights away from the nest often end with the hawk below nest level, after overbalancing on landing and fluttering down another branch or so in the adjacent trees. However, after practising on branches in the nest tree, the youngsters are adept at sidling up the branches until they can jump to a higher one and thus reach a height from which they can flutter back to the nest. For the first week after fledging, they still tend to sleep on the nest, and return there for food, but as their feathers grow and their prowess in flight improves they spend more time at greater distances away. By the end of the second week after fledging the hawks are travelling up to 300m from the nest, but they seldom exceed this distance until about three weeks after fledging (Figure 53). McClaren *et al.* (2005) obtained a very similar movement pattern for Goshawks on Vancouver Island.

Detailed analysis of their sociality showed that the young were cohesive, in that their movements tended to bring them together more often than expected if they moved about at random. In the second and third weeks after fledging, they were still quite often seen together, even perched shoulder to shoulder on a branch several tens of metres from the nest. If they were apart and one screamed at the sight of a returning parent, the others tended to move towards it and on to the nest and often screamed too.

During 65–80 days of age, most young hawks reached the peak distances from the nest that were recorded before their dispersal: up to about 1,000m. The 65-day threshold for a second phase of spreading coincided with completion of growth of their main flight feathers. However, the young hawks were still recorded within 300m of the nest during 74% of checks. Moreover, young females were still within

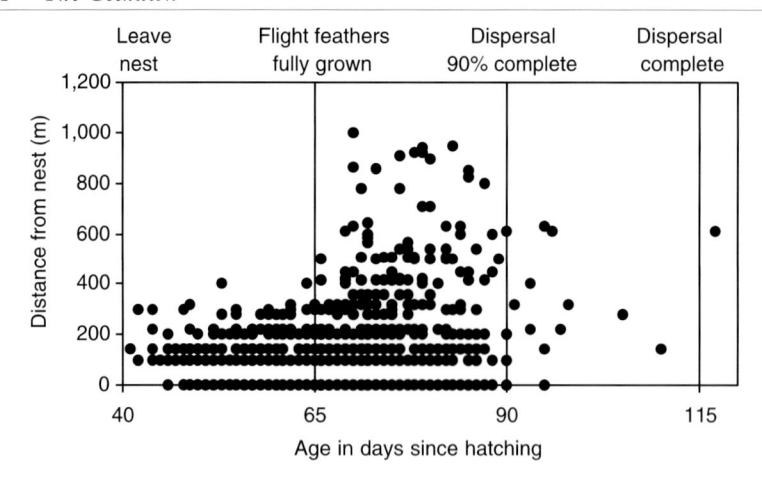

[Figure: scatter plot showing distance from nest (m) vs age in days since hatching, with phases marked: Leave nest, Flight feathers fully grown, Dispersal 90% complete, Dispersal complete]

Figure 53. Distances from the nest of 107 radio-tagged Goshawks on Gotland during the post-fledging dependence period, from fledging at 40 days of age. Data from Kenward *et al.* 1993a.

50m of the nest 20% of the time, and males 10% of the time (Kenward *et al.* 1993b). At other times, they perched near feeding sites, or wood edges from which the adults tended to approach with food. They flew out to meet incoming adults, sometimes taking food from them in flight (Brüll 1964). This tendency to meet the adults may take the youngsters to a separate stand from the nest. Their choice of perching areas may also be influenced by human disturbance.

However, cohesion was now weaker than in the first phase of spreading, and remained significant only for females. Behaviour of young Goshawks contrasted with young of the similar-sized Common Buzzard, which were not recorded at nests after their 65th day but remained strongly cohesive with each other (Tyack *et al.* 1998). Compared with the Buzzards, Goshawks were focussed more on the nest and less on each other. This was probably because the male at least continues to deliver food mainly to the nest. He is at some risk from the young, especially the females, and is now spending so little time in the nest area that he is rarely recorded there.

Adult females too are seldom recorded near the nest once young are fully developed for flight. This was shown elegantly for North American Goshawks by Dewey and Kennedy (2001), who compared distances from the nest of seven adult females that were fed near their nests with females at seven other broods. After the ten days from hatching in which the female broods chicks, the artificially fed females remained significantly closer to nests than the others and generally in nest stands until fledging (Figure 54). This was consistent with females maintaining a role of nest defence unless provisioning was inadequate.

Attendance of artificially fed females within 200m of nests remained at 60% of the time even in the phase when fledged young had incomplete feathers (and at 40% for other nests). However, after the young were 65 days old the adult females were

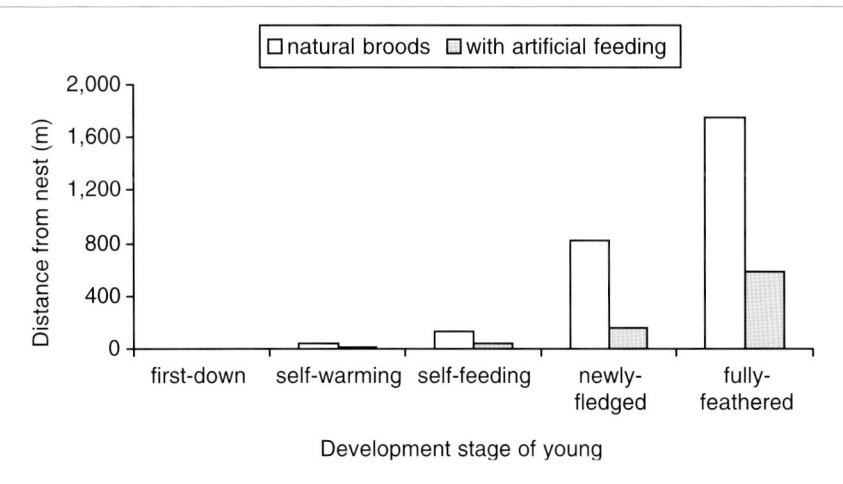

Figure 54. Distances from the nest of 14 radio-tagged adult female Goshawks in Utah were consistently small for those fed artificially. Data from Dewey & Kennedy 2001.

near 'treatment' nests only 18% of the time (and 0% for 'control' nests). In urban pairs, where breeding females remain near nests well into the post-fledging period, there is no evidence that they tend to avoid their young (C. Rutz, pers. comm.).

During at least the first three weeks of the post-fledging dependence period, young hawks on Gotland appeared to be much more interested in dead carcasses than in taking live food. Traps set for them at this time were most effective if baited with carrion as well as the usual live pigeons (Karlbom 1981). As well as completing their feather growth at this time, the young are putting on weight to a level well above the average for independent juveniles trapped in the autumn (Chapter 2). Their lack of interest in live prey may serve to prevent them wandering away and squandering their reserves in fruitless attacks before their powers of flight are developed to the full.

DISPERSAL

For the 212 young radio-tagged Goshawks that were monitored through the post-fledging dependence period on Gotland, dispersal was often abrupt and involved a movement of several kilometres without subsequent return to the nest. Repeated traverses of the same area then indicated that the hawk was establishing a home-range (Figure 55a). However, there were variations on this theme. Nearly 10% of young hawks (22) were recorded making excursive movements of up to 10km and then returning to the nest before finally dispersing. They might also move home-range again after initial settlement (Figure 55b). Others settled in areas adjacent to the nest and occasionally visited it during the following weeks (Figure 55c). Others made

126   The Goshawk

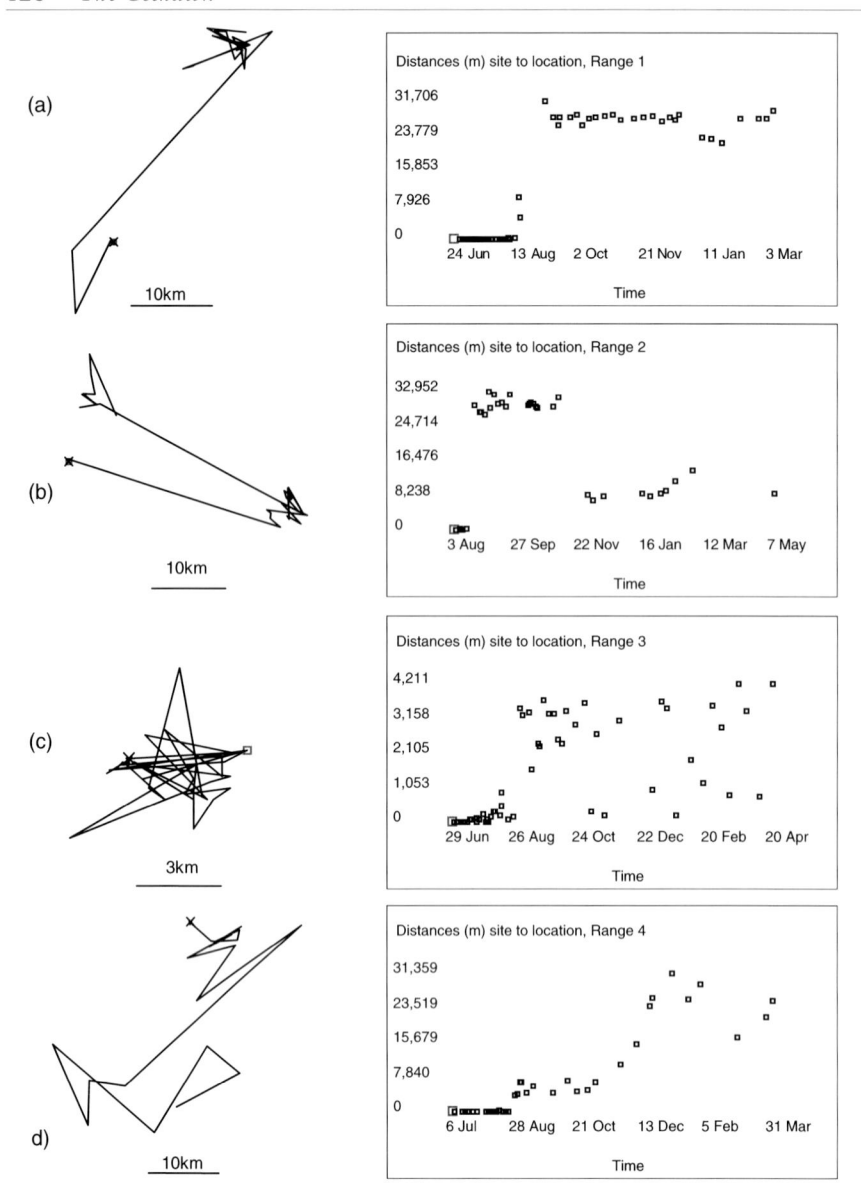

Figure 55. The paths of hawks radio-tracked on Gotland, showing (on the left) how they moved away from the nest (X) and (on the right) the distances of their locations from the nest site as months passed. Movement patterns differed for individuals that (a) dispersed abruptly and settled 30 km from the nest, (b) shifted range again in November, (c) made excursions and then settled near to the nest with occasional visits during winter and (d) moved in a nomadic fashion throughout the winter.

return visits to nesting areas in the following spring. A small minority drifted away in an ill-defined fashion (Figure 55d).

To avoid classing excursions as dispersal, we defined dispersal as exceeding 1.5km from the nest and not returning for at least two days (Kenward *et al.* 1993a). In the studies organised by Pat Kennedy in the USA, dispersal was defined as maintaining a distance of two kilometres for a week (Ward & Kennedy 1994, Kennedy & Ward 2003). That would have defined later dispersal dates for a few of the hawks on Gotland. Less subjective dispersal detection is now available as software (Kenward *et al.* 2002).

Dispersal occurred on Gotland mainly in the fourth to seventh week post-fledging for males and in the fifth to eighth week post-fledging for females (Figure 55). Overall, 90% of the young dispersed between 65 and 90 days after hatching, and 98% by 95 days after hatching. Males tended to depart before females at the same nest, and earliest from nests in areas with the least good food supply. The coastal areas on Gotland have relatively thin soil overlying the limestone fundament of the island, and support rough pasture, scrub and pine woodland with abundant rabbits. In the 136 km^2 of this habitat, in the east of the 846km^2 intensive study area, the bag of hunted rabbits on sample plots was one per hectare, compared with 0.01/ha on inland areas with fertile soil and intensive land-use. Within the area with abundant rabbits, there was one Goshawk nest per 10.5km^2, double the density elsewhere. Within the rabbit-poor inland area, Goshawks dispersed about a week earlier than on the coastal area, except when they were fed artificially at 12 nests (Figure 56).

Although an abundance of food tended to delay dispersal, the artificial feeding did not result in the young Goshawks on Gotland remaining for more than 90 days before they dispersed. Therefore, lack of provisioning was not the ultimate reason for dispersal. Nor did parent hawks drive young away: Figure 56 includes hawks from four broods where parents were live-trapped and removed to test that possibility. Indeed, radio-tagged adults on Gotland and in Alaska (Zachel 1985) were still bringing food to the nest area when dispersal occurred. The young hawks were either leaving early if parents had problems with adequate provisioning, or otherwise when they were ready to leave.

Differences between males and females were further evidence that dispersal reflected a 'biological decision' on the part of the young rather than by the parents. Not only were males less socially cohesive than females, but male dispersal was most accelerated by being in the rabbit-poor area. Males also left earliest when they were in large broods. The tendency for males in large broods to leave early depended on numbers of siblings, not numbers of sisters, so males were not being driven out by females.

Young hawks did not seem interested in hunting before they dispersed. We did not observe any prey capture attempts by young during 52 hours of watching four broods from hides, and young were more interested in dead meat in the traps than in live pigeons. However, two young that I tracked during dispersal each made their first kills within two days. This raises the possibility that dispersal is triggered by development of an increasing tendency to hunt. Such a tendency might well be accelerated by food shortage. It might also be stronger in males than females, as a

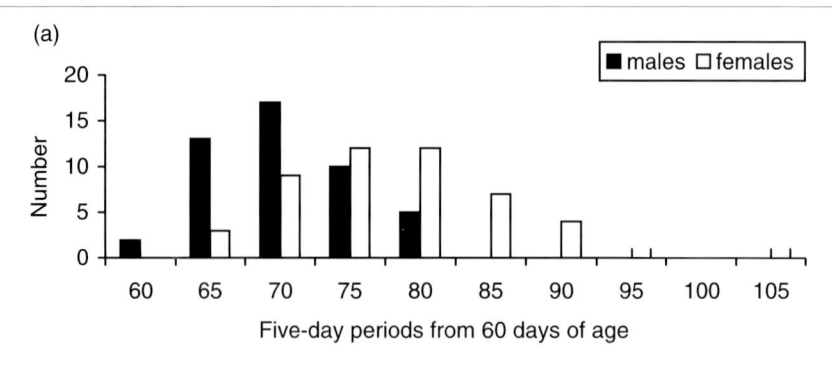

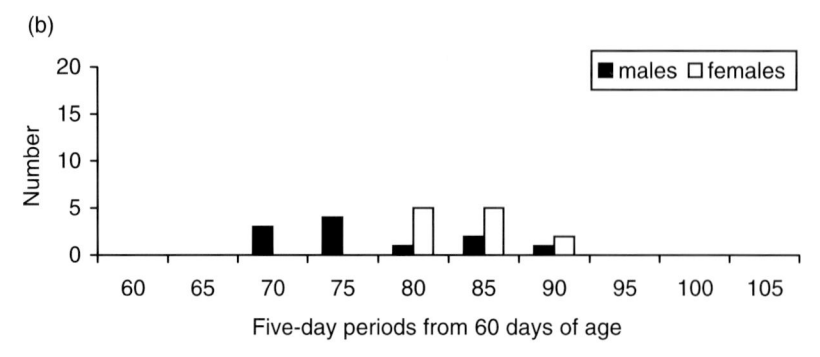

Figure 56. Dispersal of 119 Goshawks on Gotland that (a) were fed only by their parents or (b) also fed artificially, in an area with few rabbits. Data from Kenward *et al.* 1993a.

mechanism evolved to encourage finding a good feeding area in which later to provision a family. Those males that disperse early are most likely to find any unoccupied areas with a good food supply.

A 'hunting triggers dispersal' mechanism could even explain why young hawks in New Mexico tended to delay dispersal to as late as October when fed artificially (Kennedy & Ward 2003), whereas dispersal had ended by September on Gotland. Perhaps hawks whose initial hunting is unsuccessful tend to return after several days to their nests, and on Gotland their initial hunting was rarely unsuccessful.

POST-FLEDGING BROOD-SWITCHING

On Gotland, a relatively high density of Goshawk nests also provided another option for dispersers that did not kill quickly. Among 70 young males and 70 females that dispersed with tail-mounted radio-tags, seven males were recorded joining young at nests 3.3–22.7km away for up to ten days. These 'brood-switchers' had similar sociality scores at the nests they visited as at natal nests, so they were not being marginalised at the nests they joined.

Another 15 male and one female brood-switchers were recorded. These were trapped at nests in the post-fledging period and recognised because 12 lacked rings and four had been ringed in other nests. The female had moved the shortest recorded distance, just 2.5km to a neighbouring nest. All ten males of known origin came from the rabbit-poor area. Of the 22 male brood-switchers, 16 joined the 34 resident young at nests where there was hawk trapping in the rabbit-rich area and only six joined the 105 young at nests in the rabbit-poor area. Thus, the brood-switchers were almost always male, from relatively early nests in the rabbit-poor area, and they joined broods in the rabbit-rich area.

This behaviour is not unique to Goshawks or to Gotland, but is probably prevalent in many accipiter populations. One young stranger was trapped at a Goshawk nest in Germany (Kollinger 1975), and unringed youngsters have been trapped at Sparrowhawk sites where all the brood was ringed (Newton 1986). However, the radio-tagging on Gotland showed how frequent such behaviour can be, and confirmed that brood-switchers were not simply passing through but staying at the new sites.

Why did young hawks swap parents in this way? Typically, these were hawks that left their nest areas almost as soon as their main flight feathers were full-grown, rather than dispersing in the fifth or sixth week after fledging. They might then find a nest with more food, but there was a risk that they would not. If the broods they left were short of food, early dispersers improved the chances of survival of their sibs, with which they shared half their genes. By improving their nest-mates' chances they promoted the survival of their own genes, provided that the increase in their risk was not greater than half the risk reduction for their sibs. It is hard to put values on these risks, but the early dispersers did seem adept at finding new broods. One tactic they may have used was to give contact calls as they moved about, as a stimulus for other Goshawk broods to announce their presence by answering. I heard one young Goshawk screaming as it flew by in this period, the nearest nest being some 2km away.

But why did parents and young of adoptive broods tolerate the presence of strangers? Even if there was plenty of food, brood-switchers must have reduced its availability for the adoptive brood, and there must be a small risk for the adults associated with every extra hunting flight. An extra youngster reduces the chances of a passing predator killing one of the original brood. However, any dilution of risk may be offset by the tendency of predators to return to the site of previous kills. Goshawks themselves do this to broods of their prey (Schnell 1958, Sulkava 1964). Perhaps it is simply too costly to evolve a defence against the cheating. It would be costly for parents or young to kill or drive away any strangers if genuine kin were ever likely to be rejected by mistake. Goshawk parents certainly seem not to distinguish between their own young and strangers which are deliberately released near their nests as a means of returning them to the wild, and will also rear young Peregrine Falcons successfully from nestlings to independence (Trommer 1981, 1996).

FIRST-WINTER MOVEMENTS

The distance at which ringed hawks are recovered from their natal nest is strongly influenced by latitude. Young hawks travel further in the north (Figure 57).

Both the average distances travelled and the proportion of hawks recovered at longer distances increased strongly to the north. Thus, recovery distances of first-year hawks declined with latitude from an average of 377km in northern Sweden to 70km in the south (Fransson & Pettersson 2001). However, there is probably little further decline in distances for hawks south of 55°N, as studies in central Europe have recorded a mere 5–8% of Goshawks moving further than 50km from their nests (Glutz von Blotzheim *et al.* 1971).

The age of hawks at recovery provides some information on the timing of long distance movements. In the Swedish data reported by Höglund (1964a), only 3% of recoveries before the hawks' first September were at more than 100km in the south (and 17% in the north), but the 12% (and 55%) recorded during October and November was not much less than the 15 (and 67%) for December–February. Thus, most of the long distance dispersal was completed in the autumn. The proportion of long distance recoveries for hawks three or more years old was appreciably less than for juveniles in central and northern Sweden, and for juvenile males in Finland (Byholm *et al.* 2003). Shorter recovery distances for adults than for juveniles could mean that hawks were returning to their natal areas to breed after wintering elsewhere as juveniles, but could also result from the furthest moving juveniles dying before reaching adulthood.

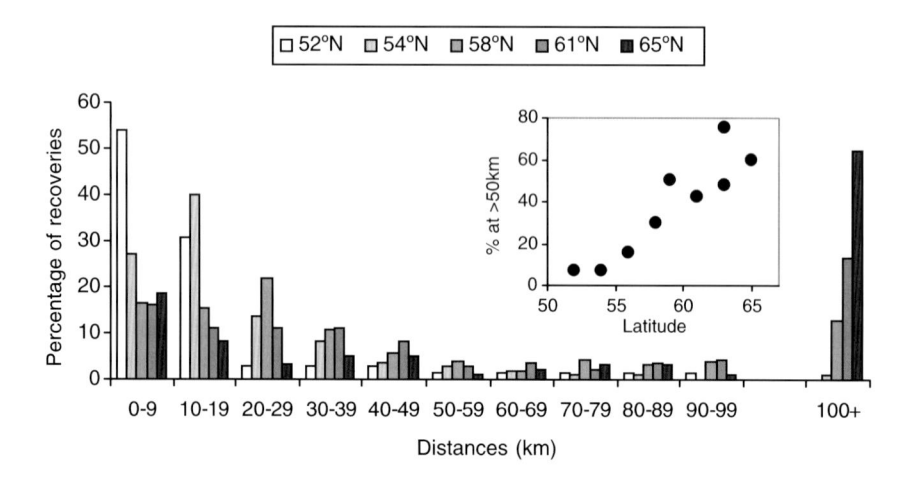

Figure 57. Distances from nests of origin to points of recovery for Goshawks ringed in Holland (Bijlsma 1993, *n*=189), Schleswig-Holstein (Looft & Biesterfeld 1981, *n*=110), Denmark (Nielsen & Drachmann 1999a, *n*=206), Norway (Sollien 1978, *n*=347), Finland (Sulkava 1964, *n*=158) and three regions in Sweden (Höglund 1964a, *n*=398).

Another complication in ringing estimates for the north of Scandinavia is the low density of humans. Goshawks from northern Sweden had about a third of the recovery rate of those in the central and southern areas with more human inhabitants, and a high proportion of the recoveries were to the south (Höglund 1964a). In Norway too, 40% of 404 recoveries were in directions SSE-SSW compared with 14% in NNE-NNW (Halley 1996). However, there was no such southward recovery tendency for hawks ringed at nests in Denmark, Germany, the Netherlands, southern Sweden or anywhere else south of Scandinavia. So was the tendency of northern Scandinavian Goshawks to be recovered further south simply a result of a greater reporting rate for hawks that travelled south?

The answer appears to be no, because there were nearly twice as many southward as northward recoveries among 34 northern Swedish hawks recovered within 50km of their nests, and thus still in the thinly populated north (Höglund 1964a). Thus, a preponderance of southward movements among northern Swedish hawks, with 19% of them recovered in central and southern Sweden (several hundred kilometres from their natal areas) does seem to reflect a tendency to move south. However, this raises another question. Are the long movements a one-way trip or do the travellers that survive return eventually to the north?

This question was answered by the ringing in autumn and winter of 1,213 Goshawks at stations run by game wardens in central and southern Sweden (Marcström & Kenward 1981b). Fifty-three hawks marked as juveniles and recovered after their first January had developed a strong northward tendency in their long-distance movements (Figure 58), despite the reduced likelihood of recovery in the north. These northward movers may not all have originated in the north. However, at least one hawk ringed at a northern nest was later retaken as an adult, after being recorded wintering far to the south (Nils Höglund, pers. comm.). Moreover, Halley *et al.* (2000) recorded a Goshawk radio-tagged in the nest that moved 75 km south-west to Trondheim for three winters, twice returning north past its nest area in spring to nest eventually about 100km northeast of Trondheim.

These records distinguish two types of movement shown by first year Goshawks. In central Europe and southern Scandinavia, most hawks disperse no more than 50km, in all directions from the nest. In northern Scandinavia some hawks are recovered after similar, short, omnidirectional dispersal at all stages until adulthood, but others make long southward movements. In some cases at least, these long distance movements qualify as migration between separate summer and winter ranges. We will now examine these movements in more detail.

HOW DISPERSAL MOVEMENTS OCCUR

In view of the gradual decline in ringing recoveries at increasing distances from the nest (Figure 57), it might be assumed that hawks drift gradually away from the nest until they settle or die. Radio-tracking provides a different picture, in that initial

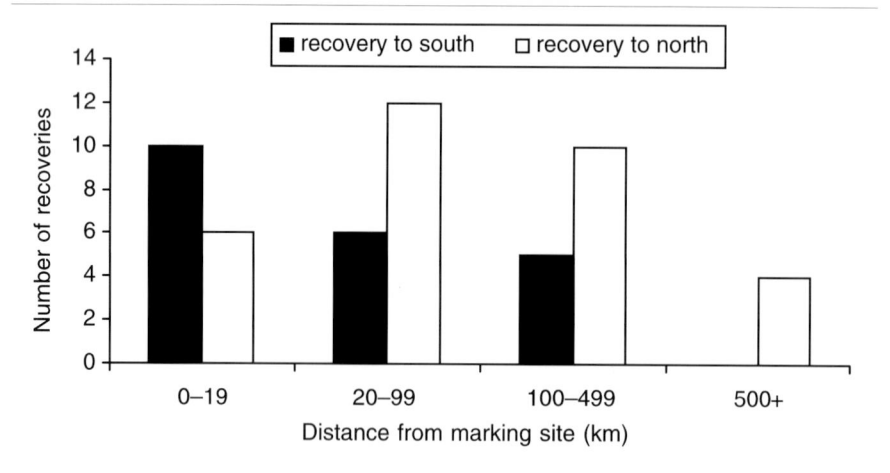

Figure 58. Distant recoveries of hawks marked in winter in central and southern Sweden tended to be to the north. For trend, p=0.02. Data from Marcström & Kenward 1981b.

movements away from the nest area tend to be abrupt. One day the young hawk is a few hundred metres from the nest, and the next it may be 20–30km away.

During abrupt dispersal of this type, young animals have three decision processes, affecting when to leave, how to travel and when to stop. Nest-mates may sometimes simply follow others, which could explain how five recoveries of two or more nest-mates have been in exactly the same direction, two of the hawks being caught in the same trap 35km from the nest (Lüders 1938, Höglund 1964a). However, such synchrony was not found among a further 53 nest-mate recoveries reported by those authors, nor among the 77 broods monitored by radio-tracking on Gotland (Kenward *et al.* unpublished), so it must be rare.

Moreover, lack of synchrony between nest-mates, which are all at a similar stage of development and are similarly fed, indicates that any tendency to leave in the same direction or wind conditions is weak. Nor is there any sign of a general tendency for dispersers to move with the prevailing wind. Many dispersal flights may therefore be made without the aid of the wind, except as a source of updraughts. Hawks in hilly country tend to move along valleys and ridges, often using the topographic wind currents to their advantage. Radio-tagged Swedish hawks also made use of thermals for dispersal flights, which were mainly at a time of year when warm weather created many such upcurrents.

Although there is no systematic information on how Goshawks move during dispersal, there are some data on when they move and plenty on how far they go. For Scandinavian Goshawks, radio-tracking on Gotland is complemented by data extracted by Patrik Byholm, Pertti Saurola, Harto Lindén and Marcus Wikman from vast ringing effort in Finland (Byholm *et al.* 2003), where 44,562 Goshawks were marked between 1913 and 2000, with 7,641 recoveries (Saurola 2001). A ring recovery shows a single distance and direction for an individual. Radio-tagging can show travels in more detail, and indicates that individuals often make more than one substantial movement in the course of dispersal. The two types of data permit inferences about the influences of sex, age, sociality, habitats and food supplies on dispersal.

Early analyses of recoveries from ringed nestlings in Finland found that juvenile male Goshawks tended to disperse further than females (Haukioja & Haukioja 1970). This is unusual in birds, among which females typically disperse further, possibly as an adaptation to avoid inbreeding (Greenwood *et al.* 1978). A 'juvenile-males-furthest' tendency was confirmed among juvenile Swedish Goshawks trapped and released in autumn and winter (Marcström & Kenward 1981b), but the trend differed for adults. Among adults it was the females that were most likely to move beyond a 20km radius (Figure 59), which reflects the maximum home-range span of settled hawks. The same effects were found among 11 radio-tagged juveniles and nine adults by Kenward *et al.* (1981a) and among 65 radio-tagged adults by Per Widén (1985).

Among recent recoveries of hawks marked as nestlings in Finland, not only did males travel furthest among those recovered as juveniles, but adult males were recovered closer than juvenile males to natal nests and adult females further than juvenile females, so that distances for adult recoveries were shortest among males (Byholm *et al.* 2003). This must result either from males showing philopatry by tending to move homeward after their first year, or from low survival of males that travel furthest (so that the adults' recoveries are mainly hawks that remained closest to home).

The timing of movements of radio-tagged hawks reveals more about what is happening. During the course of the autumn, many hawks that had settled into a home-range after dispersal made further movements and established new home-ranges (e.g. Figure 55b). Most of these movements were completed by December, and they involved mainly young males (Figure 60). December and January were then months with little movement, and in spring it was mainly the females that moved, especially the adults that had been radio-tagged in autumn and early winter.

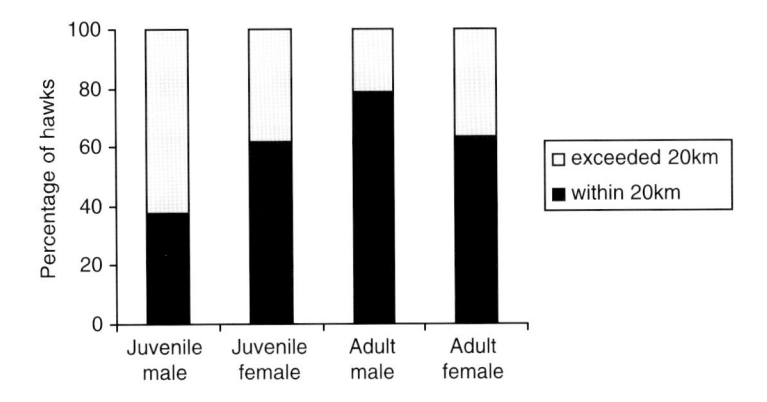

Figure 59. Male hawks ringed in winter had the greatest tendency to leave an area as juveniles, but the least as adults. Data from Marcström & Kenward 1981b.

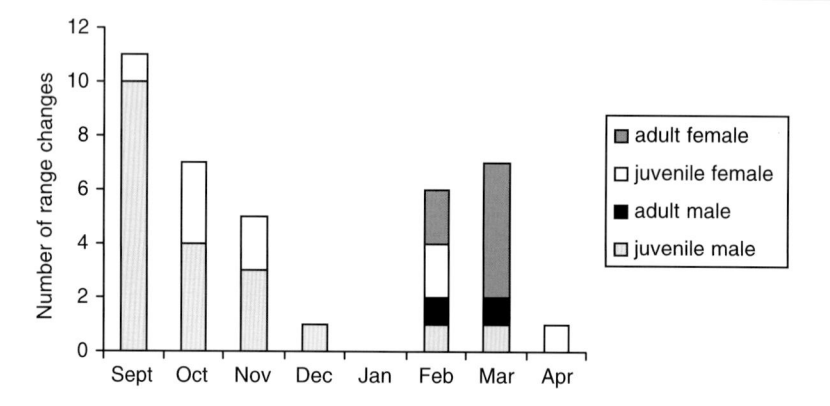

Figure 60. Changes of home-range of 38 hawks that had settled after the initial dispersal movement on Gotland mainly involved juvenile males and adult females.

In Finland, Sulkava (1964) noted that young hawks were recovered furthest from the nests in years when squirrels and tetraonids were scarce, and food seemed to be the key to movements of the young hawks on Gotland too (Kenward *et al.* 1993a). When young hawks dispersed, they tended to move in all directions until they reached the coastline, and then to move along the coast until they found a rabbit-rich area, as in the east of the intensive study area (Figure 61). A strip on the north-west of the intensive area was also soil-poor and rabbit rich, but was a military exclusion area and was therefore omitted from the road-bounded area in which all nests were sought and young marked each year.

As the rabbit-rich part of the study area was so attractive, hawks of both sexes that were raised there had moved shorter distances by September, after their initial dispersal, than hawks from the rabbit-poor area (Figure 62). However, rabbits that reach full size are twice the body mass of male hawks, and therefore a challenging prey. By the mid-winter period of minimal movement, males from the rabbit-rich area tended to have moved again, and to have travelled as far as hawks that started in the rabbit-poor area. In spring, when the relatively dense breeding in the rabbit-rich area would have engendered strong territoriality, the juvenile females that stayed until January tended to move away too.

In Finland, where the movements of large numbers of ringed hawks were studied in relation to counts of woodland grouse in 50-km grid squares, the tendency to move and the distances travelled were linked in the same way to food supplies (Byholm *et al.* 2003). Males tended to travel far unless counts of woodland grouse were high. Hawks were most likely to leave their natal grid square, and males were especially likely to travel far, if they hatched late, with hawks that hatched first in broods the least likely to leave a square. Conditions in nest areas, such as brood size and local habitat, also had effects on distances travelled by Common Buzzards (Kenward *et al.* 2001b), which raised the possibility that young birds may be 'primed' to travel far by their local natal conditions, irrespective of what they encounter on the way.

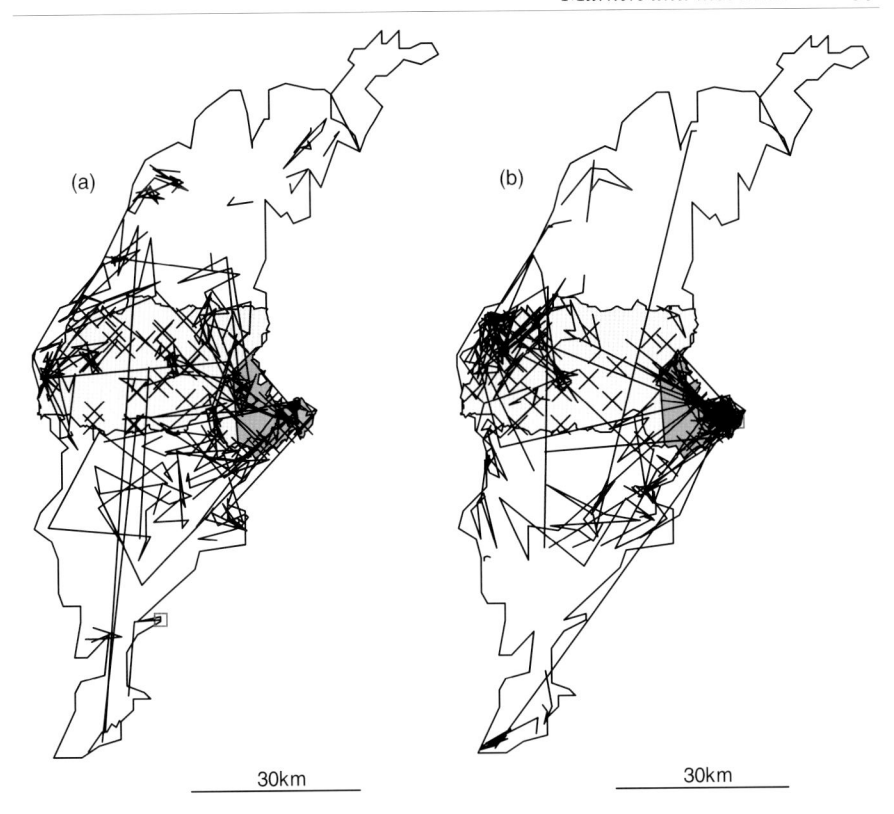

Figure 61. Tracks during October-April 1980–87 of juvenile Goshawks from nests (crosses) in the area studied intensively (grey) on Gotland. The 63 males (a) were less likely than the 65 females (b) to stay for winter in rabbit-rich coastal areas to the east (dark grey) and northwest of the main study area.

Other conditions, such as weather, may affect hawks during travels across wide areas. For example, poor weather may initially make prey vulnerable but in the long term reduce prey numbers. Thus, the distances travelled by Goshawks from winter marking sites increased with the duration of snow cover (Marcström & Kenward 1981b).

Although food availability seems to be important for dispersal movements, it is not clear how hawks are cued by the food supply to leave one area or settle in another. They certainly do not wait until the lack of food has greatly reduced their body mass before they leave an area. The masses of hawks that left an area shortly after radio-tagging in Sweden were, if anything, slightly higher than those that stayed (Kenward *et al.* 1981a). It makes sense for an animal to leave an area before loss of reserves reduces its ability to travel, and then to survive while it becomes familiar with a new range. Goshawks which moved from an autumn or winter home-range tended to travel increasingly outside the range, but without managing to kill, for two to three days before leaving completely. This would not greatly reduce their weight,

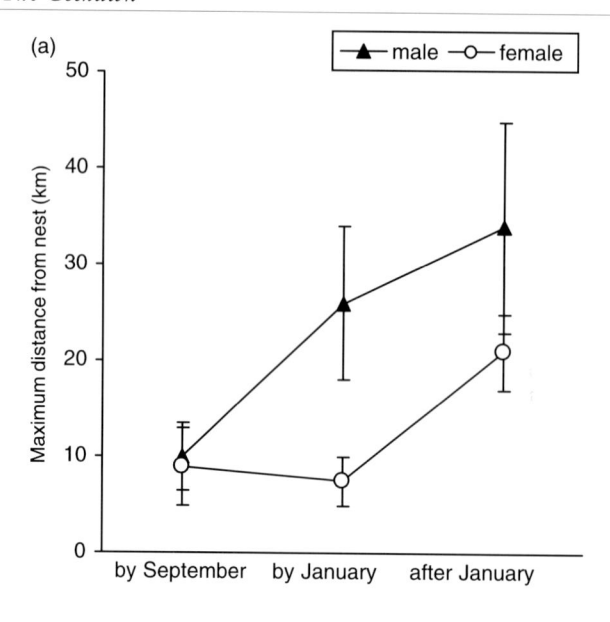

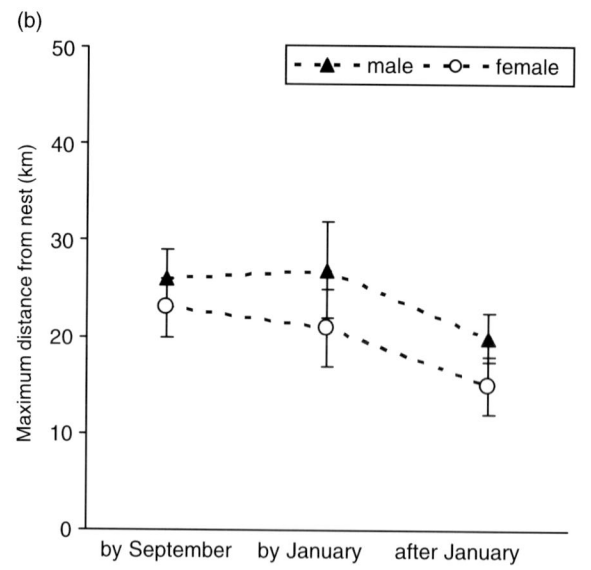

Figure 62. Maximum distances (means ± standard errors) from natal nests of male and female Goshawks that fledged in rabbit-rich (a) and rabbit-poor (b) areas on Gotland.

but they would certainly be hungry. Hunger during several days might be the cue to move, but failure to see prey in familiar areas, or after several hours of hunting, might also stimulate a hawk to seek a new range. Hawks might also respond to encounters with conspecifics, as competitors, or potential mates or merely as an index of good feeding conditions.

There is also evidence that the decision to settle in an area does not depend on the making of a single kill. Four hawks radio-tagged in their first autumn made substantial kills (three pheasants, one leveret) at 1–8km from the trap site, and fed for two or more days on the carcasses, but then left the area after hunting again for less than a day. Similarly, four hawks released in north Oxfordshire made several single kills without settling, but established ranges in areas where they made two or more kills in rapid succession (Kenward 1976a). A hawk may need to make another kill rapidly in an area, or at least to see much potential prey, before it will settle. This would reduce the chance of settling in a poor area as a result of one lucky kill.

On Gotland, although some philopatric excursions towards natal areas were recorded in spring, especially for male hawks (e.g. Figure 55b), range centres of males did not shift significantly homeward. On the whole, males completed their travels in the autumn, and were more responsive than females at that time to food supplies. There was a slight tendency for females that had remained close to natal nests through the winter to move outwards in spring. However, male range centres remained the furthest from natal nests (Figure 62). The tendency for male recovery distances to decline after the first year in Finland is therefore probably at least in part due to males that travel furthest being those least able to find adequate food. If these long-distance travellers are the hawks most likely to die young, then adult recoveries will include more of those that stayed at home.

One curious aspect of Swedish Goshawk movements is that so few hawks cross the five-kilometre Kattegat straits into Denmark and central Europe. Less than 1% of Swedish recoveries are from Denmark or further south (Höglund 1964a). Similarly, only six (1%) of 404 Norwegian hawks were recovered in countries south of Norway and Sweden (Halley 1996). Somewhat more (5–7%) of Finnish juveniles are estimated to reach southern Sweden, some perhaps by island-hopping and others when the Baltic freezes in winter (Marcström & Kenward 1981b); another 4% may have crossed water before recovery in east-central Europe (Saurola 1976). However, maps in Saurola (1976) and Halley (1996) show that many hawks accumulate on southern coasts, including the Swedish side of the Kattegat, and Goshawks are rare compared with Sparrowhawks at hawk-watch sites like Falsterbo (Roos 1974). During 1986–97, the mean annual count of Goshawks over Falsterbo was 47, 96% of them juveniles, compared with 14,653 Sparrowhawks (Kjellen 1998). Just 3% of progeny from Gotland crossed the 50km to the mainland (Kenward *et al.* 1999). Dispersing Goshawks seem not to like crossing water.

HOMING

Movements of hawks ringed and released by game wardens in Sweden showed some 'homing' ability, which could help emigrants return to natal areas (Marcström &

138   *The Goshawk*

Kenward 1981b). Among 286 released within 10km of the original capture site in autumn and winter, 37% were recaptured there, which was no great surprise as hawk home-ranges span up to 20km (Chapter 6). However, among 50 hawks released at least 20km from their capture site and retaken at least 20km from the release site (to eliminate those that may have settled where released), there was a significant orientation of recoveries towards the original capture site (Figure 63). This 'homeward' orientation remained significant even if hawks re-taken at the origin site were omitted (because those that went home were more likely to be re-trapped than those that moved to areas without traps).

The homing tendency was not very strong among these hawks, which were 84% juvenile, at distances greater than 30 km. Only three (2%) of 180 were recaptured at the original site after release more than 30km away. Goshawks probably only 'home' if their movements happen to cross ground they covered between their nest and the original capture site. On the 120 × 30 km island of Gotland, where movements are restricted by the sea so that many hawks travel over much of the island during their first year, four radio-tagged adults returned immediately to their original ranges after release 40–60km away. There are three reports of Goshawks homing from more than 200km in Germany (Rüppell 1948).

MIGRATION

The southward movement of Goshawks reared in northern Scandinavia does not involve the whole juvenile population. Risto Tornberg and Alfred Colpaert (2001)

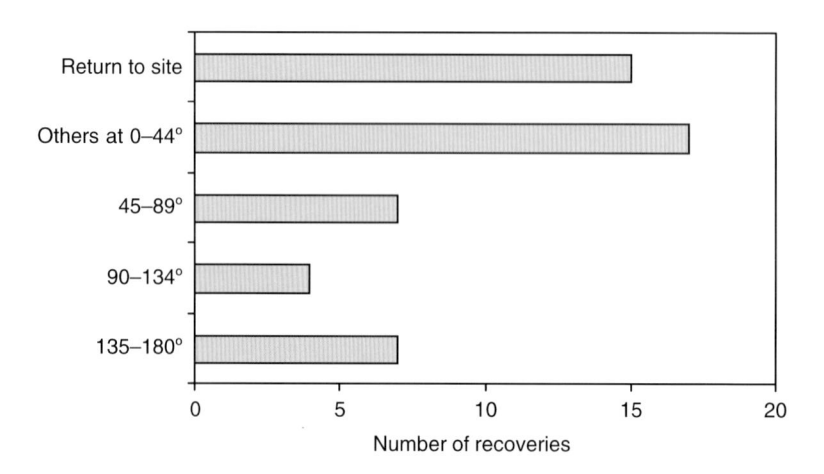

Figure 63. Among Goshawks released more than 20km from their capture site, even after discounting those re-trapped at the original site (top bar) there was still a significant excess of birds with recoveries in the homeward quadrant of 0–44°. Data from Marcström & Kenward 1981b.

trapped seven juveniles, seven second-year and 12 older hawks during November-February in a northern Finnish study area in winter. The birds had stable winter ranges and showed no tendency to move south. The southward passage is therefore no more than a partial migration of juveniles, and does not often (if at all) involve adults that have bred.

Radio-tracking of breeders tagged at nests has shown a stronger migratory tendency among adult Goshawks in the upland forests of the western United States. John Squires and Leonard Ruggiero (1995) tagged two males and two females at altitudes above 2,000m in Wyoming, and found that three moved quite strongly south in September. Two were lost at distances of 70km and 140km. The fourth travelled 65km WSW in November.

Three further studies with VHF tags tracked from the ground and aircraft showed that adults did not always migrate. Thus, Kirkley (1999) found five of 11 Goshawks in Montana near nest sites in January. Boal *et al.* (2003) recorded 25 of 26 adults within 12.4km of their nests during three winters in Minnesota, with mean distances from nests of 6.8km for 13 males and 6.9km for 12 sedentary females (one female settled further away in winter). Stephens (2001) found that all five males breeding above 1,400 m in Utah remained within 20km of nests, although winter ranges of two of them were at a lower altitude than their nests. Only two of 10 females in this study did not go more than 20km, but only one went beyond 100km. Their emigration was primarily in November and December, later than the usual autumn dispersal period, and may therefore have been a response to poor weather.

Sarah Sonsthagen and Jared Underwood, working with Ronald Rodriquez and Clay White, tagged adult females in six National Forests at 1,200–3,300m in Utah with UHF tags tracked by satellite (Sonsthagen *et al.* 2006a, b). Of 36 hawks monitored in 1999–2001, 13 (36%) left mainly in September and October to winter 100–613km away. Three others dispersed lesser distances in November. Only a single (100km) movement had a marked northerly component, whereas eight abrupt long movements were close to due south. The birds typically returned to natal areas in spring, and six of nine birds tracked for two winters exhibited site fidelity in both summer and winter (Stephens 2001, Underwood *et al.* 2006)

Combining data from all hawks that left their home-ranges, only two of 27 movements were to the north (Figure 64). Long movements tended to be before November, with two VHF values for September underestimated by tag loss. A similar seasonal decline in distances travelled by dispersing Buzzards was explained by deterioration in weather conditions suitable for prolonged flight (Walls *et al.* 2005).

All these ringing and radio-tracking studies show that Goshawks do make long distance movements with a southward orientation that qualify as migration, but it is partial migration. It affects part of the juvenile population in northern Fennoscandia and part of the adult population in upland western USA. A regular passage of Goshawks, along with large numbers of other raptors, is channelled past recorders by line features such as lake shores and mountain ridges in eastern and western USA (Mueller & Berger 1968, Bednarz *et al.* 1990, Bildstein 1998) and also in Eurasia.

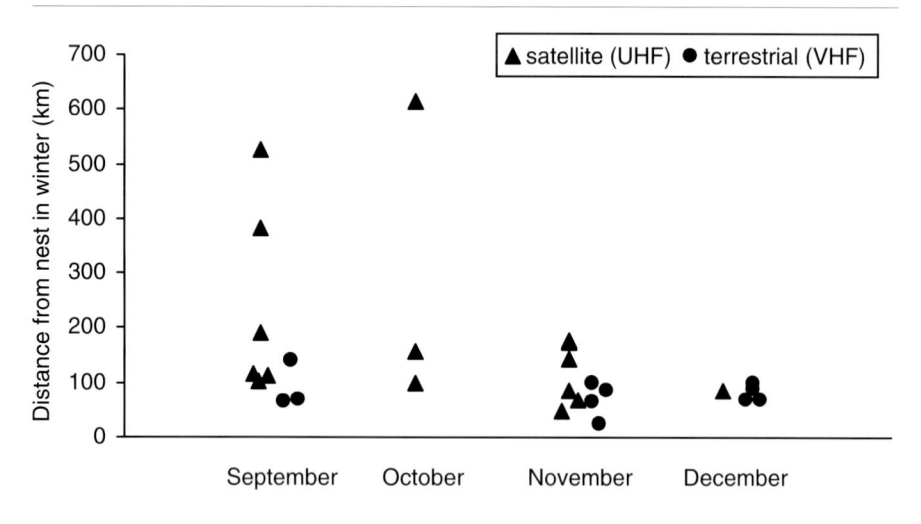

Figure 64. Radio-tagged adult Goshawks that moved >20 km from nests in Utah and Wyoming tended to travel further if their date of dispersal was early. Distance decline $p = 0.05$. Data from Squires & Ruggiero 1995, Stephens 2001, Sonsthagen *et al.* 2006a.

Although the Baltic tends to block migration from Fennoscandia, appreciable migration is observed at more easterly sites, such that Goshawks were 19% of 106 raptors observed during five days passing Vitebsk at 55°N in Belarus (Bashkirov *et al.* 2001). Proportions of Goshawks were lower further south, with 1–5% in counts of 1,100–2,900 raptors near Kiev, at 50°N in the Ukraine (Domashevsky 1995, 2001) and less than one per thousand among totals of 40,000–120,000 raptors passing towards Turkey at Adhzaria (44°N) in Georgia (Abuladze 1999, Abuladze & Edisherashvili 2003). Goshawk migration extends further south in *buteoides* country east of the Urals, with 9–17% among counts of 7,000–20,000 raptors skirting Lake Baikal at 50°N (Ryabtsev *et al.* 2001, Krasnoshtanova 2001, 2003) and 16% among 900 at Kungei-alatau (43°N) in Kirhizstan (Beishebaev 1984). In south-western Europe, fewer than one in 15,000 raptors crossing the Strait of Gibraltar are Goshawks (E.F.J. Garcia, pers. comm.).

IRRUPTIONS

At intervals of about a decade, a low level of Goshawk migration in America becomes an irruption of Goshawks into the northern states of the USA. The periodic 'invasions' have been well documented at a banding station on the shore of Lake Michigan by Helmut Mueller, David Berger and George Allez (Mueller & Berger 1967, 1968, Mueller *et al.* 1977), and by Lloyd B. Keith, who is well known for his explanation of North American predator-prey cycles (Keith 1963, Keith & Windberg 1978, Keith & Rusch 1988).

The key to these cycles is the growth of Snowshoe Hare populations to a level at which they over-browse their food supply, with the result that their numbers crash and few young are produced. Predator populations have been building up as the hares increase, and exert such heavy pressure on the smaller hare population that, although the hares are breeding well, their numbers are held down while the vegetation recovers from over-browsing. Predators turn increasingly to other prey species (including other predators) at this time, sometimes reducing the numbers of those species in phase with the hares. However, shortage of prey eventually reduces predator numbers too. The hare population then escapes from the temporary 'predation trap' and increases again to over-browse its food and start another cycle (Keith *et al.* 1977, Krebs *et al.* 2001).

When the hare population crashes, mobile predators such as Goshawks emigrate from northern areas. Much larger numbers than usual are then caught for one or two years at more southerly sites. Lake Michigan serves to funnel dispersing hawks along its shore to the Cedar Grove Ornithological Station (Figure 65). The majority of hawks are juveniles, except in the 'invasion years' of 1962–63 and 1972–73, which were associated with a marked increase in second-winter hawks in the first winter and in older hawks the following winter. However, most hawks were juveniles in the irruption of 1965–68.

Studies of Goshawks at Kluane lake, in the heart of Yukon 'hare-cycle country', show what probably happens north of Cedar Grove. In the year following a Snowshoe Hare population crash, there is little Goshawk breeding and therefore few juveniles. There is also increased mortality and dispersal of hawks, including adults (Doyle & Smith 1994, Doyle 2000). Under these circumstances one would expect the least experienced adults, the second-year hawks, to move out first. If the

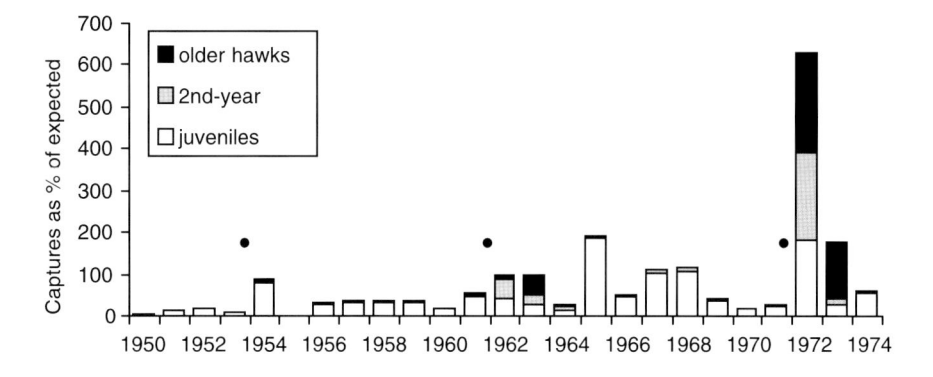

Figure 65. The annual abundance of goshawks, shown as a percentage of expected captures in order to correct for changes in trapping effort, at Cedar Grove Ornithological Station in Wisconsin. Black circles show the years (1954-5, 1962-3, 1972-3) when Snowshoe Hare populations crashed. Data from Mueller *et al.* 1977.

depression in populations of hares continues for a second year, and other prey are affected by predator switching, there may be few second-year hawks left and some of the older hawks are forced to move too.

This interpretation implies that the irruptions are simply movements of the type found in Fennoscandia and the western USA, but with stronger annual variation as a result of more extreme prey population fluctuations. Indeed, there are further similarities between the movement patterns registered at Cedar Grove and in southern Sweden. For example, males predominate among birds trapped as juveniles in non-invasion years, with a more normal juvenile sex ratio during irruptions (Mueller *et al.* 1977). Likewise, the excess of juvenile males in southern Sweden is less pronounced in colder winters (Marcström & Kenward 1981b). Similarly, the sex ratio among adult hawks, which does not differ significantly from unity in normal years, tilts in favour of females during North American irruptions and hard winters in Sweden.

In North America, there has been too little marking in the north to show whether there is a pronounced southward orientation to the movements of irrupting Goshawks, or whether they move in all directions with only the southward component being recorded. However, from the southward orientations recorded by ringing in Fennoscandia and radio-tracking in the western USA, such orientation is likely. It is also unclear whether an appreciable proportion of the wanderers later return north to breed. Some may do so, since three of four hawks banded in invasion years were later recovered 130–425km in northward directions.

There may also be irruptions from the Siberian taiga, but they are not well documented. Although more than 15 years of counts for two sites at Lake Baikal are now available for the *buteoides* heartland (Ryabtsev 2001), the only counts in *albidus* country east of the Verkhoyanski range are in single years at different sites at the Anadyr basin and northern Sea of Okhotsk (Enaleev 2001, 2003). There are occasional visits of *buteoides* hawks to Western Europe (Glutz von Blotzheim *et al.* 1971, Fischer 1980) and rare records of *albidus* hawks reaching China (see Chapter 9) and the Kurile Islands (Nechaev 1969).

TO MOVE OR NOT TO MOVE?

Throughout the Goshawk range, some birds disperse short distances in all directions, either drifting across country or shifting their home-range ten or more kilometres. Those hawks living in the northern taiga or the upland forests of the western USA face harsh winter conditions that vary in severity from year to year. Some of these emigrate southward and establish winter ranges, typically at lower altitude, from which they return to breeding areas in spring and to which they may return in later winters. But what determines whether an individual hawk flies south?

One possibility is the presence in some but not all hawks of a gene complex that obliges migration, at least in juveniles, such that individuals with genes for migration always move in autumn. A hawk which lacks the gene and remains in the north might disperse a short distance in adverse weather, but would avoid the risks of not finding a suitable area in the south and could be at a breeding advantage through familiarity with the local food supply in a good year. However, in a poor winter a bird with merely the 'disperser' gene complex might be much less likely than the migrant to survive and breed.

The disadvantage of a gene complex for obligate migration is that its proportion in a local Goshawk population might be expected to rise sharply following harsh winters, possibly leading to severe problems for offspring of such birds if southern areas also varied annually in their suitability. A more flexible genetic strategy would be for the gene complex to be facultative, triggering migration only if conditions started to deteriorate, perhaps to the level at which a hawk without migration genes would disperse. Indeed, the same facultative trigger could serve both dispersers and migrants.

With a facultative trigger based on food shortage, juveniles would presumably be more likely than adults to move as food availability declines, due to relative inexperience in hunting. Similarly, second-year hawks would have less experience than older hawks, and adult females less experience than males that have bred and have therefore successfully met the increased requirements of feeding a family. If genes were generally present to trigger dispersal or migration facultatively, the addition of genes for migration (present at higher frequency in hawks at higher latitudes and altitudes) could explain the different movement patterns observed across the global distribution of the Northern Goshawk.

PRE-NUPTIAL MOVEMENTS AND SITE FIDELITY

The detailed study of movements by radio-tagging also provided information on pair formation. The checks on hawks at intervals of one to two weeks on Gotland gave an indication of when males and females arrived at breeding sites (Chapter 3, Figure 28), but other studies have provided more detail for individual birds. For example, a radio-tagged German male spent 66 of 92 winter nights at the site where he subsequently nested, flying back after hunting up to 10km away (Ziesemer 1983). We noted similar long dusk flights back to nesting areas by radio-tagged male Goshawks in Sweden. In contrast, two German females and others in Sweden spent relatively little time at their subsequent nests during winter, but did roost there increasingly during the two months before egg-laying.

Records of courtship calling (Penteriani 1999, 2001) also show that both sexes of established pairs *are* sometimes present at sites in winter, which might be expected from the high site-fidelity of goshawks between breeding seasons. Change of nest site (and presumably of partner) was estimated for only 2–4% of females from feather markings in three European studies (Ziesemer 1983, Link 1986, Drachmann & Nielsen 2002), but this method is likely to underestimate movement between sites. An early study by banding in North America recorded 80% change for females

(Detrich & Woodbridge 1994), but two larger subsequent studies have each estimated 89–90% fidelity for females and 96% for males. Fairhurst (2004) recorded one male change of partner (to a nest 2.1km away) for 27 pairings without change, and eight changes of females (1.3–10.6km) for 62 without change, while Reynolds & Joy (2006) reported two male and five female site and partner changes in 50 opportunities monitored with bands and radio tags.

However, the question of which sex is first at a site when a pair forms will remain a matter of inference until Goshawks are tracked systematically to record life-paths from fledging to first breeding. One might expect the male to be first, because of the difference between his role and the female's during breeding. The female becomes increasingly dependent on her mate for food as her eggs develop within her and as she moults during incubation, such that her own survival, let alone her investment in the breeding attempt, is at risk if he is not a proficient hunter. As surmised by Ian Newton (1979), it therefore behoves a female to choose her mate carefully, whereas he should pick a site at which he is best able to provision a female.

On this basis, a male's best strategy for pairing would be to pick a site in his range within easy reach of his best hunting areas, and spend as much time as possible there when the courtship period starts, perhaps doing some preliminary nest building with conspicuous greenery to serve as a 'calling card' when he is absent (Newton 1979). A female on the other hand, should move around, perhaps even somewhat outside her winter range, to find her ideal male. If she has bred successfully before, she should go first to that site, which would explain why pairs often stay together, and also fits the finding that Sparrowhawks are more likely to breed again at a site if they have bred there successfully than if they failed (Newton & Marquiss 1982).

A female is not short of information for assessing whether a male is a competent hunter. First of all, was he at a breeding site to meet her? The more time he had to spare from hunting, the more likely he is to be there. Secondly, is he bringing her adequate food? Otherwise she may drift off hunting for herself and perhaps meet another male. Finally, is he attentive at the nest, engaging in a calling duet before he departs to hunt for that food?

Pair formation in this way predicts (1) that males should be first at the breeding site, (2) that sites should tend to be more central to the ranges of males than of females, (3) that whether a female lays eggs should link to male movements during the courtship period and (4) that in the year after a breeding failure, females should be more likely to try another site within their ranges (and thus a new mate) than males (who may well already be using their optimum site).

As Goshawks first bred in their second year on Gotland, the only data on prenuptial movements of definite first-breeders were those for second-year hawks. These were tracked too infrequently to detect differences between sexes in arrival records (Figure 28), and winter range centres could be estimated for only six hawks. The geometric mean distance of range centres from nests was 5km for four males, comparable with 3km for seven older male breeders and 4km for five female breeders in full adult plumage. However, two females in their second year travelled 13km and 61km to nest, at sites well outside their winter ranges.

There is therefore evidence from a variety of techniques and studies that female Goshawks are more likely than males to move in search of a suitable partner. Chapter 8 provides further information on how male and female movements link to breeding performance. In Sparrowhawks, male foraging during courtship is a good predictor of breeding success. Mick Marquiss and Ian Newton (1982a, Newton 1986) found that radio-tagged males in areas with the best food supply foraged closest to the nest, and that the foraging distances of females were reduced by providing food artificially. It is worth noting that if a female selects a competent male with whom to breed, she is also selecting for hunting ability in her offspring, inasmuch as this is based on genes. It is less easy to see how a male can select a female with whom he is most likely to father successful young.

IMPLICATIONS FOR CONSERVATION AND MANAGEMENT

Information on how animals disperse and settle is important for modelling colonisation. Raptors have at times become absent from large areas due to past persecution or pollution (Chapter 10). The development of domestic breeding techniques (Chapter 9) has been followed by many programmes that have successfully re-introduced small raptor populations (Cade 2000). Such programmes should benefit from modelling of dispersal movements to estimate rates of spread (South *et al.* 2002). For example, if spread is fast, it may be a waste of scarce conservation funds to arrange releases in an area that would be colonised naturally within a few years. On the other hand, if spread is slow, many small releases may be better than a few large ones.

One way of modelling dispersal distances is to use the distribution of distances from ring recoveries. This approach was used by Lensink (1997) to make estimates for British raptors, from which he suggested that unexpectedly slow recolonisation by Goshawks and Red Kites was due to continued illegal killing. Future models will need to allow for variation in distributions of dispersal distances (Figure 57). Reliable prediction of colonisation requires an understanding of the factors, such as variation in habitats and food supplies, which are likely to affect the movement decisions made by individuals in each area.

Radio tracking helps greatly to determine when dispersal has occurred, in order to analyse for effects of social and habitat conditions at departure from relatively few marked birds (Kenward *et al.* 2001b). Knowledge of weather at the time can help make inferences about its importance (Walls *et al.* 2005). However, it remains difficult to track many individuals systematically while they move, in order to assess what conditions influence settling, because dispersal flights are so infrequent.

Fortunately, the Northern Goshawk has been restored successfully to Britain and it remains abundant elsewhere. However, in case of future problems (and because the Goshawk may be a good test for techniques before application to rarer species) it would be useful to learn more about how juvenile dispersal differs from that of adults.

Do juveniles move more than adults merely because they are more vulnerable to food shortage? Do juveniles move differently, perhaps exploring routes more gradually so that if adults need to leave summer areas they can travel more directly to the best winter areas? Is apparent migration by some adults really a rapid re-enactment of a juvenile dispersal movement? Is there a dispersal/migration dichotomy at all, or merely variation along a continuous scale, as in so many other aspects of Goshawk behaviour? Will the current movement patterns of Goshawk populations help or hinder adaptation to changes in land-use and climate? With its ability to carry sophisticated tags, without travelling too far, the Northern Goshawk may be a good species in which to investigate genetic and other factors influencing migration.

Other important questions concern the immediate cues for dispersal and settling. Is it the failure to make kills in rapid succession? An understanding is important not only for individual-based modelling of colonisation, but also for release projects. During releases in Britain, it was my impression that a hawk would survive unaided if it had made two kills.

In assessing the risks to individuals through encounter with areas of pollution, or stock protection or other hazards (e.g. wind-turbines), it may also be important to understand how individuals move during their daily foraging or territoriality. These daily movements are described in the next chapter, after considering the diet that influences that foraging.

CONCLUSIONS

1. Rings, radio tags and other markers can conveniently be attached to young hawks in the nest, or to older ones after capture. The hung nets and nooses used to capture adults at nests must be closely attended, whereas regular checks suffice for live-baited box traps and kill-baited spring-nets left elsewhere.
2. Nesting Goshawks can be identified with colour rings, by logging micro-transponders or from moulted feathers by using natural markings, dyes or DNA analyses. Attaching radio tags to legs or tail-feathers is probably safe for hawks, but use of tags on harnesses, to gain power and life or for automated location by satellites, requires careful training.
3. After fledging to neighbouring trees, Goshawks remain within 300m of the nest for about three weeks until their flight feathers harden, at 60–65 days old, after which they routinely move within 1km of nests but with occasional excursions up to ten kilometres away. Nests remained a focus for brood care and adult females only moved far from them after young were flying strongly if provisioning was good.
4. On the Swedish island of Gotland, 90% of young Goshawks dispersed in the fourth to eighth week after fledging. Dispersal was delayed by good feeding, but all were independent by early September even if they settled nearby and occasionally revisited the nest.

5. Dispersal movements were abrupt and may have been triggered by initiation of hunting. Some 10% of young males, typically early dispersers from food-poor areas, joined fledged broods at other nests, typically in food-rich areas, for 1–10 days.
6. Young hawks tended to disperse early and further if they were male, from large broods and in years with poor food abundance. In contrast, adult males were less likely to change home-range than females, which may have been travelling to mate. The difference in natal dispersal distances of male and female Goshawks is brobably better explained by a need for males to find a good brood-rearing area than by avoidance of inbreeding.
7. Ringing indicates that the proportion of juveniles dispersing more than 50km from nests in Europe increases sharply beyond 55°N, from <10% to >60%. Some young hawks from beyond 60°N make long distance movements to the south in winter and then return. Among hawks released away from capture sites, there was some evidence of homing, perhaps through landscape recognition. Few Goshawks crossed expanses of water.
8. Radio-tagging of adults breeding at high altitudes in the western USA also showed return from wintering areas 100–600km to the south, especially among females, with some hawks (especially adult males) also dispersing short distances to lower altitudes. Site (hence mate) fidelity of breeders was 96% for males and 89–90% for females.
9. When Goshawks irrupt southward from northern areas after cyclical crashes in Snowshoe Hare populations, a juvenile passage in normal years is supplemented at first by second-year hawks and older females and later by older males.
10. Further information on the decision mechanisms and genetics that may underlie Goshawk movements for wintering and pairing would be useful for management projects and for modelling how populations may respond to changes in habitats and climate.

CHAPTER 6

Diet and foraging

I have checked from three points along the shore of the Mälaren and therefore have no doubt that the bearings to the tag cross on Björnö (Bear Island). The options are to drive 12km round via a bridge, or to cross 500m of ice, which the winter has now made safely thick. My winter boots have flanges at the front that slip into the bindings of the narrow wooden skis that Rolf Brittas chose for me in Uppsala. I check that the pair of short spikes on wood handles hang securely round my neck, in easy reach to bang into unbroken ice and drag myself out if necessary.

The best place to cross is a kilometre south of the hawk, where I have already formed tracks on the snow-covered ice while crossing to ring Bridget from a phone kiosk. This detour gives speed in the smoothed tracks, plus the safety of a tested route and an unseen approach to the hawk along the shore. She is no longer making hunting flights and is probably eating. My objective is therefore to stalk close enough to find the kill quickly when I disturb her from it, so that I can take measurements and leave her in peace.

This hawk is unusual, because she is routinely crossing two or three kilometres of water to forage on other islands and across this inlet of the Mälaren. Fridtjof

Ziesemer subsequently tells me of a hawk in Schleswig-Holstein that has a winter range with a 4-km estuary across it. Perhaps Goshawks are shy of crossing water only if they see no woodland on the other side. There is little but woodland on this island. The bears are long gone.

The radio signal is weak as I cross the ice, but as I approach her estimated location its strength increases steadily in my earphones, so she is definitely close to the ground. However, bearings and volumes along the steep slope to the shore can be fickle, due to bounce from trees and the steep slope of the terrain. Before the signal gets even stronger, as I come into line of sight from the tag and she might see me through the trees, it is important to detour in from the shore and take bearings down the slope. Fortunately there is a track along which I can move fast at a crouch until my Yagi antenna, with elements horizontal to reduce reflection from the trees, points straight down the slope. As I pass where she must be, the strongest signal is increasingly back towards the way I came. She seems to be near the shore. Without my skis I can use a gully to get very close.

At the mouth of the gully, I know from the bearing that I am just beyond her again. The noise of the beeps rises rapidly as I now move slowly straight towards her, then wavers and goes faint again. She has flown away low along the shore, and I didn't see her go. It is so much easier for a hawk to see my movement, as it sits small and inconspicuous, and I seldom see the wing-beat of a discreet departure that is low with the weight of a kill.

From an easy stalk with such good cover, I have only five metres to search to the sides of my final approach and I rapidly find this kill. On a fallen birch trunk, and beneath, is a circle of reddish hair and grey guard fur plucked from a squirrel. The intestines lie to one side, near the end of the tail. She has taken the rest with her. A week later, I pass the same way and stop to look again at the kill site. In the meantime, a little fresh snow has fallen, and melted. With difficulty, I find a few squirrel hairs pasted to the top of the fallen birch. Had I not known the story, I would have been blind to the evidence of this past meal.

STUDYING DIET AT NESTS

The diet of Goshawks has been studied in several different ways, including the collection of prey remains at nests and elsewhere, the direct observation of prey brought to nests or killed elsewhere, the analysis of stomach contents from shot hawks, and the finding of fresh kills by radio-tracking (Rutz 2003 and Lewis *et al.* 2004 provide concise recent reviews in Goshawk studies). All methods have their pros and cons. This chapter first describes the study methods and then the understanding that they bring, not only about the Goshawk diet but also about how they catch their prey and their foraging in general.

The collection of prey remains at nests, either as pellets or bones in and beneath the nest, or as remains at plucking posts, is the quickest and thus the most common technique. By the middle of the 20th century, Üttendorfer (1939, 1952) had amassed 9,022 such remains from German Goshawks, not to mention thousands more from other raptors. However, Seppo Sulkava (1964) found several inadequacies in this method when he compared it with two other techniques at nests in the forests of central Finland. Even collecting the remains every two days, he found that they represented only 30–40% of the prey recorded when he sat in hides and watched the parents bring kills to the nest. At some nests there were hardly any remains to be found on the ground, probably because foxes or badgers had removed them. Moreover, the method biased against recording small prey such as woodland grouse chicks, which tended to be eaten completely. There was also a bias against recording the largest prey, whose remains were carried from the nest area by the female, so that medium-sized prey species were over-represented in the samples. The only time the prey remains gave a good picture of the prey spectrum was during courtship and incubation, before prey species had young of the year.

An alternative method for use during the nestling period, once the young Goshawks were old enough to be left unbrooded, came from the USA (Selleck & Glading 1943) and was first used for Goshawks by Nils Höglund (1964b). This involves placing a wire mesh cage over the young, so that they cannot eat the prey brought for them. The kills are counted and the young fed at the end of each day. The female parent must be trapped and held captive; otherwise she may either 'tidy' the kills from the nest or feed prey to the young through the mesh, morsel by morsel. The 'cage technique' thus records only prey taken by males, but little bias is likely because the male would normally contribute the majority anyway (Chapter 4). Sulkava (1964) found that the cage method gave results similar to those obtained by direct observation, but was more time-consuming than collecting prey remains, was interventionist and could only be used for a small part of the breeding cycle.

Unfortunately, although watching from hides gave the best data, it was also the most labour-intensive method and it was not always possible to identify the smallest prey items before they were consumed. A solution in a Sparrowhawk study by Tim Geer (1978, 1979) was to use a long grasping tool to reach into the nests from a hide on a nearby scaffolding platform, which sometimes produced a tug-of-war with the adult female. However, hides can seldom be built close enough to Goshawk nests. Early attempts to automate nest observation, as pioneered for Goshawks with camera traps on bait in winter (Tømmeraas 1977, 1980), failed to overcome issues with triggering. Modern continuous video-recording is more satisfactory (Grønnesby & Nygård 2000, Lewis *et al.* 2004).

Stephen Lewis identified 1,541 prey from 5,834 hours of recording at ten nests in the southwest Alaskan archipelago during 1998 and 1999, also collecting the prey remains and pellets at visits with two- to three-day intervals. He found that a miniature camera at an average distance of 40cm, focussed on the nest bowl, identified 96% of prey deliveries to class (bird or mammal) and 82% to genus (Lewis 2001). Whereas video-recording indicated that mammals were 22% of prey deliveries, they were only 9% of prey remains compared with 41% of prey identified in pellets. Small birds and nestlings were strongly under-represented in prey remains, as noted also by Grønnesby & Nygård (2000).

This contrasts with a finding by Christian Rutz (2003) that large prey recorded by radio-tracking were underestimated among prey remains scanned for at nests (though only by 10–15%). This was because large prey had been plucked most extensively before being transported to nests, perhaps to reduce drag and mass carried. Taken together, these studies confirm Seppo Sulkava's finding: the occurrence of the largest and smallest species in the diet is underestimated by collecting prey remains. A compromise solution might be to combine remains and pellets (Simmons *et al.* 1991), but the use of prey remains alone gives greatest comparability between studies, with similar bias throughout. Video-recording seems very promising for accurate assessment of provisioning, but will underestimate predation rates on the largest prey (Rutz 2003).

STUDYING DIET IN WINTER

Raptors which hunt in the open can sometimes be observed well enough to record their kills at a distance. However, Goshawks are so elusive that few kills are seen by chance. The analysis of pellets from roost sites of radio-tagged hawks is also impractical, because roosts vary from night to night, and even if a roosting hawk is found without disturbance by radio-tracking, there never seems to be a pellet under the tree. Perhaps the pellet is not cast until after the first flight in the morning.

When many Goshawks were being killed legally, stomach content analyses were used to study winter diet (von Bittera 1916, Bent 1937, Höglund 1964b). High proportions of domestic hens or pheasants in stomachs indicate that hawks were not sampled randomly but instead were biased towards hawks foraging near human habitation.

Another traditional technique has been to search study areas for prey remains. This method was used in scouring large tracts of countryside by students under Heinz Brüll in Germany (1964), and by Paul Opdam and co-workers in Holland (Opdam *et al.*1977), while Göran Göransson (1975) and Fritjof Ziesemer (1983) searched more restricted areas. Assuming that all habitats are searched equally thoroughly, so that there is no bias towards recording the sort of prey killed where it is easiest to search, there is a risk of underestimating predation on species with inconspicuous remains. It is much easier to spot the pale feathers from a pigeon kill, which can remain conspicuous for several weeks, than the hair from a mammal. Ziesemer (1981b) showed that visual searches could find all the pigeons killed by radio-tagged Goshawks, but only one in three pheasants, which are relatively drab-coloured, and one in eight rabbits.

It is also important to be able to attribute kills to the correct species of predator (Opdam 1975). Goshawks usually pluck part or all of their prey quite thoroughly before feeding, so that a circle of fur and feathers is spread round the carcass. If the prey is large, the hawk will then feed for one to two hours, daylight permitting, resting at intervals, and there will usually be one or more long white mutes extending radially from the kill. A recently fed hawk may have lifted its tail and squirted a 40–100cm streak up to 1.5m from a kill on the ground, even further from a carcass on a plucking post. Buteos and owls are not usually as profligate with either their plucking or their mutes, while falcons drop their faeces instead of squirting them. Large falcons are also more likely than Goshawks to bite notches in the sternum, while eagles also damage or eat bones which Goshawks would leave whole. The distance between the beak-edge crush marks on the shafts of plucked feathers further distinguish Goshawks' kills from those of smaller accipiters. What, however, if a Goshawk has killed shortly before dark, eaten rapidly with little plucking, and left no mutes before a fox found the kill in the night? This is not a trivial problem in some areas, where foxes scavenge more than half the kills (Ziesemer 1981b).

Radio-tracking is the most precise technique for studying predation away from nests. It provides information not only on what hawks kill, but also on when they kill, how much each hawk kills and eats, and on other aspects of their foraging behaviour. It is not biased towards finding kills in particular habitats, or when remains are most conspicuous. On the other hand, radio-tracking is biased against recording the smallest kills, to an extent that depends on how the technique is used.

If the radio-tagged Goshawks are tame enough to be followed fairly closely, either through use of falconry techniques (Kenward 1979) or habituation to people in towns (Rutz 2003), they can be monitored closely enough to record any stop on the ground. This probably records all kills down to thrush size, as well as the death of an occasional vole. Following single hawks continuously gives the most accurate assessment of their diet obtainable without random killing to record stomach contents. However, it is a time-consuming way to gain data in winter, averaging at best only about two kills in three days and half that rate for large prey. During the breeding season, 3–6 kills per day can be recorded (Rutz 2001).

To gain winter data more rapidly, continuous monitoring of individuals can be replaced by interval-sampling several at a time. Kills weighing above 250g (e.g. pigeons, corvids, game birds, squirrels, lagomorphs) provide a hawk with a good meal that normally takes at least an hour to eat. On large prey the hawk will feed more than once and probably rest nearby between meals. Checking hawks at one- to two-hour intervals therefore results in most kills of at least 250g being recorded. However, data from radio-tagging in Sweden, compared with stomach contents of Swedish hawks and their estimated food requirements, suggest that only half the smaller kills are found (Kenward *et al.* 1981a).

Posture-sensing tags are essential for efficient intermittent sampling, to help detect feeding (Chapter 5). Otherwise, much time can be wasted stalking hawks that are resting on the ground without kills; a young female once did this for an hour after struggling in vain with a hare. Given a good road network, four to six hawks can be monitored at a time to record three to four daily foraging locations from each and most of their kills. Mats Karlbom once recorded 24 kills in ten days on Gotland. Although this method tends to use much fuel in travelling from bird to bird, costs can be reduced by recording kills only when hawks enter an area of a few hundred hectares (Kenward 1977, Ziesemer 1983).

FORAGING BEHAVIOUR

Colonel Meinertzhagen, a keen observer of raptors who travelled widely in the early part of the 20th century, considered the Goshawk to be primarily a 'still hunter' that found prey mainly from a perch rather than in flight (Meinertzhagen 1950, 1959). A number of authors have described occasions when Goshawks have stayed on a perch for some time before launching an attack, often using their great acceleration and agility to intercept or overtake prey. The short wings and long tail of the Goshawk suit it for such flights in woodland and edge zones (Brüll 1964, Wattel 1973).

Other writers have described Goshawks diving on prey from a soaring height, with a stoop like a falcon (Kollinger 1962, Erzepky 1977, Rutz 2001). Eberhard Hantge (1980) recorded 150 such flights in a total of 186 observations of hunting Goshawks, at 50–599 m above the ground. However, he realised that the high visibility of these flights might over-emphasise their importance. He suggested that between February and November, when thermals aid soaring, up to 30% of hunting flights might be of this type.

Nick Fox (1981, 1995) noted a tendency for inconspicuous flights in prospecting for prey. These might be low and contour-hugging, or from tree to tree, as when a hawk stalked ducks that could be heard calling on the other side of a wood. Hawks sometimes also make indirect approaches to sites in which prey has presumably been encountered previously. Christian Rutz (2001) described urban Goshawks flying along below the roof-line of housing blocks and flipping abruptly over the roof to dive on birds in courtyards.

To these strategies of searching for prey may be added various styles of attack that occur after prey has been detected (Fox 1981), some as direct flights or dives and others with a degree of concealment. Attacks may start with an inconspicuous approach, for example as a prolonged glide without wing-beats, or following a flying bird in the blind spot below its tail until it slows to land, or taking an indirect route that uses cover for as far as possible. An example included a hawk flying along the surface of a stream to a point at which Moorhens, feeding in a nearby field, could be approached from the direction in which they would seek cover (Fox 1981).

Attacks on distant prey such as Woodpigeon flocks tend to have two phases (Kenward 1978b). The initial approach in low 'wing-beat-and-glide' flight typically turns into a 'chase' phase as the hawk nears its goal. In the chase phase, the wings beat continuously except for possible course adjustment in the moment before impact, when the feet shoot forward of the head (Goslow 1971) as so acutely depicted in paintings by Bruno Liljefors (Marcström *et al.* 1990). This final speed increase is especially likely if the prey bolts or takes flight when the hawk is close. In contrast, if prey responds before there is any chance of capture, the hawk abandons the attack at the end of the approach phase and heads for a nearby perch.

The chase phase of an attack can become quite complex. Madsen (1988) describes a hawk that had apparently learned to dive into a flock of Barnacle Geese as they flew off, to break them up and destabilise individuals. Westcott (1964) described a hawk harrying an Abert's Squirrel for 15 minutes, screaming at it and apparently trying to exhaust it in the initial stages. Occasionally Goshawks are observed harrying prey by hunting in pairs (Berndt 1970, Fischer 1979).

Alternatively, there may be no approach or chase phase at all, as Goshawks are great opportunists. Christian Rutz (pers. comm.) watched a Carrion Crow mobbing a soaring male Goshawk. After avoiding five or six persistent swoops, the hawk turned round on its back and grabbed the crow in mid-air, and they fell together into the forest underneath.

Whether hunting from perches, while soaring or in low flights, the Goshawk is a typical 'pause-travel' predator (Andersson 1981, Tye 1989), like Sparrowhawks and Hawk Owls (Newton 1986, Sonerud 1992). But how important for Goshawks are the different search strategies and styles of attack, and how successful are they? Answering such questions requires systematic observation, rather than chance observations that may favour the most visible techniques. Systematic study has required continuous monitoring with radio tags, which began with Goshawks released in lowland Britain.

STUDYING GOSHAWK BEHAVIOUR IN LOWLAND BRITAIN

Goshawks were released in north Oxfordshire during 1974–5 to study their behaviour and predation on Woodpigeons. The radio-tracking started with radio-surveillance, in which the equipment was mainly an aid for making visual observations of the foraging hawks. The countryside in the study area was an undulating series of ridges and valleys. Most of the woodland was in the valley bottoms, which also contained streams, ponds and more open water, as the occasional lake in parkland associated with large country houses. My aim was to watch hawks from suitable distant vantage points, using radio to find them again when they went out of sight, to get as unbroken as possible a record of their activities. The land was 8% wooded, the average for Britain as a whole at that time in the early 1970s, and the trees were almost entirely deciduous. Lack of leaves on trees in winter was of great importance for keeping hawks in sight, while the ridges provided excellent vantage points for watching their movements in the valleys. They spent most time in the valleys and only occasionally crossed the ridges, sometimes then being lost from sight for a while.

Tracking Goshawks in that way was fun. At first I was unfamiliar with the equipment, and relatively unfit, so that it was a mental and physical struggle to keep in touch with a fast-moving hawk. Then, gradually, a practical technique developed, as I learned the best routes across country and patterns started to emerge in the hawk behaviour. I was already familiar with Goshawk hunting as a falconer and began to 'think like the hawk' predicting the next moves with increasing success. Such a subjective approach might be undesirable when sampling data from many hawks, for fear of focussing on hot-spots and missing excursive activity, but was helpful during continuous monitoring of individuals.

New discoveries were frequent, little facts which hardly count as quantitative data for scientific papers but which illustrate how hawks live. For instance, when I used the approach in Sweden, a hawk had gone to roost after eating a pheasant on a molehill-sized mound in parkland. Sometime that night the snow started to fall, blanketing the ground to depth of about 8cm by morning. Before the snow, a fox or maybe a cat stole away with the hawk's pheasant carcass. Nothing remained above the snow to mark where the kill had lain, but the hawk's first footsteps in the snow showed that she alighted at exactly the right spot next morning to continue her meal. When she found nothing under the snow she dug again a wingspan away and then flew off. I had often wondered what cues a hawk used when returning some distance to a kill, and imagined that sight of the kill would be of prime importance. In this instance it was clear that the hawk had a spatial memory at least as effective as my own.

As well as these small discoveries, there is the joy of watching the hawk in flight and the satisfaction of finding its kills. These become satisfying not only in the sense of data acquired but also because one shares in a hawk's success no less than when the home team scores. There is compensation for the inward creep of cold on a frosty morning as one waits motionless for a nearby hawk to idle away an hour, and to lighten the five-mile muddy tramp at dusk to a car left at the other end of a bird's range. Yet, at the end of a long field season one begins to long for the warmth of an office and laboratory as much as one yearns for the open air after a few months of computing and writing indoors.

FORAGING IN LOWLAND BRITAIN

In the Oxfordshire countryside it soon became evident that hawk flights tended to follow particular paths. Two different birds would follow the same line across country, perhaps because from any point there was one distant perch which stood out as more inviting than others. The birds tended to move along linear features, such as streams or old railway lines, probably because these were associated with rough ground, holding trees and cover for both hawks and their prey. Since these features and woodland were in the valleys, it is hardly surprising that hawks were there more often than on the exposed ridges, and therefore spent their time at below average altitude. They also perched in woodland about four times as often as expected from a count at all the 250m grid intersections within convex polygons round all their locations (Figure 66). Moreover, these polygons contained higher levels of woodland and parkland than in north Oxfordshire as a whole.

156 *The Goshawk*

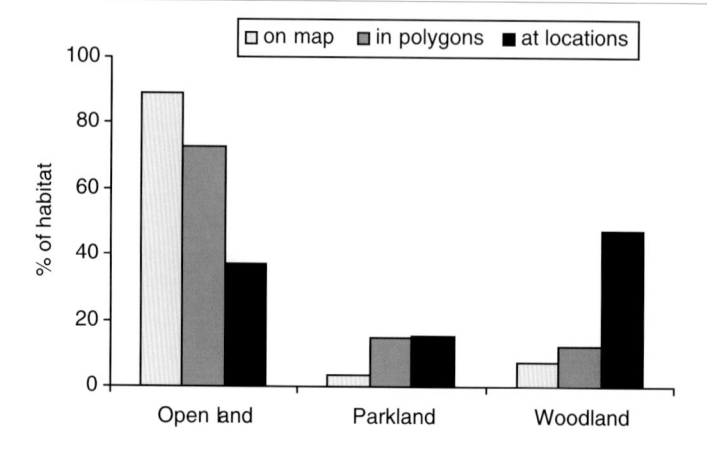

Figure 66. Goshawks in Oxfordshire foraged in areas with more woodland and parkland than available generally in the county, and were located most of the time in woodland. Data from Kenward 1982.

Hawks tended to loop back on themselves as they moved across country, often returning to near the start of the day's travel. Over several days these movements might result in a 'drift' to a different area, or might tend to be superimposed such that a hawk settled in a particular range, reusing particular hunting areas and roosts. A drifting bird would seldom visit a place twice, but after a week in a range a hawk would have visited most areas at least twice and would not be greatly increasing the area that it was covering.

When hawks were in sight, flights could be described accurately, using a pocket recorder to avoid looking down at a notebook. A typical transcription reads: 'Take off, 10.32, low flight to six metre perch in oak at field intersection SW of barn. Rouses (shakes thoroughly) and warbles (stretches leg and wing downwards simultaneously on one side, then the other), then flies high, 10.46, straight across valley to 16m south-most ash by pond. Dive at Moorhen in rush patch at pond side, 10.47, 15m from tree, misses, Moorhen runs/flies two metres to pond and submerges, hawk returns to ash perch. Sits watching pond until 10.54, then flies low west to 12m perch in mid hedge elm, north side of second field. 10.56, moves $1\frac{1}{2}$ fields west to five metre perch in corner tree.'

I transcribed the notes in the evenings, also plotting the routes on acetate sheets overlying the maps while my memory was fresh. Flight distances and directions could then be measured accurately. The 1970s technology would now be replaced by a Geographic Information System (GIS) on a computer, but immediate transcription remains advisable.

Inter-perch distances flown by the four hawks released in north Oxfordshire differed slightly between individuals, perhaps reflecting their different origins. The flights of a Finnish nestling trained for falconry (an 'eyass') tended to be slightly shorter than those of two wild Finnish hawks, one trapped in its first autumn (a 'passager') and one caught as an adult ('haggard'). A German passager had the longest flights (Kenward 1982). The feature most obvious for all hawks was that the average length of inter-perch flights that began and ended in open country averaged twice as great a distance, around 200m, as those completed in woodland (Figure 67).

Long flights in open country may at least partly have reflected the greater tree spacing there than in woodland, although hawks seldom merely flew to adjacent trees. Another difference between flights in open country and woodland was that birds tended to reverse the direction of consecutive flights twice as frequently in woodland. However there was no difference between hawks or habitats in the average time interval between the start of flights. This time interval was most commonly three to four minutes (the median time was four minutes), with rather few rests greater than six minutes between flights. The fact that hawks perched four times as much in woodland as expected (Figure 66) was explained by flights being half as long in woodland and by reversal of direction in woodland happening twice as often as in open country (Figure 67).

Eighty per cent of the heights at which the hawks perched were between 2–11m above the ground. The birds usually settled in the lower two-thirds of a tree and nearer to the trunk than to the outmost twigs. They thus tended to select inconspicuous positions, appropriate for a skulking predator. Although hawks occasionally flew at tree-top height or directly from one side of the valley to the other, flights from perch to perch usually followed a pattern which was inconspicuous and conservative of energy. On take-off the hawks tended to drop rapidly, almost to the ground, turning the potential energy of their elevation into kinetic energy of flight speed. Typically, each flight was then repeated sequences of several wing beats followed by short glide. At the end of a flight, the hawks shot steeply up from low level to a perch, often almost parallel with a tree trunk and between its main branches, turning kinetic energy back into potential energy. They thus avoided wasting their own energy on acceleration or braking, while also reducing their chances of being silhouetted against a skyline.

Once on a perch, a hawk could be seen looking around. It seems reasonable to assume the bird was scanning its vicinity for prey before moving on to a new perch. On this basis, the longer length of flights in open country may have reflected an ability to scan a greater area than in woodland. A combination of perched-hunting interspersed with inconspicuous flights has been termed 'short-stay perched hunting' (Kenward 1976a, 1982).

In the 62 days during which hawks were monitored between October 1974 and March 1975, they were foraging for about 200 of the 600 daylight hours and were otherwise feeding, resting after a meal or bathing. They were in sight for 138 hours of foraging, during which they made 946 flights. These included 39 soaring flights which averaged six minutes and a horizontal displacement of 780m (with a maximum of 2,090m in 17 minutes). Some four hours of soaring was only about 3% of the foraging time, so this was a minority foraging strategy in winter. Hawks flew at 10–15 m/sec, so with half their flights in woodland (averaging 100m) and half in open country (averaging 200m) they also spent a total of up to 4.5 hours flying

158   *The Goshawk*

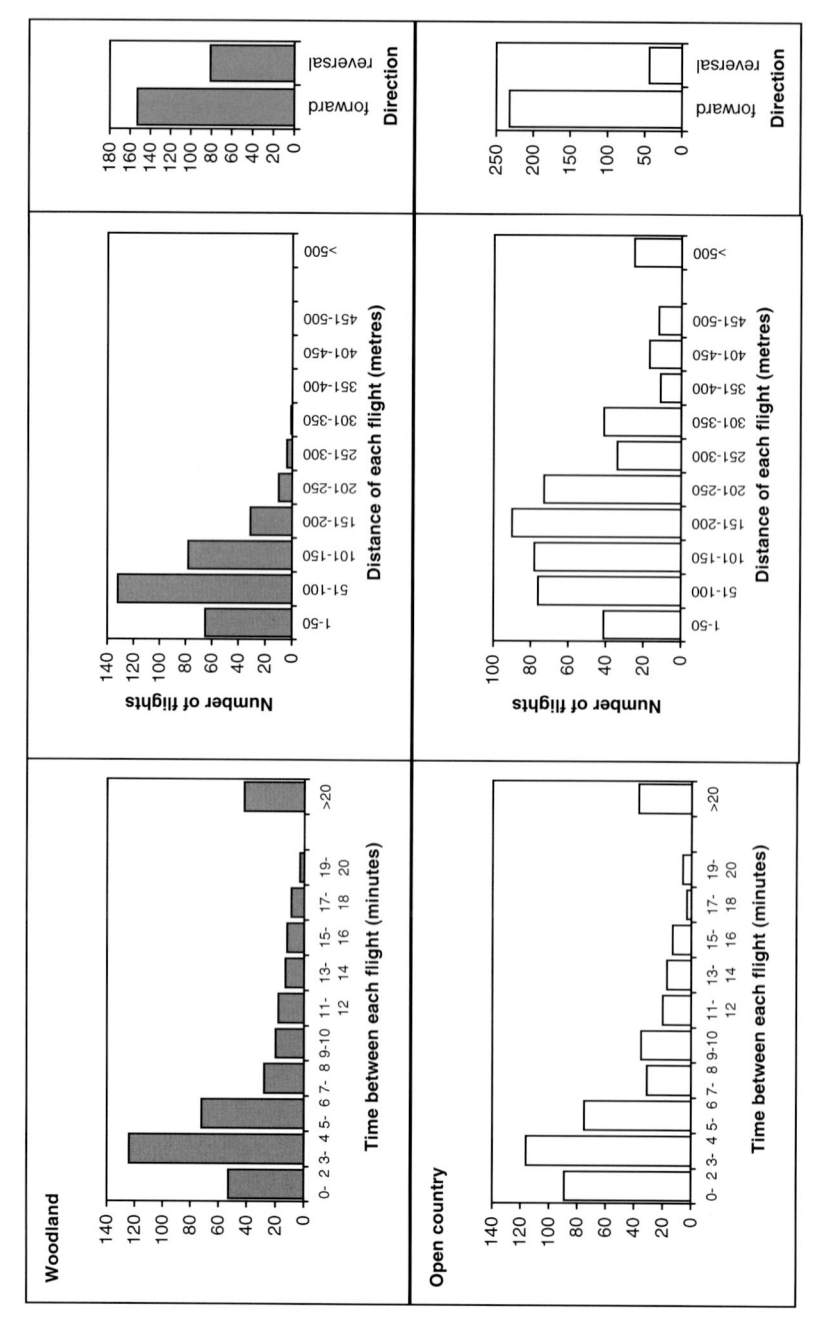

Figure 67. Goshawks foraging in Oxfordshire usually rested between flights for up to 6 minutes in woodland and open country (left), but flew half as far in woodland (centre) and with a much greater tendency to reverse direction (right) from the previous flight. Data from Kenward 1982, pooled from 4 hawks.

directly between perches. They were therefore perched for about 94% of their time while foraging and they perched for about 98% of the daylight hours in total.

ATTACKING, FEEDING, LOAFING AND BATHING IN LOWLAND BRITAIN

The 946 flights also included 79 that were attacks, of which the majority (75) were straight from a perch. Attacks were launched at prey up to 280m from perches, but the average perch-to-prey distance was about half that of the average inter-perch flight. Thus, 35 attacks in open country averaged 103m, with 54m for 28 attacks in woodland. The hawks typically dived or glided all the way to nearby prey, but otherwise used a 'wing-beat-and-glide' approach. Three attacks not launched from a perch were consecutive stoops at rabbits by a hawk that was soaring along a ridge, and another hawk checked at pigeons which flushed as it flew past. This latter attack and one at a pheasant flushed by a passing rambler were the only attacks initiated at flying prey, although birds that took flight as a hawk approached were sometimes pursued in the air.

After successful attacks, the hawks normally fed where they came to rest with the prey. Feeding was not recorded systematically because of disturbance to examine Woodpigeon kills. These were eaten relatively slowly, after prolonged plucking. Rutz (2003) noted an average 66 minutes plucking Feral Pigeons '. . .(before taking them to nests in Hamburg), compared with only 14 minutes for lighter prey. After making kills and feeding, hawks wintering in Oxfordshire typically rested in a tree nearby and usually roosted without moving far, but I did not keep systematic records.

Hawks seemed to bathe merely as a result of encountering water during foraging. This happened on seven occasions, all between one and three hours after sunrise in dry weather. All but one of the four hawks (the Finnish haggard) was seen to bathe. Shallow streams were used five times, while the edge of a ford, and a marsh, were each favoured once. Bathing did not occur with food in the crop, and the chosen sites were those surrounded by cover rather than with a good view of approaching dangers. Settled hawks may have favourite bathing places that they visit at times when they are loafing.

For bathing, hawks usually flew down to the water's edge from a nearby perch and waded in up to their ankles (i.e. to tarsus depth). Typically, a hawk then took two or three gulps of water, bending forward with its head held horizontal to fill its mouth with water and lifting its head up so that the water could run down its throat. This latter motion was often accompanied by 'beak-smacking' that may have helped the drink on its way. Captive hawks normally drink only when they bathe, unless they are ill, which raised a question: was drinking merely stimulated by presence in the water for bathing or were hawks approaching water primarily to drink, with bathing secondary?

After drinking, a hawk fluffed its feathers out in the slow way that occurred before rousing, and lowered its body partly into the water before shaking vigorously, spraying water everywhere. The wings were held loosely out from the sides, and the tail shaken in the water. Often the head was dipped into the water and lifted out so that water trickled down the back, but it was mainly the tail, wings and underparts that were wetted. The observed hawks spent 3–7 minutes in the water, after which they hopped onto a low perch, shook vigorously, and then usually flew to a high location. They remained in the vicinity, presumably drying, for up to three hours before moving on.

FORAGING IN FENNOSCANDIA

In Sweden, radio-tracking was first used to study Goshawk foraging in 1976–8, as part of an investigation of their predation on pheasants. Detailed observation of flight sequences was prevented by predominance of conifers and flatter woodland-farmland landscapes than in Oxfordshire. However, activity could be deduced from variation in radio signals for three female hawks. The most frequent inter-flight time was in the two- to three-minute category, as in Oxfordshire, so that short-stay perched hunting (SSPH) again was their main strategy.

At Frötuna Estate, where large numbers of pheasants were released annually, Goshawks would frequently kill a pheasant one day, and return early the next to feed at the carcass before loafing nearby until the afternoon, when they would feed again. Hawks abandoned 14 of 66 kills at Frötuna when they made another near the previous kill (Kenward *et al.* 1981a). Observations of Goshawks waiting for a long period on a perch, and then attacking when prey appeared, probably do not represent a 'still hunting' strategy. The hawks may have fed recently but nevertheless have been quite prepared to exploit easy quarry.

A juvenile male that had moved on from Frötuna frequently soared up above 100m during August, and was three times seen stooping at pigeon flocks. He was never seen to kill in this way, although he was found eating a pigeon far out in the open where he had been soaring. Soaring flights were often associated with movement away from an established range, and soaring may often have been for low cost locomotion rather than hunting.

Both topographic and thermal soaring were used in Sweden, the former along ridges or where updrafts occurred at the windward edges of dense woodland. The middle three-fifths of the day, from late morning to early afternoon, contained 82% of 71 soaring flights. Soaring thus occurred mainly in the warmest parts of the day, when thermals were most likely. Soaring also tended to occur in the warmer parts of the year: 13 of 16 radio-tracked hawks in Sweden soared during August to October, whereas no soaring was recorded for 17 hawks tracked later in the year. Hawks sometimes soared too high to be easily noticed. A female watched and photographed through a 400mm camera lens in August could not easily be spotted without the lens. Her peak altitude of 310m could be estimated by measuring her image and comparing with her known wingspan and length.

Whereas in Oxfordshire only 8% of the landscape was woodland, 41–61% of the Swedish study areas were wooded. The proportion of woodland in polygons drawn round all the radio locations was similar to that on the map, but hawks tended to perch mainly within 200m of edges (Figure 68).

A rather consistent 73–76% of perch locations in the three Swedish study areas were in woodland within 200m of an edge. The majority of kills were made in the edge zones or in open country within 100m (one flight distance) from an edge, and thus most probably from perches in edge zones. This was especially the case at Frötuna Estate, where hawks killed few prey other than released pheasants. About 200 pheasants had been released a year earlier at Gäddeholm but none recently at Segersjö.

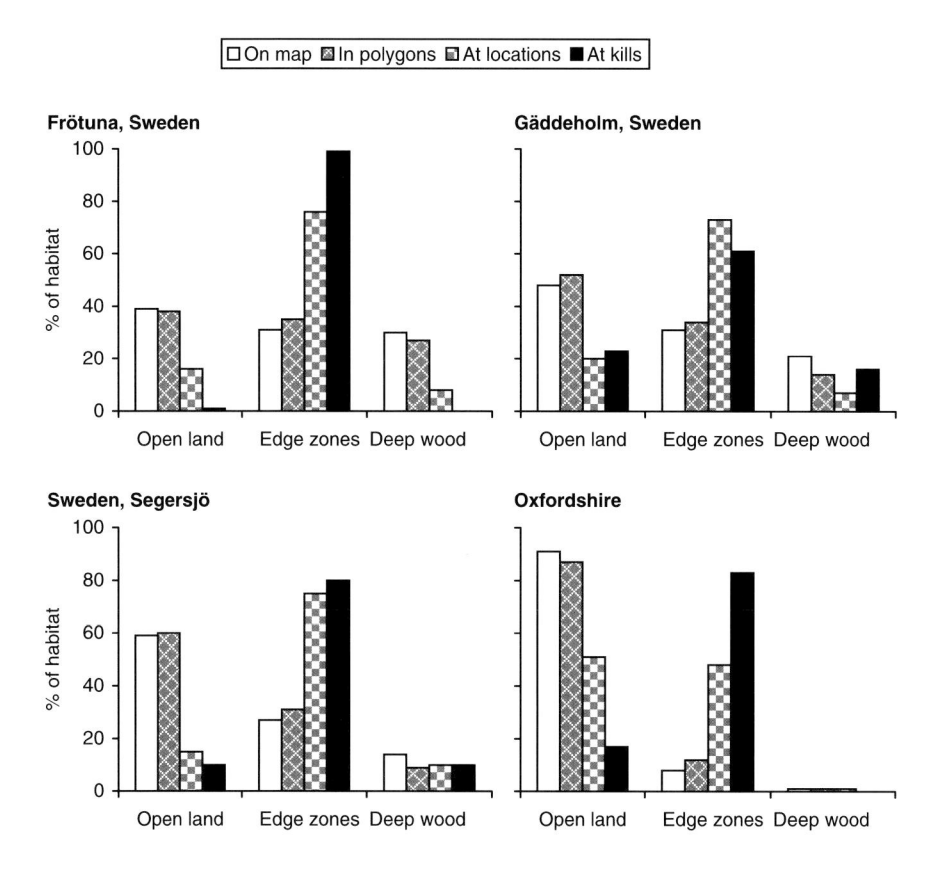

Figure 68. Goshawks in three Swedish areas and an area in Oxfordshire usually had most perch and kill sites where woodland was <200 m from edges, although these edge zones were never the most commonly available habitat on the map or in home-range polygons. Data from Kenward 1982.

162   *The Goshawk*

Goshawk foraging differed in two other areas of Fennoscandia with 75–76% forest (including clear-cut areas). In a study area on the edge of the Swedish taiga, Per Widén recorded flight times of eight Goshawks during January-March in 1980–81. He found that they moved less frequently in the 24 hours after making a kill, but that thereafter their median flight intervals were three to four minutes. This was similar to the hawks in farmed areas of Britain and Sweden, for which foraging was not recorded in the rest of the day after a kill. However, the median times in flight were 24 seconds, equivalent to flights of 240–360 m at 10–15 m/s, which was considerably longer than for Goshawks in the woods of Oxfordshire (Widén 1981, 1982, 1984b). As a result, the proportion of time spent flying increased from 3% on days with kills to 6% the next day and 9% two days after a kill, presumably as hawks became very hungry. The hawks in England made two kills every three days on average, and did not reach such high levels of flight activity as the overall 7% of taiga-dwellers.

Moreover, the habitat choice of the Swedish taiga-dwellers was markedly different from hawks in the farmland areas of Britain and Sweden. They did not avoid farmland so much as wetland and young forest up to 60 years old (Figure 69), habitats which were almost absent in the British study. These hawks tended to avoid edges, which had little to offer to them. Their prey was not edge-zone species, but was 80% Red Squirrels, which were especially abundant in 1980 and 1981 due to good cone crops on the mature conifer trees.

In the other highly-forested area, in Northern Finland, prey and habitat-use differed again. In an area on the outskirts of Oulu, the strongest preference during November-March was for deciduous areas (mainly of birch) that provided winter food for the grouse and hares that provided two-thirds of the prey (Figure 70); only a quarter of the kills were squirrels (Tornberg & Colpaert 2001). The Finnish hawks

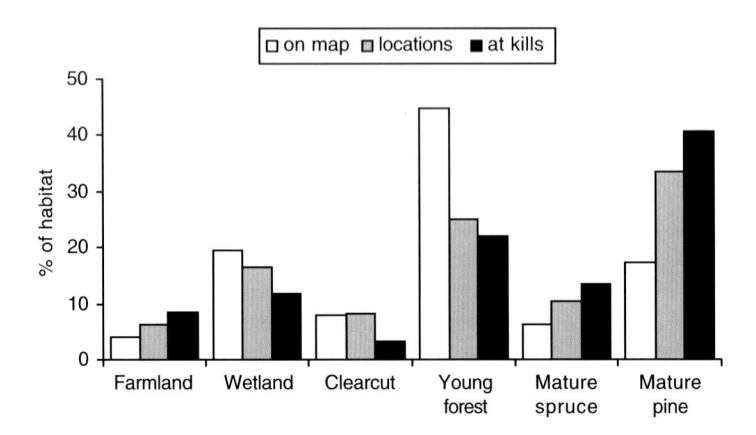

Figure 69. Although young forest was the habitat most available on maps of study areas, hawks in Swedish taiga favoured mature forest for perch locations and kills, with relatively low use of young forest and wetland. Data from Widén 1989.

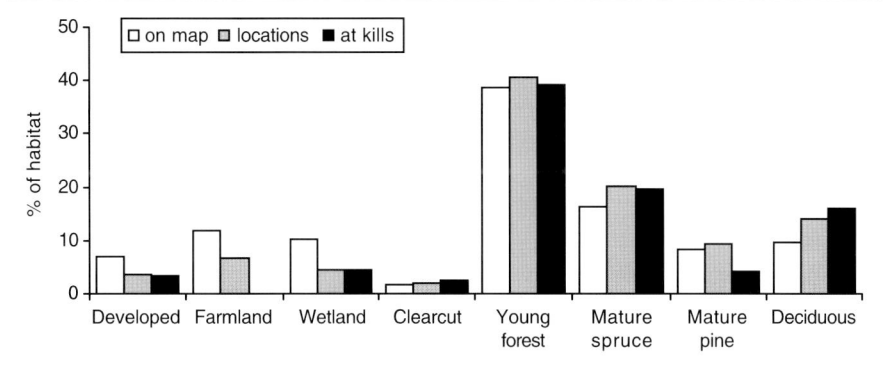

Figure 70. In northern Finland, young forest was the most available habitat and was used by hawks for perch locations and kills, although deciduous woodland was used most relative to its availability. Data from Tornberg & Colpaert 2001.

also tended to avoid the large patches of mature forest that were favoured in the Swedish taiga (Figure 71).

FORAGING IN TOWNS

Goshawks in towns have an even simpler choice, of hunting in parkland around their nests or also in the surrounding built-up areas. Christian Rutz (2001) travelled by bicycle while radio-tracking three nesting males in Hamburg for a total of 858 hours, recording 5,364 radio fixes and 122 kills during March–July 1997–99. In view of the difficulty of radio-tracking in built-up areas, where bouncing signals make direction-finding a tremendous challenge, this was an amazing feat of dedicated fieldwork.

The city-dwelling hawks spent 82–97% of their time in vegetated areas although these comprised only 13–34% of their home-ranges. Nevertheless, about half the kills of each hawk were made during short forays into the built-up areas. Pigeons comprised 37% of the prey, twice as frequent as any other species in the diet, followed by Magpies and Blackbirds which together provided another 27% (Rutz 2001, 2003, 2004).

Two of the hawks spent most time in their own parks (84% and 80% of all radio locations) and made regular but short hunting trips into built-up areas (13% and 19%), occasionally visiting other urban parks (3% and 2%). The third male hunted almost exclusively (92% of locations) in the well-forested cemetery containing its nest, and spent more time soaring than the other hawks. Soaring was much more prevalent than in the winter studies in Britain and Fennoscandia. The male hawks were flying an average 10% of the daylight hours, of which 8% (4–16% for different individuals) was soaring and only 2% comprised direct flights between perches. Daily

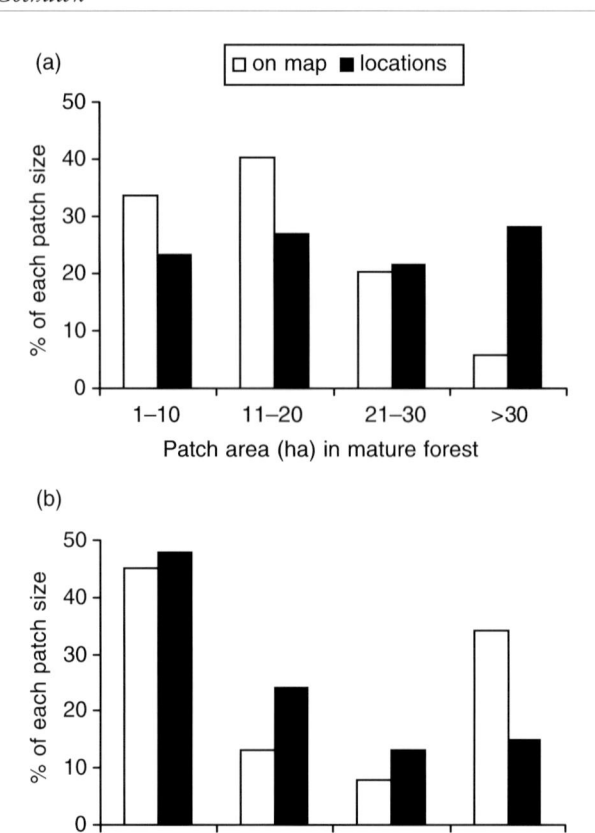

Figure 71. Hawks in the Swedish taiga (a) were located in large patches of mature forest very frequently in relation to its abundance on the map, whereas Finnish hawks (b) avoided this habitat. Data from Widén 1989, Tornberg & Colpaert 2001.

activity patterns were bimodal, with peaks in the early morning and in the evening. One hawk was observed hunting after sunset under artificial light conditions.

Moreover, the soaring flights were relatively long. As a result, although flight intervals were 14 minutes on average, which is less frequent than the medians of three to four minutes in Britain and Sweden, they flew an average 15km per day compared with 6km per day in Oxfordshire. Greater activity would be expected for hawks feeding families in summer than for hawks feeding themselves in summer, with more soaring in the warmth of summer and using the thermal generation of cities. Hawks foraging for families in rural areas may find thermals less easily, and may have less need to soar when hunting prey that skulk in trees or on the ground.

Above all, these contrasting observations show the considerable ability of the Northern Goshawk to adapt its hunting strategies and habitat-use to take advantage of a wide range of circumstances.

FORAGING IN NORTH AMERICA

Pat Kennedy radio-tracked three male Goshawks breeding in New Mexico (Kennedy 1991) and recorded 18% flight activity during the rearing and post-fledging period. This is an even higher average than the 10% recorded during March to July in Hamburg (although hunting of city hawks too increased towards the end of the breeding season). Short-stay perched hunting (SSPH), which is a 'saltatory-searching' strategy (O'Brien *et al.* 1989) was also the main technique in New Mexican forests (Kennedy 2003). The same applied in western Washington, where 72% of foraging time was SSPH alone, 13% soaring alone and 15% a mixture of the two strategies (Bloxton 2002).

In North America, most attention has focussed on comparing habitat at radio locations or kill sites with habitat at random locations during the breeding season, using very detailed assessment of habitats. The result has been more information on how foraging relates to details of woodland structure and local prey density than has been gathered in Europe. An early study found that range outlines, like nest sites, contained above-average canopy closure and density of large trees (Hargis *et al.* 1994), with a tendency also for locations to be away from open areas, where canopies were most closed and large trees were at highest density (Bright-Smith & Mannan 1994, Drennan & Beier 2003).

During the breeding season, there was also a tendency to make kills in forests not so much in the areas of highest prey-density as where large trees were most abundant, with least tall understorey to hinder flight access and with good canopy closure (Beier & Drennan 1997) or presence of dead trees (Bloxton 2002). Presumably a well-developed canopy and dead trees (snags) provide good perch sites. Good *et al.* (2001) also noted disproportionate subsequent foraging at kill sites with low ground cover, but kills did not seem to be made where ground cover was poor (Beier & Drennan 1997, Bloxton 2002).

However, some hawks changed their habitat use outside the breeding season. Drennan & Beier (2003) noted that four of five males in the Coconino and Kaibab National Forests of northern Arizona moved to areas of pinyon pine and juniper in winter, although four of six females remained mainly in ponderosa pine breeding habitat. Hawks were killing mainly Cottontail Rabbits in the pinyon-juniper habitat in winter (Stephens 2001, Drennan & Beier 2003). The hawks that Sonsthagen *et al.* (2006 b) tracked by satellite in Utah descended into similar habitat in winter (Figure 72). While hunting in this Cottontail habitat, hawks were not avoiding open areas: more than half of their radio-locations were associated with edges (Underwood *et al.* 2006).

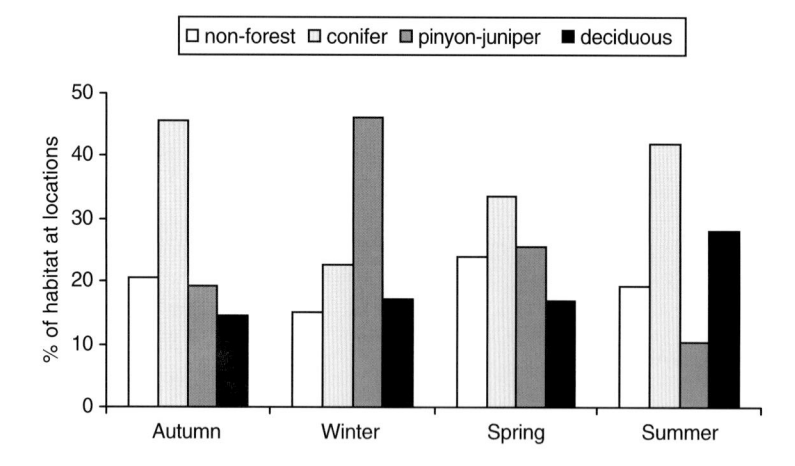

Figure 72. In southern Utah, 16 Goshawks that were tracked by satellite in 2000-01 moved from conifer-dominated areas in autumn to pinyon-juniper-rich zones in winter. Data from Sonsthagen et al. 2006b.

HOME-RANGES AND TERRITORIALITY

A home-range was originally described as the area 'traversed by the individual in its normal activities of food gathering, mating and caring for young' (Burt 1943). However, the term is more generally applied to areas used annually, seasonally or between dispersal movements (Craighead & Craighead 1956) rather than specifically for breeding. It is most easily defined as a repeatedly-traversed area, and therefore by implication starting with an arrival location after a single dispersal movement (or at an individual's birth-site) and ending with the last location before dispersal (or at a site of death). However, this concept can also accommodate the use of fixed dates to define, say, a weekly or monthly range for animals drifting across the countryside (Doncaster & Macdonald 1991).

Home-ranges have traditionally been drawn as the polygon with a minimum peripheral length around all the locations, the 'minimum convex polygon' (MCP). Unless individuals are monitored continuously, home-ranges are usually based on taking a sample of locations that is large enough for the areas covered by most of them no longer to be increasing: the animals have then been tracked throughout their current range. When three Goshawks were first tracked in Sweden, sampling three active locations and a night roost each day, there was little increase in area after the first ten days (Figure 73). Moreover, night-roost locations were not peripheral. Thereafter, standard ranges were measured by sampling 30 locations for ten days, a practice also confirmed by Tornberg & Colpaert (2001).

As the area of a convex polygon home-range depends on peripheral locations, its extent is strongly influenced by occasional excursions that hawks make from the area

Diet and foraging 167

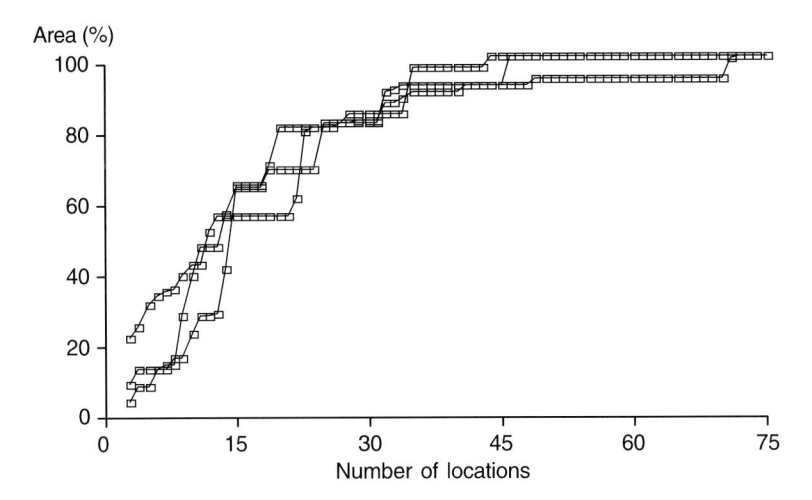

Figure 73. The size of convex polygon home ranges round all locations (100% Minimum Convex Polygons) for three Goshawks tracked at Frötuna in 1976 showed that area estimates increased little after the first 30 locations were recorded.

used mainly for hunting. Other methods have therefore been developed for representing core areas that are most used, based on contouring the density of locations (Dixon & Chapman 1980, Worton 1989) and on defining clusters of locations with minimal nearest-neighbour distances (Kenward 1987a, 2001). In the former case, a contour round 95% of the density distribution is used as an outline equivalent to the MCP round all the locations, while a 50% contour estimates the core.

In the latter case, polygons around clusters that include 85% of the locations have been found to exclude excursive locations (Hodder *et al.* 1998) that may sometimes reflect exploratory activity rather than hunting. For example, a hawk that had a pheasant kill at Frotuna made a journey of several kilometres to the north one autumn day, mostly in flight, before returning to feed at the kill (Kenward 1977). For Buzzards, which were mostly eating worms on the ground in autumn, the locations that were furthest outside their main range cores were most often for birds in flight (Kenward *et al.* 2001c).

Although contour outlines estimated for Goshawk ranges tend to ignore distant excursions, they also tend to expand into areas not visited. This is especially so for kernel techniques (Worton 1989), in which the whole density distribution is influenced by distant outliers, and least evident in harmonic mean contouring (Dixon & Chapman 1980).

Contour expansion due to outlying locations affects 50% and 95% outlines. On Gotland, there were three cases where the 50% contours for adult hawks of the same sex overlapped in winter by 25–100% whereas round 85% cluster polygons overlapped by 0–30% (Figure 74). Overlap of core areas for ten first-year hawks was more extensive (Figure 75) and also occurred at other sites (Kenward 1977, Kenward

168   *The Goshawk*

1) Minimum Convex Polygon,
for 100% of the locations

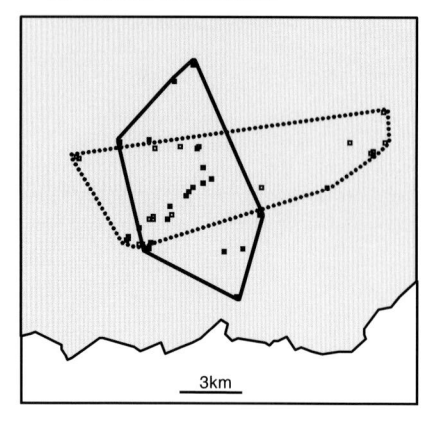

2) Fixed kernel contours,
for 95% of the location distribution

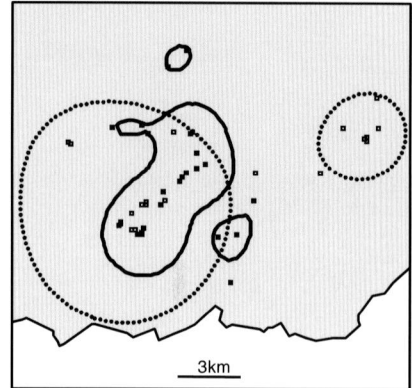

3) Fixed kernel contours,
for 50% of the location distribution

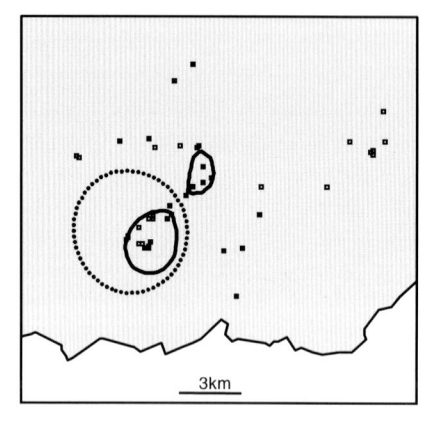

4) Cluster analysis polygons,
for 85% of the locations

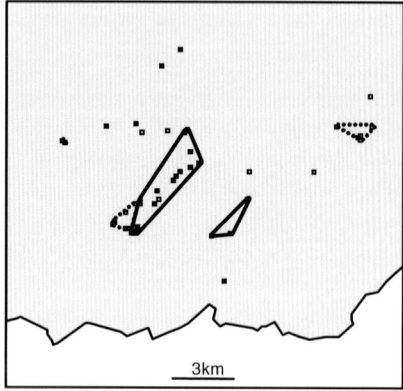

Figure 74. Expansive home-range methods showed extensive peripheral overlap of two second-year females (non-breeders) in early March 1981 in a rabbit-poor part of Gotland, although cluster polygons indicated exclusive cores.

& Walls 1994). Although juvenile Goshawks share foraging areas extensively, core areas of adults seem to be more discrete. Buzzards too tend to have extensive range overlap as juveniles, especially between siblings, but discrete cores as adults (Walls & Kenward 2001).

Among adults and juveniles of both sexes, 30-location ranges during October–March were on average smallest in the rabbit-rich area, as shown by use of a 1km scale bar in Figure 75 compared with a 3km scale for the rabbit-poor area in Figure 74. In the three Swedish mainland areas where predation on pheasants was studied, average range size increased from 20km^2 where pheasants were released

Diet and foraging 169

1) Minimum Convex Polygon,
for 100% of the locations

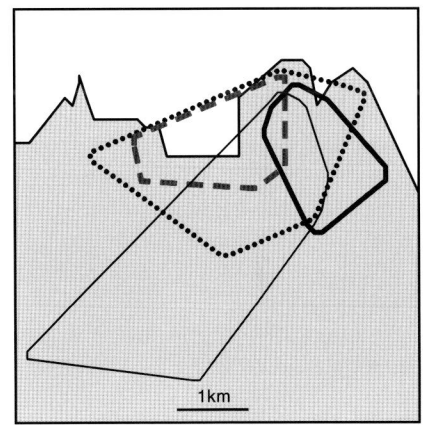

2) Fixed kernel contours,
for 95% of the location distribution

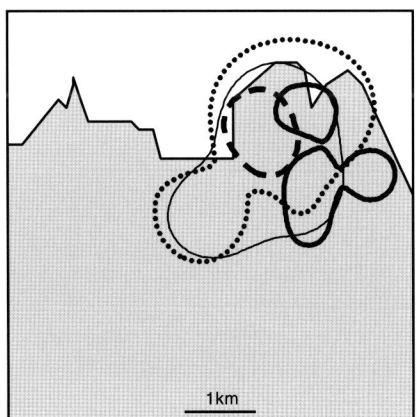

3) Fixed kernel contours,
for 50% of the location distribution

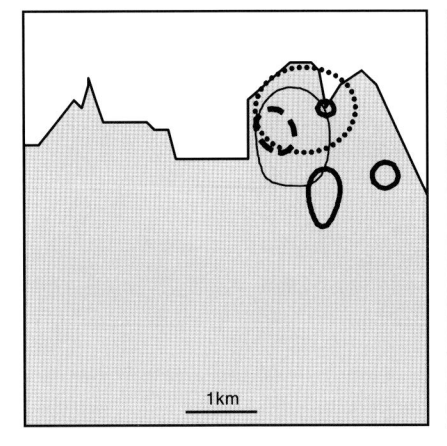

4) Cluster analysis polygons,
for 85% of the locations

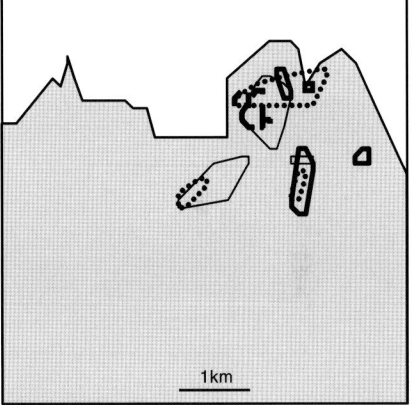

Figure 75. Expansive home-range methods showed extensive peripheral overlap of four first-year females in December 1980, and cluster polygons showed overlap persisting in the cores. Home-ranges were small in this rabbit-rich part of Gotland.

through 26km^2 with mainly wild pheasants to 54km^2 around an estate with only wild pheasants (Kenward 1982). In Hamburg, with such abundance of Feral Pigeons that breeding females seldom hunted, three breeding males covered areas that averaged only 7.4km^2 during the breeding season (Rutz 2001). Three hawks released experimentally in prey-rich lowland Britain settled for 9–12 days (until recapture) in areas of only 3.8km^2, 10.5km^2 and 13.6km^2 (Kenward 1976a).

However, food supply is not the only factor that influences range size. In order to catch prey, the landscape must be structured adequately for hunting. Although Goshawks in towns hunt from roof tops, antennas and even window ledges (Rutz 2004), elsewhere they need trees in habitats used by suitable prey. For prey that favours edge zones, woodland edge is an important hunting resource and individual hawks with most access to edge have the smallest ranges (Kenward 1982). Where mature forest is the most important area for hunting, individual hawks with most access to old forest have the smallest ranges (Tornberg & Colpaert 2001).

Indeed, the area that a hawk covers in order to obtain its food may be a rather useful indication of the prey and habitat available. Relationships between prey, habitat and range size will be considered again after examining the prey on which Goshawks subsist in different areas.

VARIATION IN DIET WITH LATITUDE AND ALTITUDE IN EUROPE

The more data that has been obtained on Goshawk diet, the more it has been found to vary. Figure 76 shows the variation represented in remains from five groups of prey species at nests from 22 studies in central Europe since 1980 (Rutz *et al.* 2006 a). These groups are shown because they all exceeded 10% of the diet in at least one study. They are what may be called 'staple prey', and represent the commonly occurring birds and mammals with masses of 50–2,000g.

In these groups of prey, the mammal species were mainly Red Squirrels and rabbits and the gamebirds mainly pheasants and partridges (with some Red Grouse in

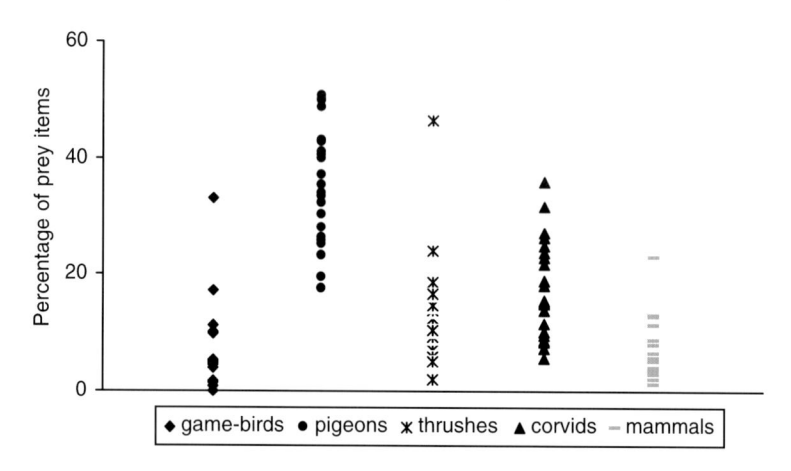

Figure 76. The variation of main groups of prey in remains at nests for 22 central European studies published since 1980. Data from Rutz *et al.* 2006a and as cited in Appendix 2. For easy comparison with birds, mammals are shaded grey in this and following figures.

Britain). There were a variety of thrush and corvid species, with both Woodpigeons and Feral Pigeons. Pigeons were the most important single group, comprising 18–51% of prey items across all studies. However, in the far north of Europe there are no rabbits and relatively few pigeons. Woodland grouse then become increasingly important prey (Figure 77).

A similar trend occurred with altitude in areas of northern Britain. The importance of partridges and mammals declined as altitude increased; grouse increasingly dominated the diet (Figure 78). Further south, Paul Toyne (1998) worked in a Welsh area that lacked grouse, Red Squirrels and upland hares. He showed that as rabbits declined in the prey remains with increase in altitude it was the prevalence of pigeons that increased.

In the south of Europe, there is a decrease in the proportion of birds in the diet, associated in part with the taking of many rabbits (as also in Britain and on Gotland) and also with increased killing of reptiles. Among 33 prey remains at a nest in the Spanish Sierra Guadarrama, there were 12 Jackdaws, six rabbits and no fewer than seven lizards (Araujo 1974).

VARIATION IN DIET WITH TIME

Especially in the northern part of the Goshawk range, the abundance of staple prey species can vary considerably from year to year. Census counts of woodland grouse have been strongly linked to representation of grouse in the diet at Goshawk nests the following spring (Figure 79). As grouse became more abundant, hawks killed them at a higher rate. Although the percentage in the diet did not increase above

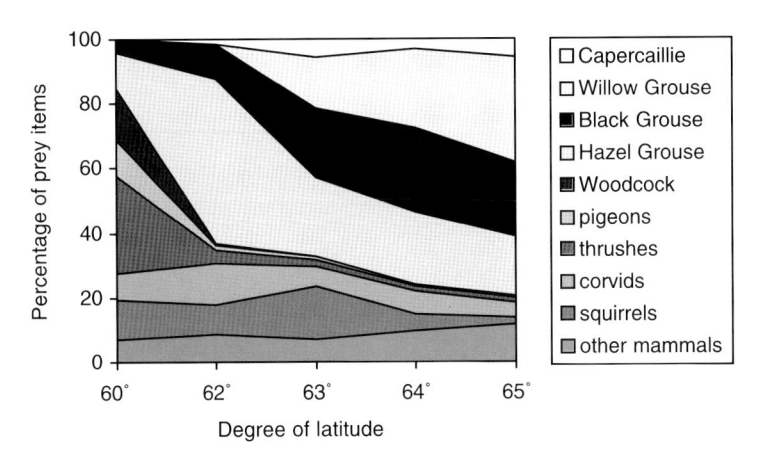

Figure 77. The increase in prevalence of woodland grouse species among prey remains at Goshawk nests to the north in Finland during 1957–81. Data from Wikman & Lindén 1981, Huhtala & Sulkava 1981, Tornberg 1997.

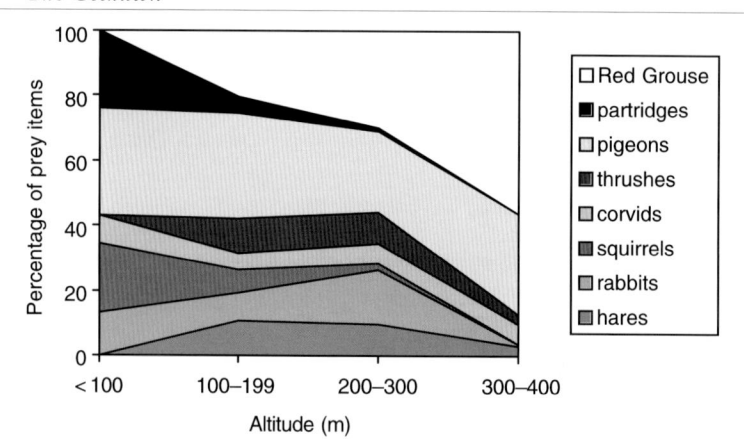

Figure 78. The increase in prevalence of grouse remains with increase in altitude of Goshawk nests in Britain. Data from Marquiss & Newton 1982b.

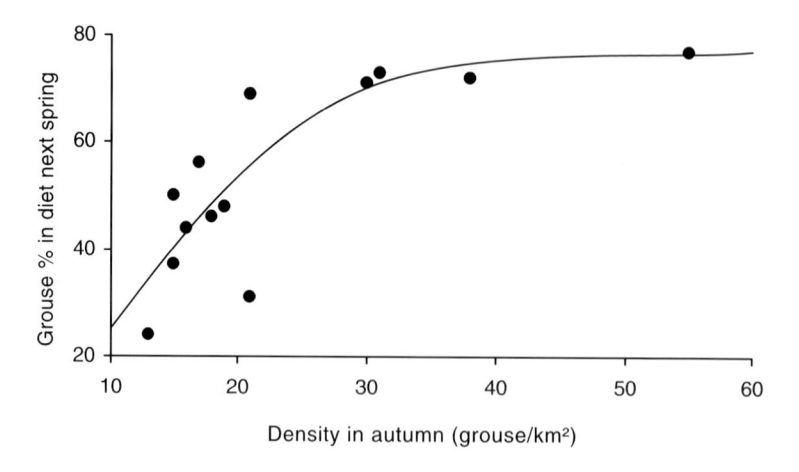

Figure 79. The proportion of woodland grouse among remains at nests in northern Finland increased with grouse density. Redrawn from Tornberg & Sulkava 1991.

80%, the kill rate must have continued to increase at grouse densities above 30/km² because clutch size continued to increase (Figure 35 in Chapter 3) and hawks therefore had more young to feed.

As well as varying from year to year, the Goshawk diet also varies considerably during the course of the breeding season. For example, prey remains collected by Tornberg (1997) showed that between courtship and the early nestling period (June) there was a decline in numbers of woodland grouse taken, offset by an increase in the proportion of other birds (Figure 80). This trend reversed after young grouse became available in June.

Diet and foraging 173

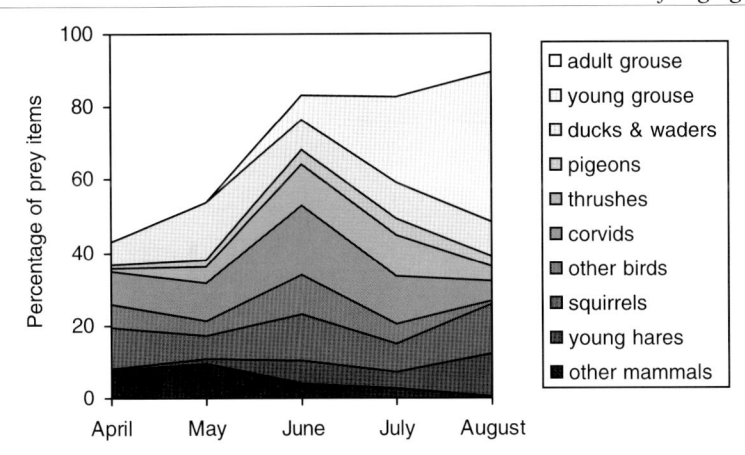

Figure 80. Changes during the breeding season in the diet of Goshawks nesting in northern Finland during 1988–1994. Data from Tornberg 1997.

The explanation for these changes had much to do with changing vulnerability of prey, as young became available in turn first for waterbirds, then corvids and thrushes, and finally for grouse. Indeed, the collection of prey remains tends to underestimate the importance of young prey, which are more strongly represented when other methods are used to assess the diet. The importance of young birds in the summer diet was noted in the hide observations in California by Schnell (1958), in the cage method data from Sweden and Finland (Höglund 1964b, Sulkava 1964) and in modern video records (Grønnesby & Nygård 2000, Lewis 2001). Not only were adults a mere 17% of all bird prey brought to nestlings in Finland, but 17 of the 57 altricial birds (30%) were pulli that had probably been taken from nests. The availability of vulnerable young prey also resulted in a marked seasonal change in the species of tetraonid killed: whereas adult capercaillie made up only 8% of 423 tetraonid remains during incubation and courtship (35 greyhens: 1 cock), 43% of 83 chicks were Capercaillie (Sulkava 1964).

An abundance of vulnerable young bird prey is very important for accipiters. However, whereas in Sparrowhawks it is egg-laying that coincides with the start of this bounty (Newton 1986), Goshawks lay before young prey become available. For Goshawks it is the peak food demands of rearing that coincide with the peak emergence of young prey (Tornberg 1997, Toyne 1998).

Further south in Finland, woodland grouse and mammals show a greater decline in the diet as the season progresses, and are replaced by corvids and thrushes (Lindén and Wikman 1983). In Germany, 36% of the prey remains were Starlings, Jays and crows in June, when young were available, but only 6–10% in March-April (Brüll 1964).

Many studies of raptor diet have concentrated on the breeding season, sometimes even assuming that the diet is similar at other times of year. However, changes

174   *The Goshawk*

similar to those during the breeding season also occur during the rest of the year. Indeed, although diet may be most easily studied in the nestling period, it may then be least typical of the year as a whole, especially where species that migrate or hibernate in winter are important summer prey.

Nils Höglund noted that mammals were always better represented in stomach contents of European Goshawks than among prey remains at nests, contributing 35–55% of the prey in early studies (Rörig 1909, von Bittera 1916, Munthe-Kaas Lund 1950, Sládek 1963). However, as the stomach contents contained many poultry and game birds, there is a danger that birds were over-represented in strongly anthropogenic environments. The radio-tracking of hawks tagged in the nest, or trapped in more natural environments, should provide less biased material.

Studies in Finland, Germany, Sweden and Britain have each recovered 38–179 winter prey by radio-tracking in nine different areas. Although kills of small mammals (and birds) may be under-recorded in such material, mammals were strongly represented in the winter diet in the six northernmost areas without pheasant releases, mainly as squirrels and lagomorphs (Figure 81). The southernmost area, in southern Germany (Kluth 1984), was more continental in climate and lacked rabbits. In the seven areas without released pheasants, a focus on pigeons was replaced by squirrels and grouse beyond 55°N (i.e. Scandinavia). Waterbirds and rabbits were also lost from the diet in northern or continental winter climates.

The proportion of mammals in the winter diet in the seven areas without pheasant release averages 54%, and is therefore much higher than the average of 6% mammals in summer diet shown in Figure 76. However, it is more robust to compare diet seasonally in just three of the Fennoscandian areas, where it was recorded at nests and in winter (by radio-tracking) in exactly the same localities (Figure 82).

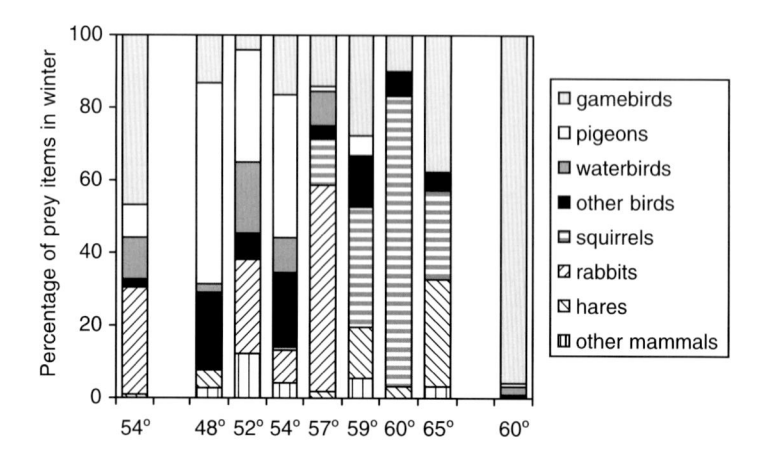

Figure 81. Diet during August-March of radio-tracked Goshawks in rural areas. The first and last columns represent areas with released Pheasants. Data from Kenward 1977, 1981b, Kenward *et al.* 1981a and unpublished, Ziesemer 1983, Kluth 1984, Widén 1987, Tornberg *et al.* 2006.

Diet and foraging 175

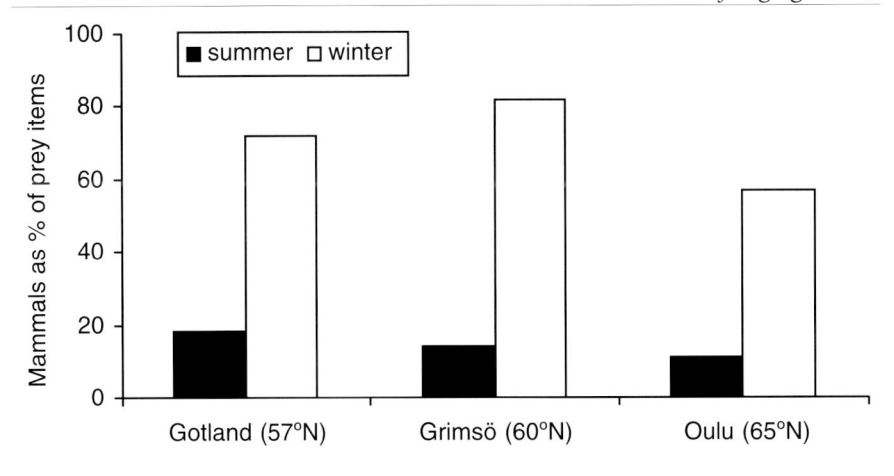

Figure 82. The proportion of mammals in prey remains at nests and among kills of radio-tagged hawks in winter from three Fennoscandian study areas. Data from Kenward *et al.* unpublished, Widén 1987, Tornberg 1997, Tornberg *et al.* 2006.

Moreover, the 34% mammals in Oxfordshire winter kills can be compared with 6–22% among prey remains at nests in Britain (Marquiss & Newton 1982b, Toyne 1998, Petty *et al.* 2003a), and the 14–30% mammals killed in winter in Schleswig-Holstein (Ziesemer 1983) with records of 1–15% mammals at nests in four studies there (Looft & Biesterfeld 1981). Thus, wherever comparisons have been based on radio-tracking, Goshawks have killed more mammals in winter than are recorded in prey remains at nests. Comparisons based on searching for kills have not given such marked differences between diet at nests and in winter (Opdam *et al.* 1977, Brüll 1984, Nielsen 2003a).

VARIATION IN DIET BETWEEN SEXES AND AGES

The difference between the prey spectrum of breeding and wintering hawks may reflect not only the abundance of young birds as summer prey, but also a dietary difference between male and female hawks. Male hawks, which provide most of the food at nests, tend to have fewer mammals in their winter diet than do females. This was noticed by Höglund (1964b) among stomach contents, and shows clearly in radio-tracking data (Figure 83).

Where hares were common in Sweden, five of 12 females took adults in winter but none of the ten males were so bold. Most of these hares were in excellent condition, and the largest weighed 3,670g, which was 2.4 times the weight of the hawk that killed it. Two males each killed a leveret, the largest (1,600g) again being 2.2 times

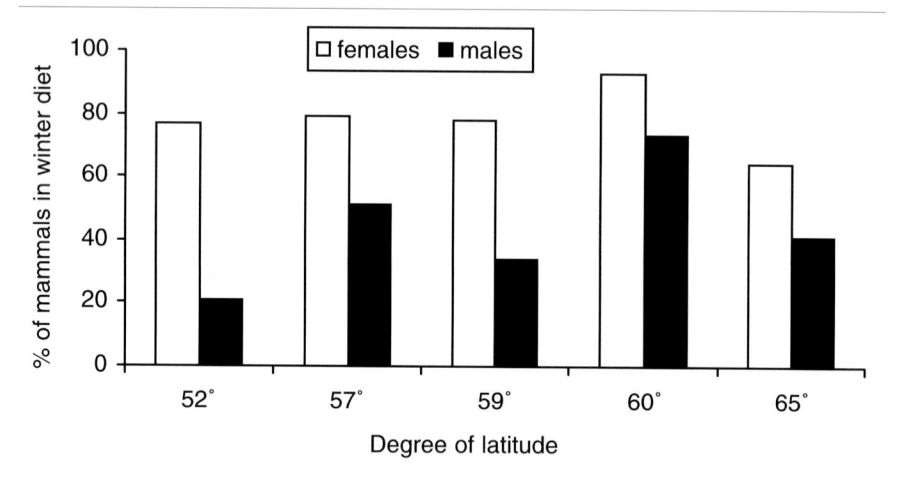

Figure 83. The proportion of mammals in kills of radio-tagged hawks showed least difference between males and females in the far north. Data from Kenward 1981b, Kenward *et al.* 1981a and unpublished, Widén 1987, Tornberg *et al.* 2006.

the hawk's weight. Both males and females took rabbits, but these were a higher proportion of the females' kills and a much higher proportion of the food biomass for this larger hawk sex. Both sexes killed red squirrels, but males released in Britain found the larger Grey Squirrel too difficult to handle.

Males compensated for lack of biomass from mammals with kills of large birds, primarily pheasants and Woodpigeons. The wild pheasant population at Segersjö provided 45% of the biomass for male hawks, whereas hares accounted for 41% of the winter food for females (Kenward *et al.* 1981a). Fridtjof Ziesemer (1983) reported that Woodpigeons were taken mainly by female Goshawks, whereas the radio-tagged males captured mainly Feral Pigeons. He noted also that Paul Opdam and co-workers (Opdam *et al.* 1977) had found mainly Feral Pigeons in the diet at nests of Dutch Goshawks, and data in Brüll (1964) show an increase in Feral Pigeon remains relative to those of Woodpigeon in summer diet records. He suggested that males, being the more agile sex, were better adapted for darting among buildings to take the Feral Pigeons, whereas the greater speed of the females was at a premium for getting close to Woodpigeons feeding in open fields. Nevertheless, male Goshawks released in Britain took more Woodpigeons than any other prey, and Woodpigeons feature strongly in the diet at British Goshawk nests.

For urban Goshawks, which have a more limited choice of prey than in the diversity of rural habitats, pigeons are a particularly convenient large prey for provisioning nests. However, their speed, agility and flocking behaviour makes pigeons hard to catch. The growth of a Goshawk population in Cologne provided an unusual opportunity to study how the diet of young male hawks changed with age, on the expectation that the killing of this challenging prey would increase with experience. Sure enough, older male hawks had a higher proportion of pigeons in their diet (Rutz *et al.* 2006 b).

CARRION

Some writers have felt that Goshawks are much too rapacious to consider eating carrion (Engelmann 1928). However, others have reported regular visits by Goshawks to carrion baits for foxes and crows (von Schweppenburg 1938, Prill 1959). Tømmeraas (1977, 1980) provides some fine photos of adult Goshawks using feeding sites for White-tailed Eagles in Norway. In America, Goshawks have been seen eating sheep carcasses in irruption years (Bent 1937) and at other times feeding on offal left by hunters (Squires 1995).

Goshawks will eat carcasses that are far from fresh. The young male that was soaring and killing pigeons near Frötuna ate half of one Woodpigeon and then killed another near his subsequent roost. It returned to the original two days later, in warm August weather, and finished the carcass which was thoroughly fly-blown. Another hawk killed a couple of rabbits after release in Oxfordshire, but then dined for a week on the carcass of a vixen that had probably succumbed because of the severe weather (Plate 21). The weather had become warmer and the fox became very 'unpleasant' before the local farmer's spring ploughing interred the remains and put an end to the unsavoury feast.

Although Ziesemer (1981b, 1983) noted that one radio-tagged juvenile fed extensively on carrion in Schleswig-Holstein, the habit is not restricted to inexperienced juvenile hawks. On Gotland, I found an adult female almost underground in the rib-cage of a long-dead pig, in a farm's charnel heap, some 5km from the nest where she was roosting in March. Two among 20 juveniles were recorded eating carrion (including the kills of other hawks) during 28 tracking periods of 10 days on Gotland, compared with 6 among 18 adults tracked for one period each. Although carrion is not a major part of the Goshawk diet (and was excluded from data used in the previous figures), it is eaten widely and may occasionally be a life-saver.

COMPARING DIET IN EURASIA AND NORTH AMERICA

At North American nests, the proportion of mammals in the diet at Goshawk nests is much higher than in Europe (Figure 84). The 33 American and 68 European studies shown (and listed in Appendix 2) are all based predominantly on prey remains or visual observations, to remove possible bias towards recording mammals in pellets.

In Europe, the highest proportion of mammals was found at nests in the west and south, which are the areas where a more maritime climate and consequently limited

▲ NW ◆ SW ■ east – central △ north ◇ west ○ south □ east ✕ urban

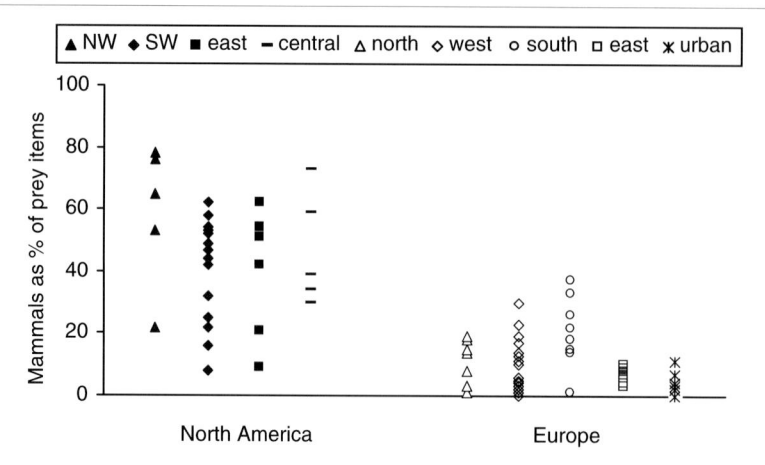

Figure 84. The proportion of mammals among prey at nests is higher in North America than in Europe. Data from Rutz *et al.* 2006a, which excludes studies based mainly on pellets, as cited in Appendix 2.

winter snow permits rabbit populations. Rabbits are not found where there is prolonged snow-cover. Summer prey are mainly birds in Scandinavia, eastern European countries and towns.

There is also no good evidence that mammals become substantially more frequent in the Eurasian diet at Goshawk nests east of Poland. Jevgeni Shergalin helped me find 15 diet studies published in Russian since 1985 (Appendix 2). In none of the 14 from west of the Urals, ranging from Belarus in the west to Ukraine and Georgia in the south and Pechora in the north, did mammals account for more than 9% of the 69–601 prey remains. Mammals were also 9% in each of two small data sets from east of the Urals, including 35 prey records from the Altai (Mitrofanov 2003) and 22 from the Anadyr basin (Krechmar and Probst 2003). Fischer's extensive knowledge of the earlier (Soviet) literature reveals no studies of Goshawk diet in the vast tracts of land from the Caucasus across Mongolia and Tibet into China. However, there are occasional observations of hawks killing eastern equivalents of the bird species and mammals killed further west (Fischer 1980).

In contrast, birds are *least* prevalent at nests in the northern and central states and provinces of North America. And whereas woodland grouse were so important in northern Europe, Snowshoe Hare remains were more common than bird remains in pellets collected during three years at Alaskan nests (McGowan 1975). Although the hare remains may have been unusually common because of a population peak (and the use of pellets), the remains of squirrels, the second most common mammalian prey, were three times as common in pellets as the remains of grouse. At more southern nests, the squirrels (including Flying Squirrels in Oregon) were more common than lagomorphs, although Snowshoe Hare bones were 8% of the prey remains in Oregon. The most frequently recorded birds were short-billed crows (45%

of all prey) in the east and thrushes (including American Robins) or jays in the west. Compared with Europe, gamebirds (mainly woodland grouse) were very much less prevalent in the diet.

The main exceptions to these trends tend to prove the rule of mammals being particularly important for North American Goshawks. One is the Tongass National Park in the archipelago of south-east Alaska, where Lewis *et al.* (2004) found mammals to comprise only 9% in prey remains, 22% in video records (used in Figure 84) and 41% in pellets. Grouse were 50% of prey biomass, and corvids another 19% (Lewis 2001), on islands that mostly lack lagomorphs and in at least one case (Prince of Wales Island) lack squirrels. A similar situation exists on Haida Gwaii and Vancouver Island off the coast of Canada, which lack Snowshoe Hares (Doyle 2006).

Goshawks at Tongass also have exceptional foraging behaviour. The mean 100% convex polygon home-ranges in the breeding exceed 100km^2 (Iverson *et al.* 1996). This is due mainly to three females with ranges of 250–950km^2, probably because they were crossing water. Even using median values for these hawks, their range sizes lie above the trend line for increase in range size with latitude in North America (Figure 85). Data for this figure are based on combining male and female ranges, because studies measured these through the whole season (i.e. into the period of female provisioning) and found no significant differences. Data from one study that used convex polygons round 95% of the locations (Hargis *et al.* 1994) were adjusted with data from another study that gave sizes for both 95% and 100% polygons (Keane & Morrison 1994). Data from satellite tracking were not used because their inaccuracy creates an upward bias in home-range estimates (Kenward 2001).

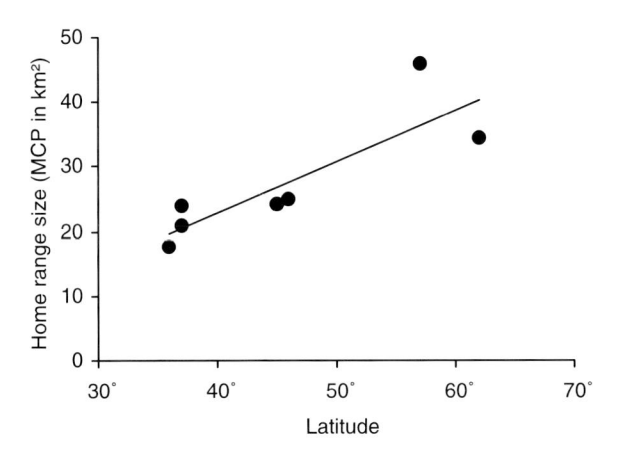

Figure 85. Convex polygons round all the breeding-season locations indicate that home-range sizes of North American Goshawks increase to the north. Data from Austin 1993, Bright-Smith & Mannan 1994, Doyle & Smith 1994, Hargis *et al.* 1994, Keane & Morrison 1994, Iverson *et al.* 1996, Boal *et al.* 2003.

180   *The Goshawk*

There are too few data from breeding Goshawks for a similar plot for Europe, but winter range estimates from rural areas show a strong trend of increase to the north (Figure 86).

These are values only for areas without released pheasants. However, the lowest estimate, of 9.3km², is for nine to ten days continuous monitoring of three male hawks that had just settled after release in Oxfordshire, and probably underestimates the area that would have been covered if tracking had continued. Winter ranges averaging 20–40km² may be more typical for western Europe south of Fennoscandia.

IMPLICATIONS FOR CONSERVATION AND MANAGEMENT

It is important to have accurate information on diet of raptors, to indicate what prey species may be important for them as food and also to assess their likely impact on prey populations (Chapter 7). The recording of a fourfold difference in the representation of mammal prey between pellets and prey remains (Lewis *et al.* 2004), and the finding of only one in eight rabbit kills by scanning areas for prey remains in winter (Ziesemer 1981b), shows just how much the result can be influenced by the technique.

On the other hand, ensuring accuracy has a cost if it requires most effort or equipment. That would not be the case if the duration of video-recording is long enough to offset the cost of equipment by a need for fewer nest visits. Radio-tracking to recover kills in winter is hard in terrain that lacks good access, but scanning for kills is not then feasible either, and the radio-tracking can give much other information. Satellite tracking is very expensive and currently unsuitable for estimating

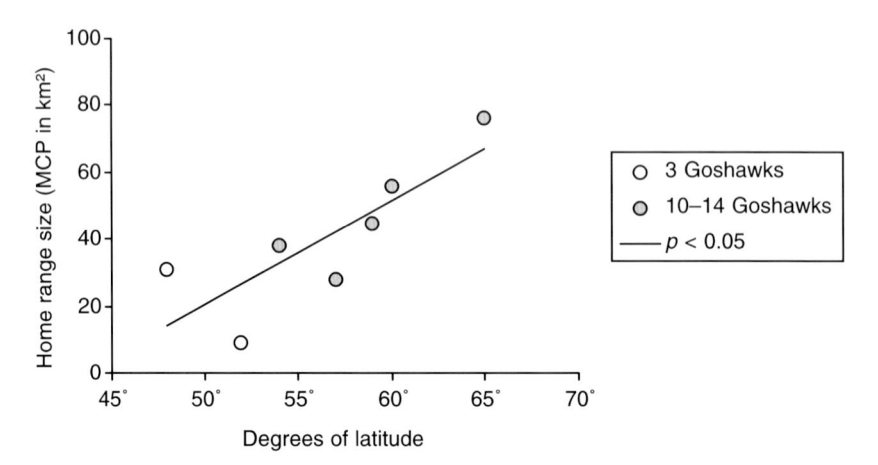

Figure 86. Convex polygons round locations show that Goshawk winter range sizes in Europe increase to the north. Data from Kenward 1976a, Kenward *et al.* 1981a and unpublished, Ziesemer 1983, Kluth 1984, Widén 1989, Tornberg & Colpaert 2001.

home-ranges or fine details of habitat-use (Britten *et al.* 1999). However, it may eventually be possible to use GPS modules to increase accuracy of transmission to satellites from small tags and sophisticated sensors to indicate kills as they occur, which would make such studies more attractive for researchers.

There is also scope for the use of trained volunteers to gather data, in roles often now filled by student projects. Video recording and VHF-tracking of hawks can be fun. For volunteers with a falconry background, it is worth noting that the behaviour of tamed hawks seemed to be similar to that of wild ones and hawks tamed at the age of fledging (eyasses) are easier to observe at close quarters. I recovered such hawks to the fist from kills after two weeks living free.

The use of data from the nestling period to assess predation year-round is inadvisable, because (i) the diet is then orientated to young prey that are available only seasonally, especially if they are migrant species, (ii) male hawks take most prey and in Europe they differ in diet from females even in winter and (iii) diet of inexperienced non-breeders may differ from breeders. However, do prey remains collected in spring represent winter diet, does diet differ as much between sexes of the less dimorphic Goshawks in North America, and does experience greatly influence diet in rural areas?

The Northern Goshawk has been a convenient species in which to show how the size of home-ranges may be influenced both by habitat structure and prey availability. This raises a question about whether such an effect scales-up from an individual to a regional scale: does increase in range-size to the north indicate poorer resources there? Can home-range or other location data be used to examine resource factors that may be important when Goshawk populations decline, or to understand differences between continents?

Examination of relationships between range size and latitude were possible because most Goshawk researchers used comparable home-range measures. Convex polygons round all the locations may not be the best measure for other analyses, so Goshawk researchers should keep in touch to ensure adequate standardisation in future. It would be advisable to obtain at least 30 locations to standardise range outlines; 50 may be wiser for defining cores (Robertson *et al.* 1998).

The success of Goshawks in European towns and citites, as indicated by small ranges and high productivity (Rutz 2005c), destroys the idea that Goshawks require forests. When I joined with Per Widén to ask 'do Goshawks need forests' (Kenward & Widén 1989), we were not asking whether they need trees, but whether they require large expanses of woodland. The *Oxford English Dictionary* defines forest as 'an extensive tract of land covered with trees'. European Goshawks do very well without forests. Moreover, some North American Goshawks breed and spend winter in areas where woodland is naturally fragmented (Younk & Bechard 1994, Underwood *et al.* 2006). The importance of forest for density and productivity of Goshawks can be examined in Europe, because most studies record how much of their area is forest, but lack of comparable data hinders investigation of mechanisms underlying Goshawk association with forest in North American (Rutz *et al.* 2006 a).

The question as to why Goshawks in America may be more dependent on forests than in Europe will be considered again in Chapter 10. However, it is worth noting here (i) that mammals seem to be especially important for American Goshawks, (ii) that very different range sizes and diet on islands of the Pacific northwest raise particular concerns about the impact there of changing forestry practices and (iii) it would be useful to standardise research techniques and approaches between North America and Europe to facilitate intercontinental co-working on this holarctic raptor.

CONCLUSIONS

1. Collecting prey remains at nests estimates diet with reasonable accuracy in spring, but underestimates mammal and young bird prey during the nestling stage. Video-recording is a promising alternative to observing from hides to record diet and predation rates during brood-rearing.
2. Tracking with VHF radio tags can record diet and kill rates accurately throughout the year, with intensive data on foraging if single hawks are monitored continuously. The sampling of locations and activity at intervals can record kills and locations from more birds, but tends to underestimate predation on species of <250g.
3. Goshawks are pause-travel hunters, using a saltatory search technique of short-stay perched hunting (SSPH) as an alternative to soaring and low prospecting flights. SSPH dominates in winter and continuous forest, but soaring becomes more frequent in warm weather and can dominate hunting of urban Goshawks during breeding.
4. Goshawks favoured woodland edge zones and made most kills there when foraging for lagomorphs, pigeons and pheasants in fragmented woodland of Britain and Sweden, but favoured large blocks of mature forest for squirrels in Swedish taiga and smaller stands with deciduous trees when also hunting lagomorphs and grouse in northern Finland.
5. In North America, hunting during breeding was mainly in forest with good canopy cover, high density of mature trees and good flight access. Hawks breeding in conifer forests at high altitude in the western USA, especially the males, tended to descent to lower altitude in winter and were hunting Cottontail Rabbits in pinyon-juniper habitat and edge zones.
6. Minimum convex polygons round all the recorded locations have proved effective for relating Goshawk range size to prey availability, habitat structure and latitude. The outer polygons and contours of same-sex neighbours often overlap extensively, as can range cores of juveniles, but cores estimated by cluster analysis tended to separate for adults.
7. In Europe, diet at nests varies with latitude and altitude, and with season as young prey become available. In northern Fennoscandia, where numbers of the dominant woodland grouse prey vary annually, so does their representation in the diet. At lower latitudes, corvids, thrushes, pigeons and rabbits become more

important, but mammals only exceed 20% of prey remains at nests in western and southern Europe.
8. Mammals, mainly lagomorphs and squirrels, are more important prey for European Goshawks in winter than in summer, especially at high latitudes, and are taken more frequently by females than by males. The proportion of elusive prey in the diet can increase with age of hawks. All ages may feed on carrion at times.
9. Mammals are generally more prevalent in diet of breeding Goshawks in North America than in Europe, a notable exception being data from hawk nests on islands that lack lagomorphs and where large home-range size may indicate foraging problems.
10. Many questions remain about factors that influence foraging, especially concerning differences between Europe and North America in the use of landscapes by Goshawks. Greater transatlantic communication and standardisation is recommended to help solve problems that may stem from general changes in land-use or differences in Goshawk behaviour.

CHAPTER 7

Prey selection and predation pressures

According to my field notes, the 27th of January 1973 was a mild and overcast day, without rain in the afternoon but with a moderate south-west wind. Just after 3 o'clock, I drove my blue Mini Countryman up the rough track south of a large field of Brussels sprout plants, 20 miles north of Oxford. The edible 'buttons' on the stems of these plants had mostly been harvested already and no doubt some had been enjoyed with turkey at Christmas. However, a few plants still had tops that had not yet been eaten by pigeons.

'Foot', so called because he has grabbed me in one of my careless moments during training, sits on a perch covered with soft brown leather, screwed to the top of the back seat. His Finnish compatriot 'Mouth' is too vocal for our work with Woodpigeons and stays at the university or on loan to friends. Foot is looking through the side window at pigeons feeding in the field, about 150m away. I take him onto my fist, check his field jesses and hold him so he can leave through the passenger window.

At 3.15 he starts a low flap-and-glide flight, almost between the rows of sprouts. He gets as close as 20m before the 30 pigeons leave as fast as they can. None fall behind and he does not bother to chase them but swings up into a convenient tree nearby. The pigeons circle too, and settle again 150m further from him. It is getting towards the end of the afternoon, and they are ready to start filling their crops for the night. They do not fill their crops until the last hour of daylight, probably because the extra loading is a risk: the success rate of Foot's attacks increases five-fold in the hour before sunset.

Foot relaxes in the tree. He fluffs out his feathers, stretches one leg down in an 'L' shape with claws closed and then tucks it up under his feathers. Two crows pass, one dipping in flight over him, but they don't break their journey to spoil his idle observation of the pigeons, which are being joined by others in a slow trickle from a wood to the east. He has already made three unsuccessful attacks earlier today and is not wasting energy.

At 3.30 his feathers suddenly tighten and both feet grip the branch. I can see that he has spotted a lone pigeon, which is flying rather weakly over the far hedge to land some distance from the flock. As it lands, he is off, wings beating fast as he skims the sprout crowns, no gliding even though the flight is gently downhill. White feathers puff into the air. I note the response of the other pigeons, which again circle and settle further down the field. The main objective of this research is to record the behaviour of the pigeons in response to successful and unsuccessful attacks, to see whether wild hawks or their effigies might help protect *Brassica* crops. Predation data are an extra benefit.

When I coax Foot back onto my fist for a small piece of beef, I can see immediately that the pigeon has an old shot wound to one wing, and that one leg is broken. This is Foot's twenty-first Woodpigeon. Although he is occasionally refusing now to fly at distant flocks, he clearly had no doubt about this victim. My subsequent *post mortem* examination records that it has the third-lowest mass of breast muscle of the 32 he takes that winter and I find two more lead shot in its abdomen.

PREY VULNERABILITY

The last chapter showed how the foraging and diet of Goshawks can vary with time, location, sex and even age of hawk. This chapter considers how that variation may influence the impacts of Goshawks on the characteristics and abundance of their prey, and how that can affect human interests. A first step in this process is to see how the vulnerability of different prey influences what Goshawks kill in different situations. This sets the stage for a look at how predation rates and the numbers of Goshawks can respond to changes in prey density and impact different prey populations.

Regional and annual variation in the composition of the Goshawk diet may relate to the ways in which the abundance of different staple prey varies with time and place (Chapter 6). However, prey abundance is not the only factor influencing what hawks eat. To show this, Jörg Dietrich and Hermann Ellenberg (Dietrich & Ellenberg 1981) plotted the frequency of different prey in the Goshawk diet against their abundance in an area around Saarbrücken, in south-west Germany. Kills of Woodpigeons, Feral Pigeons, Jays and thrush species (including Blackbirds) were recorded more frequently than expected, whereas duck, crows, Starlings and smaller passerines appeared to be neglected (Figure 87). The analysis was based mainly on pluckings and will thus over-represent the most conspicuous kills (e.g. pigeons), but there may be a similar bias towards conspicuous species in the abundance estimates, which were visual point counts and may thus under-represent skulking prey (e.g. pheasants).

Risto Tornberg (1997) ranked the divergence of percentage availability and percentage in diet to indicate the vulnerability of each species to attack.. This is in keeping with the concept of varying vulnerability of prey introduced by John and Frank Craighead (Craighead & Craighead 1956). The term has since broadened to include a variety of factors that may influence attack success (Quinn & Cresswell 2004). In general, the frequency of prey in the diet is likely to reflect not only the encounter rate (which depends on abundance) and vulnerability to attack, but also

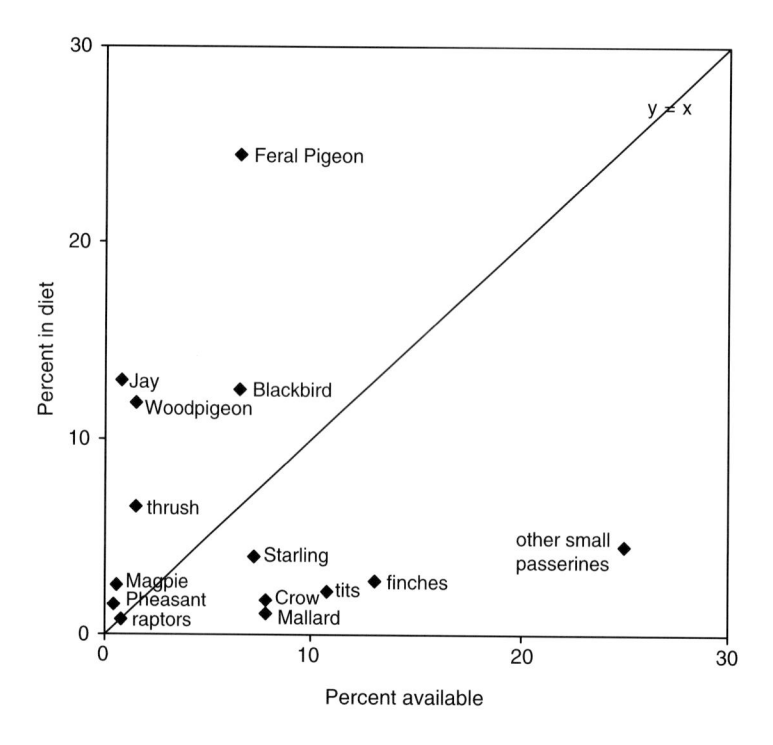

Figure 87. Goshawks near Saarbrücken killed pigeons and Jays more than expected by chance, but crows, small passerines and Mallards less than expected. Data from Dietrich & Ellenberg 1981.

the probability of detection when encountered. Whereas some species are especially hard to catch when detected (e.g. the flight-agile finches), others are hard for the predator to detect (e.g. rabbits in holes).

In principle, this means that kill-rates, and hence frequency in the diet, should be predictable from probabilities of encounter, detection and capture. Thus:

$$kill\text{-}rate = encounter\text{-}rate \times probability\text{-}of\text{-}detection \times probability\text{-}of\text{-}capture$$

with predation rates estimated by combining kill rates with density of predators and prey:

$$predation\ rate = \frac{density\ of\ predators \times kill\text{-}rate}{density\ of\ prey}$$

Prediction of predation rates from encounter-rates, detection and capture is complicated, because all the components vary with habitat structure, time of year and experience of individual predators and prey. It is reasonable to surmise that encounter rate is involved when both prey abundance and diet change between locations, and that much seasonal change in diet and predation rates within the same location can be attributed to changes in the abundance and vulnerability of different prey. Prey increase in abundance through breeding at different times in summer, and newly fledged (or fledgling) Jays and crows are weaker and less experienced than their wily parents (Tornberg 1997). Hares are sufficiently vulnerable as leverets to be taken by both Goshawk sexes, but the adults are too large for male Goshawks to subdue in Europe unless they are diseased, for example by an outbreak of tularaemia (Tornberg *et al.* 2006).

This chapter considers studies that assessed how Goshawks impacted prey populations. However, in order to predict and hence perhaps manage predation in future, rather than surmise about diet and predation *post facto*, it is important to understand the mechanisms that influence encounters, detection and attack success. In Oxfordshire, radio-tracking of free-living Goshawks let me assess how often they encountered fields where pigeons fed (Kenward 1976a), at which 100% detection of this conspicuous prey could be assumed. Encounter and detection rates could in principle have been predicted from foraging movements (Chapter 6). However, it was much harder to measure attack success.

ATTACK SUCCESS

The problem with assessing attack success is that, even with the assistance of radio-tags, any form of direct observation seems to be biased against recording successful attacks. The four hawks released in Oxfordshire (Chapter 6) made 79 attacks in the 6,638 minutes during which they were in sight, but only five of the attacks were successful. The five out of 79 represented a success rate of 6% and a kill every 1,328 minutes of foraging. On the other hand, the hawks made 41 kills during the 5,432 minutes when they were out of sight, which is a kill every 132 minutes. If attacks occurred at the same frequency when hawks were in sight, they must then have been 63% successful, or 32% overall. Quite possibly attacks were both more frequent and more successful when hawks were less visible to me, and also presumably to their prey, so their attack success could not be estimated reliably. The overall rate of 262 minutes per kill is a reasonable estimate of their foraging success, but lacks the detail needed to understand the underlying mechanisms.

Chance observations are likely to be even more biased. Christian Rutz (2001) saw 26 of 176 attacks by radio-tagged Goshawks in Hamburg to be successful (15%), but there was not one success in 89 flights of untagged hawks observed by chance at the same time. Hantge (1980) recorded a similar low success rate (5%) in chance observations of Goshawk attacks. Accurate estimates can only be obtained when hawks remain in sight all the time, as can be the case when they hunt by soaring. Thus, Rob Bijlsma (2005a) was able to record 11 kills from 50 attacks (22% success) by adult hawks on homing pigeons from such 'waiting-on' flights (although only 0.2% of the passing pigeons were killed).

On this basis, a success rate of 15–30% across all attacks may be a reasonable estimate, but reveals nothing about factors affecting individual attacks. More revealing information is that Goshawks in Hamburg took only 35 minutes of active hunting to secure each prey (Rutz 2001), which is dramatically better than the 262 minutes per kill in Oxfordshire.

ARRANGED ATTACKS

Another approach to measuring attack success is to use an older technology than radio-tracking. Hawks trained for falconry can be used to investigate many aspects of raptor biology, including their safety on power lines (Nelson 1978) and mechanisms that may help to increase or decrease their predation (reviewed by Lindsay 1981).

For example, Nick Fox (1981, 1995) has nicely illustrated how trained hawks can vary in the success of their attacks on different prey species. He found, for instance, that Goshawk attack success was only 11% in 18 flights at hares, compared with 50% in 693 attacks on rabbits. Success was intermediate, at 24–25%, in attacks on ducks (17), pheasants (84), and partridges (12). None of five attacks on Blackbirds were successful but 67% of 104 attacks on Moorhens resulted in kills, as did 11 of 14 attacks on Herring Gulls. These are the most extensive data published on capture risk of different species, although some prey may have been distracted by the human presence and were hence at greater risk. Prey that are taken mainly by surprise in the wild may be less able to escape when flushed out by a human or dog. However, Nick Fox points out that hawks were not assisted deliberately when attacking non-target species, such as protected species and in eight attacks on chickens (of which seven succeeded). At least, the results suggest that hares and Blackbirds are at less risk of capture, once detected, than Moorhens, rabbits, gulls and chickens.

Attacks arranged to see whether Goshawks would scare Woodpigeons from crops of Brussels sprouts and rape (Kenward 1978b,c) showed how success at a single prey species can vary with habitat, flocking, time of day and the physical condition of the prey. The overall success rate in 92 attacks on Woodpigeons was 30%, but that included a very high success of 78% in 18 attacks on single pigeons.

In the 74 attacks on pigeon flocks, success was 19%, comparable to that of soaring hawks (Rutz 2001, Bijlsma 2005a). However, there were no kills from 16 attacks at flocks on grassland, probably because it was more difficult for a hawk to approach prey undetected across open land than against a *Brassica* crop that varied in height. Only three of 58 attacks (5%) succeeded when the pigeons took flight before the hawk was within an estimated 20m of them, compared with 20 of the 24 attacks (83%) with a shorter flight distance (Figure 88).

Attack success was not only greater at single pigeons than at birds in flocks, but also decreased from small to large flocks (Figure 88). Single birds were attacked most successfully because many of them were too weak to keep up with flocks or out-fly the attacking Goshawk. However, when more than one pigeon was present, the decline in attack success with increasing flock size owed much to the tendency for larger flocks to take off when the hawk was farther away, which indicated that the large flocks were best at detecting the approaching predator (Figure 88).

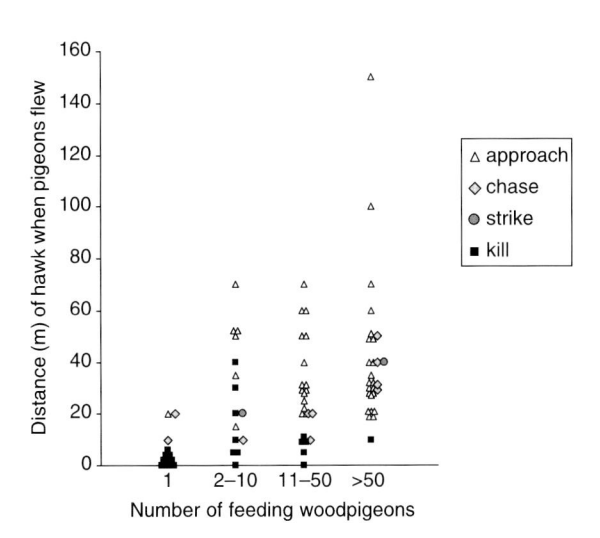

Figure 88. In arranged attacks on Woodpigeons, success was greatest at single prey. Attack success decreased with flock size, as a hawk was seldom able to get closer than 20 m before the largest flocks flew off. Data from Kenward 1976a, 1978b.

A hawk might in theory be confused by a large flock (Hamilton 1971), and this may also have contributed to the poor success of attacks on them. When a flock took off, on 52 occasions the approaching hawk swung away in a glide but on 22 occasions it kept flying powerfully in a chase, and the proportion of successful chases decreased sharply with flock size (Figure 88). Thus, seven of nine chases resulted in kills during attacks on flocks of 2–10 pigeons, compared with one of six chases after the largest flocks (with one pigeon struck but not held in each case). However, chases at pigeons in the largest flocks often occurred when flight distances were long, perhaps because confusion sometimes also worked to the hawk's advantage. Toni Lutz told me about flying Goshawks at flocks of crows, which did not take off in a cohesive group like the pigeons but as a straggle of smaller (family?) groups. Attack success was greatest at large flocks, probably because it was hard for crows still on the ground to spot a hawk when so many other birds were moving.

In the 59 attacks at pigeons flocks on *Brassica* crops, nine of 20 were successful in the last hour before sunset (45%), compared with only five of 39 (13%) earlier in the day. The pigeons might have been least alert while feeding hard to fill their crops, but the hawks did not get closer than earlier in the day before the pigeons took off. Alternatively, the attacking hawk might have been trying hardest to get a meal shortly before dark, but that would not explain why the proportion of Woodpigeons in the diet of free-living Goshawks doubled in the last fifth of the day. Increased vulnerability was probably due to reduced airworthiness when these 450g pigeons filled their crops with some 60g of food to digest overnight (Kenward 1978b).

Another factor which influenced the vulnerability of Woodpigeons was their physical condition. Mid-winter is a time of food shortage for them (Murton *et al.* 1963, 1966), and there were often diseased, starving or shot-damaged pigeons in the study area. The body mass of pigeons captured from flocks was strongly related to how close the hawk came before they took off (Figure 89).

Pigeons taken by the hawk before they took off, or very shortly afterwards, were similar in mass to the average of birds shot from flocks as they flew from woods to the same *Brassica* fields. But if the hawk was further away when the pigeons flushed, so that it had to choose them in the air, it made a kill only if there was a weak pigeon that lagged behind. These laggards were in condition as poor as those that hawks took singly. The three in poorest condition came from the smallest flocks, so presence of weak pigeons in small flocks also contributed to high attack success on small flocks (Figure 88).

SELECTION

When attack success is influenced by a physical characteristic of the prey, the result is selection of prey that diverge from the average. The arranged attacks showed that

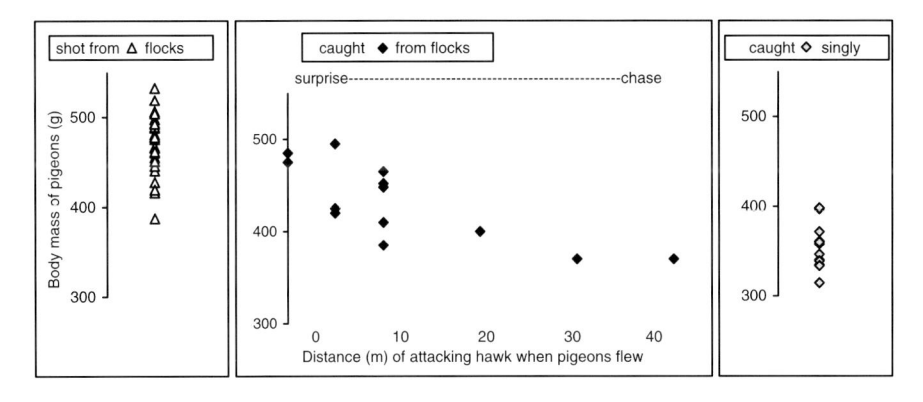

Figure 89. When pigeons were taken by surprise in arranged attacks, their body-mass was similar to pigeons shot from flocks at *Brassica* sites. Pigeons taken after a chase had masses as low as those caught singly. Data from Kenward 1976a, 1978b.

selection for poor pigeon condition could be the result of a chase, but was less likely when a flock was surprised. The hawks then apparently grabbed a pigeon at random and there was no tendency for the weakest to be selected. Single pigeons that had left flocks were also in poor condition. The radio-tracking of free-living Goshawks in Oxfordshire showed how selection occurred in the same way when there was no human participation. In this case the pigeons were often part-eaten when examined, but their body mass could be estimated very accurately from the mass of the small *supracoracoideus* breast muscle, which underlies the much larger *pectoralis major* and was usually uneaten on one side (Kenward 1978b, 1986).

Although there was a marked selection of pigeons with below average mass, some of those taken by the free-living hawks were well above the average for those shot at roosts in nearby woods (Figure 90). The plumpest pigeons had usually been caught at the edge of cover or where they roosted, while three birds killed more than 50m from cover, and therefore least likely to have been surprised, were in significantly poorer condition than the remainder. On average, however, pigeons captured by the free-living hawks in 1974–5 were in better condition than those caught at *Brassica* sites in 1972–3 (compare masses in Figures 89 and 90) because a wet harvest in 1974 had left much grain on winter stubbles.

Other physical aspects of prey can influence its vulnerability. Although Woodpigeons do not differ appreciably in appearance between the sexes, and released Goshawks did not select individuals of either sex (Figure 90), selection for one sex of the more dimorphic gamebirds is routine (Höglund 1964b, Widén *et al.* 1987). At the Swedish estate where released pheasants were very abundant, Goshawks developed a strong tendency to take females rather than males (Kenward 1977). Male pheasants are larger than females (Figure 90) and armed for aggressive defence with spurs, which probably reduced the success of Goshawk attacks

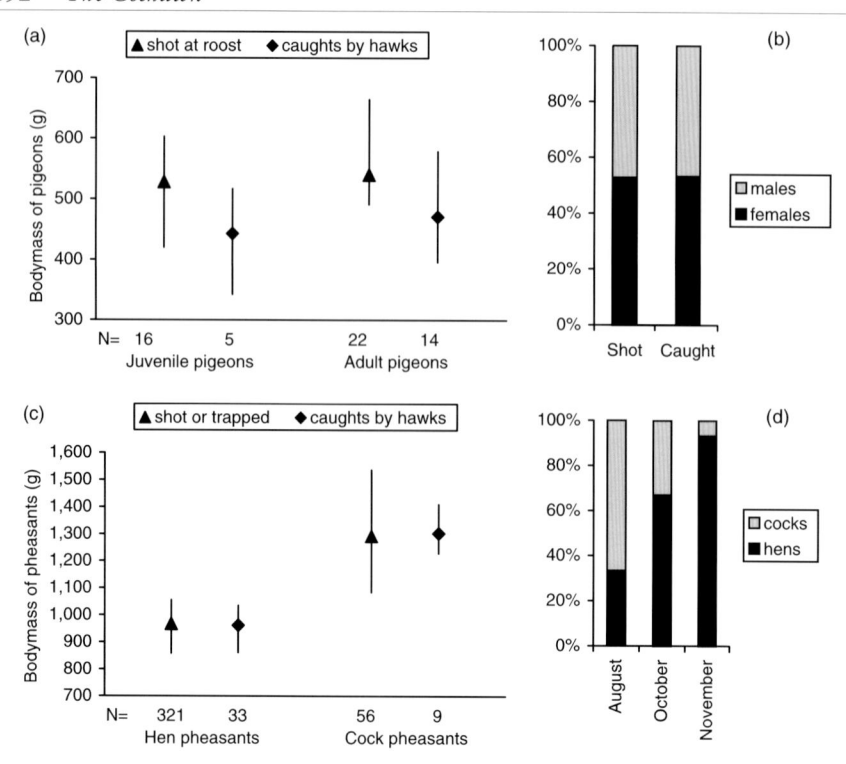

Figure 90. When Goshawks caught Woodpigeons (a) they selected prey with low body-mass (shown as means with range of values) compared with samples shot at roost but caught the same proportion of each sex and age class as in shot samples (b); Goshawks did not select Pheasants (c) in poor condition but the did selectively hunt females (d) as the Pheasants grew to full size between August and November. Data from Kenward 1986.

on them. Selection of female pheasants also occurred during snow-cover in an area with wild pheasants. In this case, snow would have reduced the value of female camouflage and their tendency to skulk in cover.

Absence of selection for pheasants in poor condition probably reflected their difference from pigeons in tactics to escape hawks. Pheasants seldom fed more than 10 metres from cover and only 4% of 79 kills were more than five metres from cover. If a hawk did not surprise the pheasants, they could run immediately into cover, with no chance of a chase in which selection for weakness might occur. In the escape flight of pigeons there was plenty of opportunity for weak stragglers to be taken in a chase, as shown in Figure 90.

Although Woodpigeons vary little in colour, one finds both dark and pale varieties of Feral Pigeons. In attacks by wild Goshawks on flocks of Feral Pigeons, Zygmund Pielowski (1961) observed that pale birds, which were uncommon in the flocks, were caught unusually often. Since the pale pigeons might simply have been the weaker birds, he created some flocks in which dark pigeons were a minority. In attacks on these flocks the Goshawks took more dark pigeons than expected, so Pielowski was able to conclude that the hawks were selecting the 'odd' coloured birds, probably because hawk attention was drawn to the unusual pigeons during a chase. Christian Rutz (2005d) showed that such selection of odd-coloured pigeons increased with Goshawk age, and that males showing the strongest selection also had the earliest broods and young in best condition.

Attack success is influenced by differences in vulnerability within prey species due to behaviour, habitat and weather, which can change from hour to hour, and to physical condition and age, which change more slowly, and to sex, which does not change. There is also variation in vulnerability between species. Effects of variation in vulnerability of prey can include (i) variation in Goshawk diet (Chapter 6), (ii) reduction in the impact on prey populations (see below) and (iii) trends in Goshawk demography (see Chapter 8).

HOW GOSHAWKS CHOOSE PREY

Selection is not only a passive result of greater success resulting from attacking prey with particular characteristics. There can also be a reinforcing element of hawk choice. As illustrated at the start of this chapter, hawks may choose to attack certain prey, or to avoid them, for various reasons. When trained Goshawks have flown at many pheasants, they often start refusing to seize the males but will still take females. Wild hawks apparently do the same when there are abundant released pheasants or when snow increases their visibility. In other words, the hawks may direct their attacks at the females if there are many pheasants of both sexes available, but if pheasants are sparse the hawks take what they can get. This would explain an absence of marked selection for female pheasants in other studies (Marcström & Widén 1977, Ziesemer 1983). Very small prey also seldom occur in the diet of wild hawks (even among stomach contents), probably because the energy spent hunting them is usually greater than the energy gained by eating them, unless they become very abundant or vulnerable.

A predator's ability to learn to avoid certain prey can be 'exploited' by the prey. Although gulls often occur in areas with Goshawks and are highly vulnerable to attack (Fox 1981), Nils Höglund noted that they are relatively seldom taken, and suggested that Goshawks do not like their taste. The same might be true of adult Rooks and crows, which are seldom killed by the hawks. Falconers who hunt corvids substitute pigeon meat when the raptor first kills, so that the taste will not discourage it (Woodford 1960). Young corvids are less acrid-tasting and passerine taste also varies with colour (Götmark 1994), but no temperate prey are as deterrent as the batrachotoxin-laced Pitohuis (*Pitohui* spp.) from New Guinea (Dumbacher *et al.* 1992). Taste may be less important for Goshawks than other deterrent techniques employed by gulls and corvids, which usually forage in the open where they are hard to approach unseen. The success of the trained hawk at flocks of more than ten was only 9% and, after making some 90 attacks on pigeons, it refused several flights at large flocks of them late in the winter.

Rooks, crows and gulls have a further behaviour which may teach hawks to avoid them: they mob raptors (Plate 6). The mobbing of Goshawks by corvids continues with great persistence right through the winter, long after young have left the nest. Zahavi & Zahavi (1997) suggest that this sort of behaviour signals to the predator that the prey is aware of its presence, and is therefore not worth chasing. An extension to this idea is that the predator may become so disheartened by the prey, especially if the mobbing is unpleasant, that it loses interest in attacking the prey at all. Predators can also be driven away by mobbing, or worse. A juvenile female Goshawk in St Petersburg was killed when caught in the open by a flock of some 200 Hooded Crows (Savinich 1999).

In the wild, the process of becoming an experienced hunter probably leads to shunning of situations where attack success is low, and hence to improved success on particular prey species (Fox 1981, Rutz 2005d). Perhaps Goshawks approach a 'giving-up-threshold' for situations in which their attack success falls appreciably below 10%. It would be interesting to know whether decisions to attack are also influenced by the size of prey, and hence the potential reward of capture, so that small species are most readily shunned.

SPECIALISATION

The learning process probably results in all Goshawks developing to some extent a preference for attacking particular prey in particular situations. When some Goshawk nests contain predominantly waterfowl remains in the same area as others that do not (e.g. Holstein 1942, Rummel 1962), this may reflect hunting in different habitats rather than specialising on particular prey. Such hawks could be considered habitat specialists.

More rarely, hawks seem to have a penchant for prey which are otherwise taken very seldom. There are occasional corvid specialists. I have been to one nest in Finland and one on Gotland where there were nothing but crow remains. Sulkava (1964) and (Slagsvold 1978) made similar observations. Some hawks may merely take temporary advantage of inexperienced newly-fledged corvids, but one falconer reported a Goshawk that had been trapped as an adult and which would take any opportunity to fly at crows and Rooks (David Kent, pers. comm.). Such corvid specialists are probably no more than 1–2% of Goshawks. There were none among 40 adult Goshawks radio-tracked in Sweden to recover kills.

Once a trained hawk has learned to kill an elusive prey it may become very adept, and even 'cheat' on the prey's normally safe anti-predation tactics. Glasier (1963) reported a Sparrowhawk which appeared to lure mobbing Magpies away from cover, only to turn and seize them. Wild Goshawks probably learn similar tricks.

If specialisation is defined as 'the frequent capture of prey which are not normally taken by other hawks hunting the same habitats', then the taking of adult hares should be seen as specialisation. Six of 10 females tracked in central Sweden all took at least two hares, and obtained the majority of their food in this way. They had smaller ranges, indicating a better food supply than the other females, which took no hares (Kenward 1982). Trained hawks must learn to seize hares at the head if they are to subdue them, and refuse to fly at hares if they have been kicked from the back end several times (Woodford 1960). Such specialisation is likely only for prey that are relatively invulnerable to attack or dangerous for Goshawks, so that no more than a small proportion of the hawk population learns to capture them. A radio-tagged Goshawk found dead with unexplained 'blunt impact' injuries in an open area (Squires & Ruggiero 1995) might have been killed by a hare.

FOOD REQUIREMENTS

During eight periods of 9–26 days in which only one or two hawks were available to be tracked in mainland study areas, three male and three female hawks were monitored so intensively that all kills of any appreciable food value were probably recorded (Kenward *et al.* 1981a). Large kills were weighed when found, and after each subsequent meal until the hawks finished them or they were taken by scavengers. Although wastage, on carcasses that hawks picked clean, was only 25–32% of body mass for pheasant and hare kills, another 19–32% was lost by the hawks because of scavengers, which often found kills after the hawks had managed only one meal. An estimate of food from smaller kills was the average mass of that prey, less the mass of the skull and any other parts remaining after an undisturbed meal.

The kill rates of males and females in these mainland study areas appeared to be very similar, at just over three days per kill, but males obtained an average 410g food from each kill, compared with 607g for females. Höglund noted that the average weight of prey animals was 462g for male hawks, and 768g for females, which is equivalent to the 410g and 607g of food consumed after wastage of 11 % and 21% respectively.

At the observed kill rate, males obtained an average 133g of food per day, compared with 169g per day for females. Of course, hawks could eat much more than the average in a day. The largest meals of males were 375g and 569g for females. While gorging themselves on kills, males obtained up to 585g and females 838g in the first two days of feeding on large kills. All the consumption figures for males were about 70% of the values for females, the same ratios as their average body mass in central Sweden. Food requirements of wild hawks were 1.43 times (females) to 1.65 times (males) those of captive birds.

Using these consumption figures, it was possible to estimate how much food the hawks obtained from their main prey, not just during the periods in which all the kills were found, but over longer periods when almost all large prey should have been found. In the central Swedish study area with a good pheasant population, the most important prey for females were hares, which provided 37% of their food requirement, whereas males obtained most food (43%) from pheasants. Squirrels, which were the most commonly taken prey for each sex, provided only 15% of their food because there was less meat on a squirrel than on a hare or pheasant. The detected kills accounted for only 63% of the estimated food requirements of females, and 74% for males, outside the period in which food consumption was monitored intensively. The overall dietary contribution estimates are probably fairly accurate for hares and pheasants, since most such kills will have been found, but the contribution of smaller prey must have been underestimated.

ESTIMATING THE IMPACTS OF PREDATION

When kill-rates cannot be measured directly, for example by radio-tracking or video records at nests, the kill-rate can be estimated from the daily food requirements of each hawk, the biomass available from each prey species and its proportion in a diet estimate:

$$\text{kills/day} = \frac{(\textit{biomass available from prey species})^2 \times \% \textit{ of prey species in diet}}{\textit{biomass from 100 kills representing all species in diet} \times \textit{biomass required daily}}$$

Biomass values should subtract wastage values from prey body mass estimates and wastage should include large prey incompletely consumed due to scavenging as well as parts not normally consumed. Wastage can exceed 50% for large prey in areas with much scavenging (Kenward *et al.* 1981a, Ziesemer 1983), but can otherwise be taken (from kills of hawks radio-tracked intensively) as 25% for birds that provide more than one meal, 30% for equivalent mammals and otherwise 20% (Kenward *et al.* 1981a). As females need more food each day than males, so an average for both sexes will underestimate predation if there is an excess of females in the population (see Chapter 8).

The predation impact for the period in which that kill-rate applies, measured as the proportion of the prey population killed is then:

$$\frac{\textit{kills/day} \times \textit{number of days} \times \textit{density of hawks} \times \textit{mortality compensation in prey}}{\textit{density of prey}}$$

The drawback to estimating kill rates from biomass is that it depends on an accurate representation of the diet and of wastage for the period concerned (see Chapter 6). The recording of kill rates obviates these problems. Accurate estimates also require a sound knowledge of predator population dynamics, in order to estimate their density correctly, and of prey population dynamics to make allowance for when mortality through predation is not additional to other sources of mortality. The hawk density is only the same as that of families at nests when there are few non-breeders (see Chapter 8). The impact of predation on juveniles before a food-bottleneck in winter is likely to be compensated by a reduction in subsequent starvation, and hence will have less impact than when adults are taken during the breeding season.

One question in the research at Oxford was whether Goshawks, which were just becoming re-established in Britain, could kill enough Woodpigeons to reduce damage to *Brassica* crops. Alternatively, could disturbance by Goshawks drive them away from crops and thereby perhaps increase starvation among the pigeons? Ron Murton's team (Murton *et al.* 1963) had shown that Woodpigeons at clover sites might need to feed for 95% of the day in mid-winter in order to maintain their body mass. Disturbance by hawks might therefore tip more pigeons into energy deficit, by increasing energy costs as well as reducing the time available for feeding.

Woodpigeons were the main prey of the free-living Goshawks in Oxfordshire and provided 40% of their kills, with a kill-rate of about one pigeon in four hawk-days. The Woodpigeon density in late winter, which is determined by the winter food supply (Murton *et al.* 1964, 1966) was estimated at 43 pigeons/km^2 in a Cambridgeshire study site in the late 1960s (Murton 1974), and comparable densities were to be expected in similar parts of lowland Britain.

Estimating a possible predatory impact by Goshawks in lowland Britain was hindered by an absence of data on Goshawk densities. It could be estimated that if Goshawks reached a density of 0.2–0.4/km^2, they would kill enough pigeons after mid-winter to reduce their breeding population in July by 18–37% (Kenward 1979). However, using a relationship between Woodpigeon survival and body mass from Ron Murton's data, it could be shown that 28% of the predation would be compensated by the selection of pigeons in poor condition (Figure 90). This constrained the maximum likely predatory impact on the breeding Woodpigeon population to a 28% reduction. Moreover, predation between breeding and mid-winter could be fully compensated by a reduction in starvation, with no reduction in the Woodpigeon population that damages *Brassica* crops in mid-winter.

At the time it was thought that Goshawks would not reach densities of 0.2–0.4/km^2 in Britain. However, densities of 0.1–0.2 Goshawks/km^2 occur in parts of Sweden and Germany without released pheasants, rising to 0.4/km^2 in the rabbit-rich part of Gotland and even higher levels in autumn where pheasants are released. Thus, possible predatory impacts by Goshawks on Woodpigeons should not be excluded.

Other data indicated that disturbance by foraging Goshawks would not increase food shortage among pigeons. Pigeons (and pheasants) will feed close to a perched or feeding Goshawk, and thus seem to perceive little threat in a stationary hawk. Pigeons were also relatively unresponsive to hawk attacks when they were most at risk of an energy deficit. For instance, on one cold December afternoon, with sleet blowing on an east wind shortly before sunset, the trained hawk killed one of a feeding flock. The remainder resettled about 150m away, and continued feeding on sprout crowns as I walked to the hawk and coaxed it onto my fist for a tit-bit. The hawk again flew at the flock and killed another pigeon. Again the pigeons resettled elsewhere in the field while I collected the hawk. Neither of the pigeons killed was in a very poor condition. It seemed that the birds were so desperate to feed that a third attack would have had exactly the same result.

At earlier times of day, pigeons left feeding sites after an attack. However, this did not reduce their food intake. A Woodpigeon eats *Brassica* leaves faster than it can digest them, and thus accumulates food in its crop as it feeds. If disturbed, a pigeon returns to feed when its crop has emptied, which may be as much as an hour later (Kenward & Sibly 1977). Disturbance by humans was just as effective for scaring pigeons as were Goshawk attacks, and was more frequent (Kenward 1978c). From the distances flown daily by the free-living hawks, it could be estimated that an Oxfordshire pigeon flock would be disturbed by them about three times a day at a high Goshawk density, whereas pigeons in roadside fields were disturbed two to three times an hour by passing pedestrians and other traffic. Pigeons in roadside *Brassica* fields showed no significant avoidance of feeding near trees and hedges, from which Goshawks might attack, but had a strong tendency to keep away from the road, where there was frequent disturbance by humans.

Although occasional Goshawk attacks were of little value for scaring pigeons from *Brassica* crops, pigeons showed a strong reluctance to settle anywhere near a soaring hawk. Raptor shapes in the form of kites or suspended from balloons have been used successfully to scare pigeons, but are practical only in suitable wind conditions.

Woodpigeons have a number of characteristics which may have evolved to reduce the risk of predation by Goshawks. Their escape response is to take flight, in which their speed is slightly greater than a Goshawk. They do not reduce their airworthiness by loading their crops until the very end of the day, when part of the hawk population has already fed. When attacked, they take to the air virtually in unison. A remarkable uniformity in size and colour further reduces risk of being singled out. If they are seized, the hawk may well get no more than a clump of feathers, which are especially thick and easily detached from the rump, the part most likely to be grasped from behind. These were the feathers lost from pigeons that were struck but not held during the arranged attacks on flocks (Figure 88).

Although Feral Pigeons are an important prey for Goshawks, estimates of hawk impacts on their populations have yet to be published. However, there are some interesting Dutch data from rings on racing pigeons based on systematic searching of Goshawk nests and the ground nearby with metal detectors. In one study on the coast, the tracing of ownership from rings showed that 81% of 465 pigeons had over-flown their destination and could be considered lost. Choice of a coastal site may have reduced the chance of pigeons being taken 'en route' between towns, but other data showed that only 0.1% of the pigeons were ones trained at local villages (Vlugt 2002). A majority of pigeons were killed in their first year of life, which enabled another study to estimate the years of occupancy of 22 Goshawk nests for the previous 15 breeding seasons. The results of this 'ornithological archaeology' agreed well with independent observations (van Haaff 2001).

GOSHAWKS AND PHEASANTS

Although there was a healthy wild pheasant population in North Oxfordshire, free-living Goshawks killed no more than one pheasant each per month in 1974–5. Whereas attacks on pigeons were as successful by free-living hawks as in arranged attacks, attacks on pheasants were the least successful observed (Figure 91). None of 16 attacks resulted in kills, the hawks being hindered by brambles, nettle stems and other ground cover. When I started work on Goshawks in Sweden in 1976, it is not surprising that I expected to show that Goshawks had little impact on pheasants there.

However, although the release and feeding sites at Frötuna provided very thick cover, the 4,000 pheasants released annually there were by no means immune to Goshawk predation (Plates 18 & 20). Goshawks were trapped near nine pheasant feeding sites during autumn and winter and 17 were radio-tagged and released on site. Up to six were tracked at a time, checking each at one- to two- hour intervals to record kills. On average, each hawk killed a pheasant every 1.7 days in August and every 2.3 days in October. The kill-rate declined to one pheasant every 3.6 hawk-days in December, and one every 8.7 hawk-days in January, after the prey population had been substantially reduced by shooting, the trapping of breeding stock and hawk predation (Kenward 1977).

The ranges of the radio-tagged hawks, mainly juvenile males, were extensively superimposed, with core areas overlapping at the pheasant release sites (Kenward & Walls 1994). An area in which convex polygon ranges of three juvenile and two adult males all overlapped in October encompassed eight of nine pheasant sites (Figure 92) and was visited three- to five-times as frequently as expected if they used all parts of their ranges equally. Occasional mildly aggressive encounters were seen between hawks, but they never resulted in radio-tagged birds leaving an area. It was often the

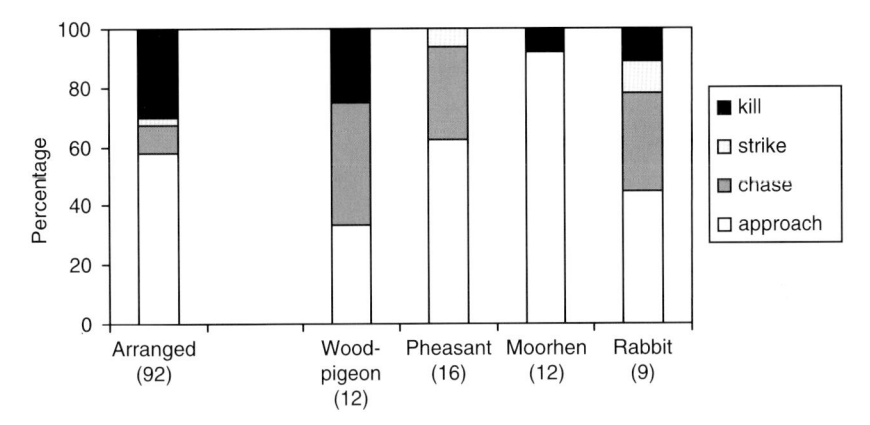

Figure 91. The success of arranged attacks on Woodpigeons, and of all attacks by free-living goshawks in Oxfordshire for which >8 were seen on each prey species. Data from Kenward 1978b and unpublished.

first-year birds which stooped at adults, as they soared together in local updraughts, and there was no aggressive screaming.

To estimate the density of hawks, radio tags were used as markers. Whenever hawks were seen 'by chance' (when they were not being radio-tracked), they were immediately checked for a radio-signal. In principle, if half the sightings were hawks with radios when six hawks with radios were around, then there would have been 12 in the area in total (Kenward *et al.* 1981a, see also White & Garrott 1990). This estimate was checked with pheasant kills found independently by gamekeepers: the ratio of kills by tracked and 'unknown' kills was close to the ratio of tagged to untagged hawks in the sightings.

In October, when the released pheasants had reached full size but before numbers had been reduced by shooting, up to 13 Goshawks were killing an estimated 40 pheasants per week, 1% of the released stock. This underestimated the possible impact, partly because 11 of 23 trapped hawks had been removed from Frötuna in the previous two months to save more pheasants for shooting, and partly because the kill-rate was 74% of that in August. The total predation impact likely without removal of hawks was estimated at 19%.

Goshawk predation on wild pheasants in Sweden was studied in the same way at Segersjö estate (Kenward *et al.* 1981a). Pheasants were counted with the help of

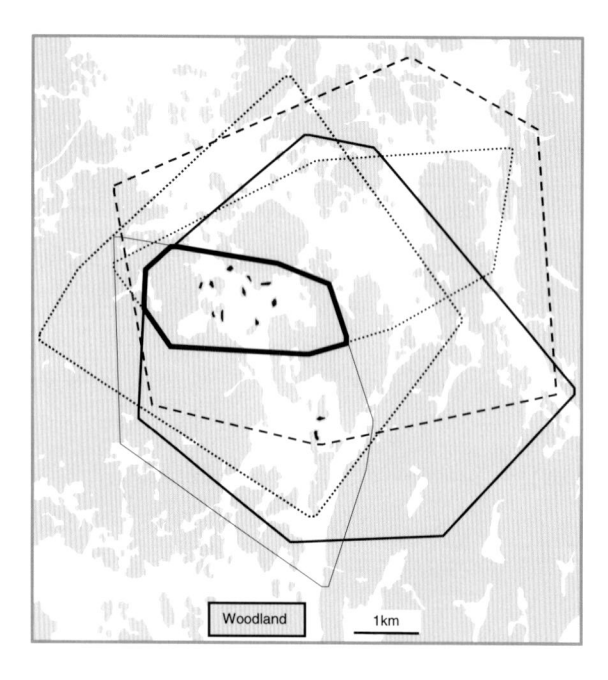

Figure 92. The October ranges of two adult males (solid narrow lines) and three juvenile males (dotted lines) at Frötuna estate in Sweden all overlapped in a high-use area (solid thick line) that included ten of the 12 pheasant release and feeding sites. From Kenward 1977.

volunteers and trained dogs in autumn 1977, and again in spring 1978 to estimate the losses during winter. Goshawks killed pheasants much less frequently than at the estate with released game, but hen pheasants in particular had a high mortality rate over the winter (64%), of which 88% was due to the hawks (Figure 93). Cock pheasants were killed mainly by shooting, which was directed at the males to preserve hens for breeding. The hen pheasants were reduced to a level at which they could only have reproduced the original autumn population by rearing four young each. Such a high reproductive rate has been recorded for pheasants in Scandinavia, but only once in three areas each studied for four years (Göransson 1975). The pheasant population was declining towards extinction at Segersjö.

Although Goshawks may take a higher proportion of a pheasant population when disease is present (Göransson 1975), the high predation by Goshawks at Segersjö was not linked to disease or poor condition in the captured pheasants. Goshawks were not selecting wild pheasants that would have died of food shortage but the sex whose loss would most affect the productivity of this polygamous prey.

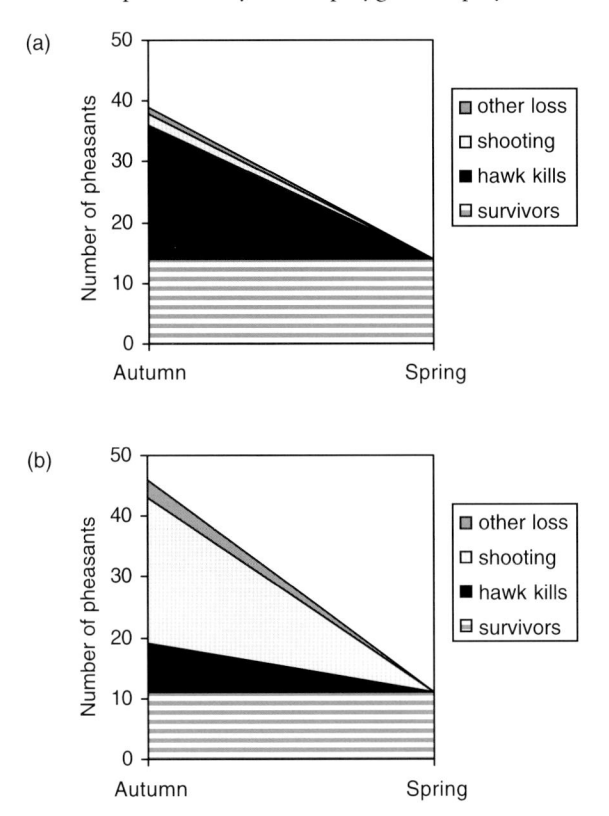

Figure 93. Most wild hen pheasants (a) lost at Segersjö estate in Sweden between October 1977 and April 1978 were taken by hawks, whereas most cock pheasants (b) were shot. Data from Kenward *et al.* 1981a.

GOSHAWKS AND WOODLAND GROUSE

Woodland grouse have been studied as important gamebirds in many northern countries. Early research in North America estimated that Goshawks killed 35% of a marked Ruffed Grouse population over the course of a winter, and found that grouse density declined with proximity to Goshawk nests (Eng & Gullion 1962, see also Gullion 1981). In Belarus, Vladimir Galushin (1970) used an index of predatory pressure to estimate the proportion of gamebird populations taken by breeding Goshawks in two study areas. With a low Goshawk population of about one pair per 100 km^2, the hawks were estimated to take 4% of all the woodland grouse, including 5% of the Hazel Grouse.

In Finland, Harto Lindén and Marcus Wikman (Lindén & Wikman 1983) used prey remains collected at Goshawk nests to estimate that hawks took an average of 12% of the adult Hazel Grouse population during the breeding season (March-July), with up to 26% per year. Per Widén (1987) estimated that breeding Goshawks killed 14% of Black Grouse males and 25% of females, and a study with radio-tagged grouse gave a very similar estimate of 20% for hens during the breeding season (Angelstam 1984). Higher estimates for the breeding season were derived by Risto Tornberg (2001) from Goshawk diet and grouse data in northern Finland, with averages (and range of values) across 11 years of 16% (4–25%) for Hazel Grouse, 22% (5–54%) for Willow Grouse and 14% (5–25%) for Black Grouse hens, but only 9% (4–15%) for male Black Grouse. Estimates for proportions of the grouse populations taken annually, using late-winter diet to represent the non-breeding season and varying numbers of non-breeding hawks, were 43–66% for Hazel Grouse, 68–100% for Willow Grouse and 29–46% for Black Grouse.

These are very appreciable proportions of the adult prey populations. Willow Grouse in particular seem prone to heavy Goshawk predation in northern forests. The grouse populations of such areas may be sustained by Willow Grouse living in more open land, where the absence of trees would hinder Goshawk foraging. In Poland, Dudziński (1987, 1990) found that the density of Grey Partridge pairs and coveys increased with distance from forest edge, with the greatest increase after 1,000m, beyond which Goshawks were seldom recorded. Of partridges killed by predators, 71% could be attributed to raptors, 'mainly Goshawk'.

GOSHAWKS AND CORVIDS

Impacts of accipiters on passerine prey species, including reduction in density near to predator nests, have been measured in several studies. For example, in Wytham Woods near Oxford, where the nesting of Great Tits and Blue Tits has long been carefully recorded (Perrins 1979), tit nesting-success decreased in the vicinity of Sparrowhawk nests, mainly due to 'disappearance' of the breeders (Geer 1978,

Perrins & Geer 1980). Recovery of tit rings from pellets at the hawk nests showed that 18–36% of young Great Tits were being killed by hawks during the breeding season. However, there was only a slight increase in tit mortality and immigration to the wood when Sparrowhawk recovered from pesticides (McCleery & Perrins 1991, Newton 1998).

Large passerine species are at risk from Goshawks. Looft & Biesterfeld (1981) noted that corvids tended to space their breeding attempts away from the nests of Goshawks, and that crow and Magpie nests closest to the hawks usually failed. This was investigated in more detail by Hermann Ellenberg's group at the University of Saarbrücken. Carrion Crow nests were never successful within 500m of Goshawk nests, compared with a failure rate of only 10–15% more than 1.5km away. Nest failures of Magpies, which are so common in the diet of urban Goshawks (Würfels 1994, Rutz 2003), were 100% within 1km of hawk nests (Figure 94). From prey remains, (Ellenburg *et al.* 1984) estimated that Goshawks could eat more Magpies and Jays than were produced annually in the study area, 43–62% of the Carrion Crows, Woodpigeons and pheasants, and also have a heavy impact on

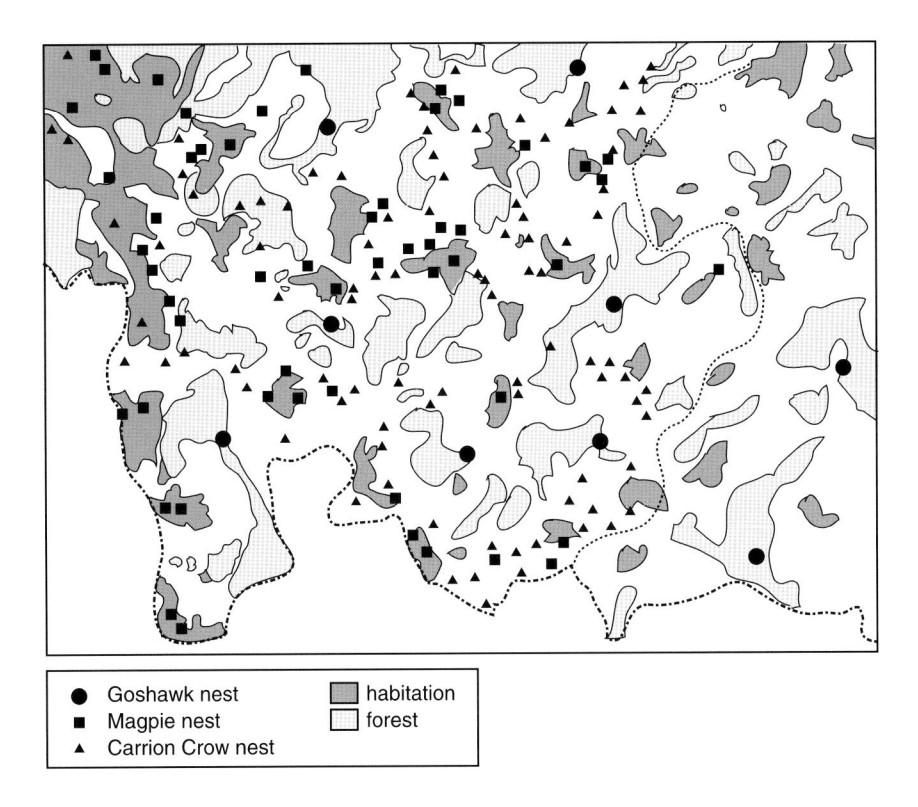

Figure 94. The spacing of successful Magpie (▲) and Carrion Crow (■) nests around Goshawk nests (●) near Saarbrücken. Redrawn from Ellenberg *et al.* 1984.

Grey Partridge, Mistle Thrush, Hobby and Sparrowhawk populations. There is a further twist to this story. Goshawks themselves may avoid nesting near Ravens, such that these large corvids provide shelter areas to small species (Ellenberg & Dreifke 1993).

At first sight, these findings seem hard to reconcile with claims that Goshawks do not hunt close to their nests (e.g. Schnurre 1934). Nesting associations of potential prey with other raptors are well known, including waterfowl (and especially Red-breasted Geese) near Peregrines (Cade 1960, Quinn *et al.* 2002) and Fieldfares and Woodpigeons near Merlins (Hagen 1947, Wiklund 1982, Bijlsma 1984). By tethering pigeons at various distances from Goshawk nests, Wyrwoll (1977) found no reluctance to attack, except that females were reluctant to leave their eggs to attack prey more than 75m away. Males were less keen taking prey within this 'female attack distance' from the nest and took kills further away to pluck them. Both Goshawk sexes are also liable to ignore prey that their previous experience has shown to be hard to catch. This probably explains why small passerines so often nest successfully in the 'lower storeys' of Common Goshawk nests. Looft and Biesterfeld (1981) recorded 14 nests of Eurasian Treecreepers, 11 of Common Starlings, two of Redstarts and one Coal Tit nest, as well as a roosting bat, in Goshawk nests in Schleswig-Holstein. Bijlsma (1993) added Stock Dove, Goldcrest and three other tit species, also two Eurasian Treecreepers in the same nest in one year.

GOSHAWKS AND OTHER UNCOMMON SPECIES

Tinbergen (1936) noted that many Sparrowhawks were brought to Goshawk nests in his Dutch study areas, and suggested that Goshawks might control Sparrowhawk numbers there. In Schleswig-Holstein, Looft & Biesterfeld (1981) found similar heavy Goshawk predation on Sparrowhawks. All 15 Sparrowhawk pluckings attributed to Goshawks were in woods that were both open enough for Goshawks and dense enough for Sparrowhawks. Ian Newton (1986) considered the Goshawk to be the main avian predator of the Sparrowhawk. He showed that whereas the median tree spacing at 89 Sparrowhawk nest sites was 2.8m in areas without Goshawks, the spacing was greater than 2.7m for only one of 43 Sparrowhawk sites where Goshawks were present. However, Goshawk feathers under intact Sparrowhawk nests show that attacks do not all succeed (Vedder & Dekker 2004).

In Oregon, Reynolds (1975, 1978) found not only that Cooper's Hawks nested in denser woods than Goshawks, with Sharp-shinned Hawks in the densest of all, but also that all three accipiters tended to space out their nests with respect to the other species. This was less obvious in the more broken woodlands of Schleswig-Holstein, where Sparrowhawks twice nested within 250m of Goshawks. However, if Sparrowhawks, Kestrels or Long-eared Owls tried to nest in a small wood that contained a Goshawk nest site, they rarely succeeded in years when Goshawks were present (Looft & Biesterfeld 1981). Buzzards, on the other hand, showed no avoidance of Goshawk nests. Although one study reported that proximity to Goshawk nests reduced Buzzard breeding success (Kostrzewa 1991), another found Buzzards nesting within 80m of Goshawks, and showed experimentally that crows were less likely to mob a Buzzard with a Goshawk present (Krüger 2002a, b).

The re-colonisation of Britain by Goshawks has created a useful 'natural experiment' for studying predation as well as nest spacing. Steve Petty and colleagues (Petty *et al.* 2003a) recorded a decline in Common Kestrel numbers as Goshawk pairs increased in Kielder Forest (Figure 95). The effect remained after correcting for a 20% decline in vole habitat, which did not cause a similar decline in numbers of Tawny Owls and Long-eared Owls. Moreover, Kestrels had not declined in a nearby area without Goshawks. Finally, diet records were used to estimate that Goshawks removed more Kestrels than were recorded each spring in the study area. There were 139 Kestrel remains among the Goshawk prey, especially in the March-April period prior to Kestrel breeding and thus when predation would have the greatest impact on Kestrel breeding population levels. Predation on young Kestrels was also heavy in July-August. It was concluded that the decline in the Kestrel population was mainly due to Goshawk predation.

A second paper (Petty *et al.* 2003b) examined Goshawk predation on Eurasian Red Squirrels, for which Kielder Forest is one of the last bastions in England against the invading alien Grey Squirrels. A similar estimation process suggested that Goshawks might kill four squirrels per home-range in summer, but take only an eighth as many in winter. This was partly because only adult males were seen

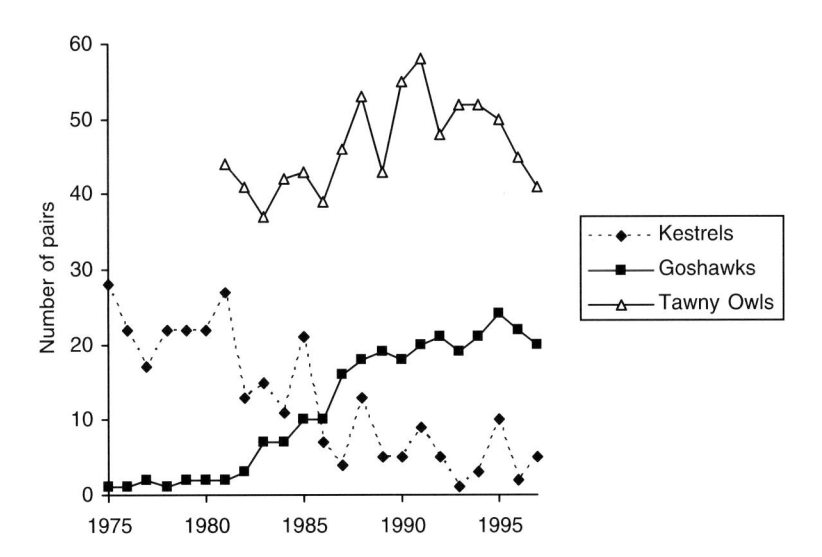

Figure 95. As numbers of nesting Goshawks increased in Kielder Forest, there was a sharp decrease in the numbers of breeding Kestrels. Redrawn from Petty *et al.* 2003a.

regularly in the forest in winter, but is surprising in view of the many Red Squirrels taken by Goshawks in Scandinavia. As prey remains contained the highest proportion of squirrels in March, radio-tracking data might be useful for confirming the winter predation estimates.

During a long-term study in Veluwe province in The Netherlands, Rob Bijlsma (2004) recorded a rising impact on Honey Buzzards as Goshawk numbers increased. The proportion of nests predated increased from 7.7–12.3% during 1981–90, to 17% in 1990–99 and rose to 33% in 2000–04. Goshawks also destroyed 25–30% of Honey Buzzard broods at a study area in Drenthe province. In both areas Goshawk populations were apparently stressed by food shortage (see Chapter 10).

HOW GOSHAWK PREDATION RELATES TO CHANGE IN PREY DENSITIES

In a few studies, Goshawk diet has been monitored for a number of years in relation to assessments of prey abundance. Thus, Tornberg & Sulkava (1991) were able to show that woodland grouse increased from 25–55% of the spring diet when there were fewer than 20 grouse/km^2 the previous autumn, to a plateau of nearly 80% when there were more than 30 grouse/km^2 (Figure 79 in Chapter 6). If the total number of prey is not changing greatly each year, this proportion is effectively the number of grouse actually killed by each breeding male hawk. Such curves are known as 'functional responses' (Holling 1959). A plateau in the response at high prey density occurs through predator satiation.

However, what happens at low prey density is more interesting, because it indicates how much predatory impact a predator that itself changes little in numbers may have when prey numbers decline. Lindén & Wikman (1983) examined how the kill-rate of individual hawks changed as the numbers of Hazel Grouse rose from 4/km^2 to 8/km^2. They found a concave relationship and, knowing the likely functional response at higher density from early access to Sulkava's data (Wikman & Lindén 1981) and from publications on other raptors (Keith et al. 1977), they predicted that the full functional response would be a 'sigmoid' S-shape (Figure 96). A plateau at something less than three times the maximum rate observed was inevitable, because at the maximum observed rate, Hazel Grouse were already 33.6% of the prey biomass.

The concave lower half of the functional response implies that the hawks did not initially increase the kill-rate as grouse numbers rose, but then switched to feed heavily on them until (at the top of the curve) they could not kill any faster. A sigmoid functional response has also been found for raptors feeding on grouse (Redpath & Thirgood 1997, 1999).

However, predation does not always switch from a prey species which is at low density. Between 1976 and 1982, Goshawk predation on pheasants was also being assessed by radio-tagging in two areas of northern Germany (Ziesemer 1983), as well

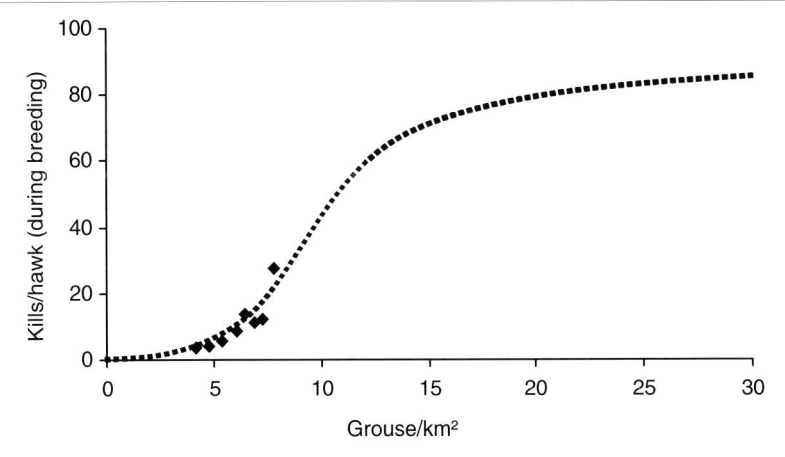

Figure 96. As numbers of Hazel Grouse increased, goshawks switched to this prey; satiation is predicted at high grouse density. From data in Lindén & Wikman 1983.

as at Frötuna and Segersjö and in three more areas of Sweden. Functional response to change in pheasant density could be drawn from seven areas of similar habitat (Figure 97) instead of from multiple years at the same site. This functional response remains convex at low prey density, as it does also for Peregrine and Gyr Falcons eating grouse (see review in Valkama *et al.* 2005). Pheasants were a preferred prey at all densities, with the hawks living on them almost exclusively if they were common enough. At the high pheasant densities, a plateau in Goshawk kill-rate might have been expected at one pheasant in three to four hawk-days, the period for which a Goshawk can obtain food from one pheasant kill. However, the hawks did not get

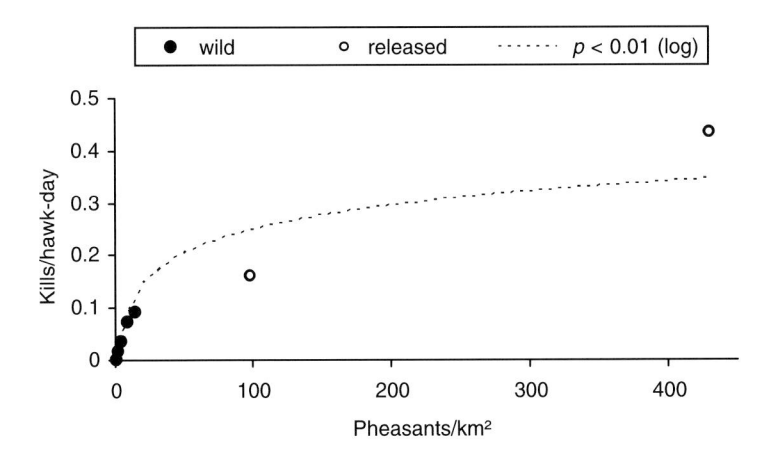

Figure 97. Goshawk kill-rate increased with pheasant density at 7 sites in Sweden and Schleswig-Holstein. Data from Kenward 1986, including Ziesemer 1983.

208   *The Goshawk*

all the food available because of scavengers (mainly foxes). To compensate for such losses, the hawks therefore killed more frequently than they would otherwise have needed for their own food requirements, especially at the two sites with greatest pheasant densities.

In all seven areas, the Goshawk density could also be estimated, and should show a sigmoid 'numerical response' to pheasant density (Figure 98), although more data are needed to be sure. A concave curve at low hawk densities is likely when other prey species maintain hawk numbers in an area that has few pheasants, and this creates the lower half of the 'S' shape. An upper limit on the number of hawks, which gives the upper half of the 'S', is a result of a limit on the number of hawks than settle, either from interference between them or because brood size constrains the number of dispersing hawks available in autumn.

The most interesting finding did not rely on the precise shape of the curve, but was a result of the high hawk density in the rabbit-rich area on Gotland. As a result of maintaining a high kill rate at low pheasant density (Figure 97) and accumulating in the area (Chapter 5), the Goshawks were killing 6% of the pheasants per month. Such a predation rate cannot be sustained by wild pheasants (it was 4% in the declining Sergersjö population). It used to be thought that game species would be 'buffered' against predation by large numbers of alternative prey (Leopold 1933). However, this may only occur when a sigmoid functional response indicates an ability to switch prey. Rabbits seemed able to aggravate Goshawk predation on pheasants by drawing in more hawks.

Was predation by Goshawks so severe because pheasants are introduced in western Europe and lack appropriate predator defences? Probably not, because pheasants meet Goshawks throughout their original range in Asia and are frequent prey where both are native, from Georgia (A. V. Abuladze, pers. comm.) to China (Z. Zhengwang, pers. comm.).

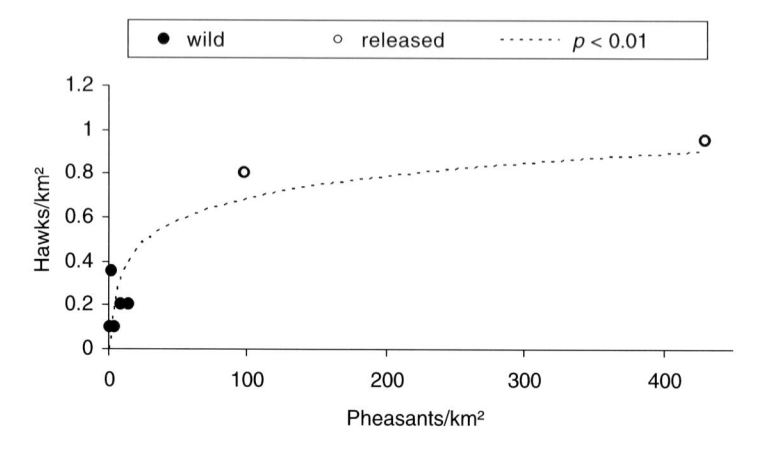

Figure 98. Goshawk density increased with pheasant density at 7 sites in Sweden and Schleswig-Holstein. Data from Kenward 1986, including Ziesemer 1983.

Predation by Goshawks seems to become most severe through accumulation of dispersers. Among hawks caught at Swedish pheasant rearing sites, 84–91% were in their first winter (Marcström & Kenward 1981b, Niedemann & Schönbeck 1990). When these dispersers are funnelled by the topography or attracted by temporary prey abundance into areas which would normally support few hawks, local extinction of certain prey is a possibility. If the young hawks then starve or move on, there will still be breeding hawks left elsewhere. Such locally severe impacts on prey may also occur during Goshawk irruptions in North America, after which the native Ruffed Grouse virtually disappear from some woods (Keith & Rusch 1988, E. Charles Meslow, pers comm).

Goshawk predation of this type has been implicated in reducing numbers of the Heath Hen prior to its extinction (Newton 1998). An even better example may be the squirrels on Mount Graham in New Mexico. The top of this southern outpost of the Rockies hosts an important astronomical site as well as being the unique location of the relatively large Red Squirrel race, *Tamiasciurus hudsonicus grahamensis*, in pine forest at risk from drought, long-suppressed fire – and Goshawks. Radio tags on the squirrels have shown an unsustainable mortality rate, twice the normal levels for this American squirrel, with most deaths due to raptors that leave neat rings of well-plucked fur (John Koprowski, pers. comm.). The peak is a typical focus for Goshawks, not only as southward dispersers (Chapter 5) but also for visits of wintering birds from the pinyon-juniper zone below (Chapter 6). Adjacent to the home state of the 43rd US President, global-warming and Goshawks may well terminate *T. h. grahamensis*. In Germany, controversy persists about the role of the Goshawk in eliminating tetraonids by moving from prey-rich areas into residual patches of heathland (Brüll *et al.* 1981, Dobler & Siedle 1993, 1994, Busche & Looft 2003).

In Northern Finland, juvenile Goshawks tend to leave in winter, but less so in years when grouse populations are high (Byholm *et al.* 2003). As a result, Goshawk numbers build up when grouse numbers are high and occupancy of breeding sites peaks two years after a peak in grouse density (Tornberg *et al.* 2005, 2006). The result is a lag in the intensity of predation, which then tends to be high when grouse numbers are low (Tornberg 2001). The effect is especially strong for Willow Grouse, but also present for Black Grouse, albeit with a lower predation impact on their denser populations (Figure 99).

As the numbers of adult grouse were measured at the end of the Goshawk breeding season, these data may in part also reflect the impact of the predation. However, the tendency for predation impact to be low at high grouse densities remained for adult Willow Grouse the previous year (albeit with marginal statistical significance), and was stronger for all grouse species combined (Tornberg 2001). As an explanation for this effect, Risto Tornberg and colleagues also showed that Goshawks were four times more numerous relative to grouse at high grouse densities than at low densities.

If predation impact is high at low prey density (negative density dependence) and in some years high enough to be unsustainable, predators are able to reduce prey breeding populations in a way that can cause local extinction, prolonged suppression

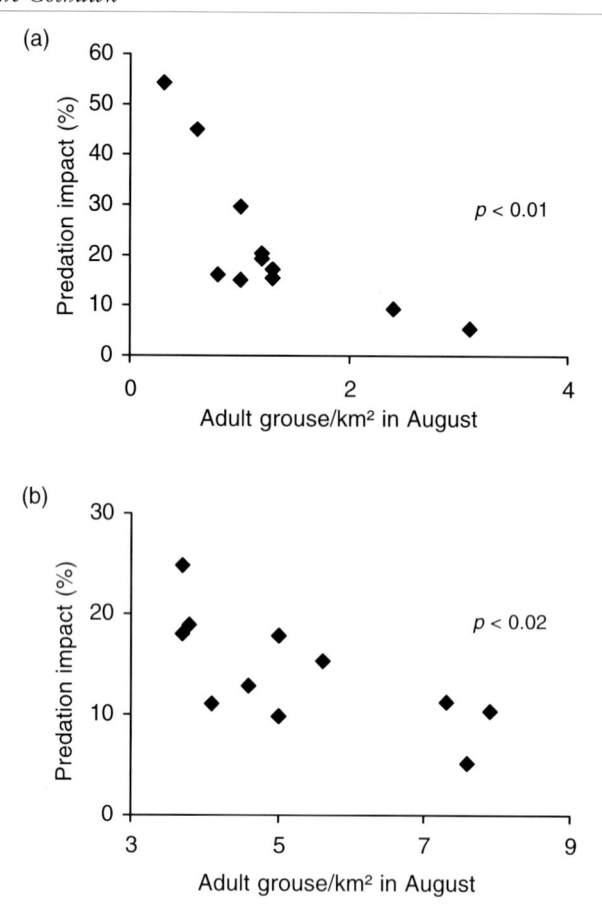

Figure 99. The proportion of woodland Willow (a) and female Black Grouse (b) populations killed by Goshawks in the breeding season is highest when there are few grouse. Data from Tornberg 2001.

in 'predation traps' or population cycles. Petty et al. (2003a) found strong negative density dependence in the Goshawk predation on Kestrels in Kielder Forest. A two-year lag in the numerical response of the hawks to grouse numbers made the Northern Goshawk a candidate for causing cycles that are observed in northern grouse populations (Tornberg et al. 2005).

IMPLICATIONS FOR CONSERVATION AND MANAGEMENT

Attitudes towards predators have changed greatly during the last century. The development of ecological science has coincided with growth of recreational interest in animals. This has put predators (and biologists who study them) into a zone of conflict between humans suffering from the predation and others admiring the predators.

Early population models indicated that predators ought not to be thought of as a nemesis for their prey, because if they did they would lack food and themselves become extinct. Paul Errington then took the scientific defence of predators a stage further. From studies of Muskrat and Northern Bobwhite, he concluded that predation was important in population regulation of herbivores because its victims were 'so often the immature, the ill-situated, the restless, the wanderers or the otherwise handicapped' and that predation therefore merely compensated for other types of population loss (Errington 1946). A similar situation was found among Red Grouse marked in Scotland by David Jenkins, Adam Watson and Andrew Miller (1964). The grouse found killed by predators in winter were those that failed to gain territories, and whose loss was therefore unimportant for the breeding population. The view became widely established that predators take mainly a 'doomed surplus' from prey populations.

The idea that raptors have little long-term impact on game populations was supported by the first extensive study of raptor communities and their food supply, which estimated that a North American raptor guild killed only 5–18% of the autumn game bird population and accounted for 18% of the total pheasant winter mortality (Craighead & Craighead 1956). Brüll (1964) concluded that Goshawks in northern Germany took too few pheasants and partridges to affect their breeding populations. Evidence was accumulating that bird populations were limited by their food supply, not by predation (Lack 1954, 1966). It was widely upheld that raptors could not reduce game populations, that they might sometimes act as beneficial 'health police' by rapidly destroying diseased prey (Leopold 1933), and that resident raptors should be preserved because they would keep others out of their territory (Brüll 1964) and even avoid taking prey near their nests (Schnurre 1935).

This was a very convenient viewpoint when predators had either been extirpated or severely reduced over wide areas by long-term persecution and more recent pesticide impacts. In many regions there was little predation on game or on species with rarity value. Poultry could be free-range in Britain without risk from Goshawks.

Figure 100. A Goshawk feeds on a freshly caught pigeon.

The information on predation in this chapter indicates that we need to re-think a previous consensus, which holds that avian predators rarely if ever have an impact on prey populations. It is being increasingly accepted that predators, including raptors, can reduce populations of ground-nesting birds, such as woodland grouse (Newton 1991, 1998). In the case of the Goshawk, the possibility of impacts on some corvid and small raptor species also needs to be taken seriously. Such impacts are likely to vary in extent in different areas, depending on local conditions that affect Goshawk abundance and prey vulnerability.

Goshawks may sometimes depress breeding populations of gamebirds and other species. However, in a review of experimental studies that showed predation impacts, Ian Newton (1998) found fewer cases of breeding populations being depressed than of post-breeding numbers being affected. Where there is concern that predation by Goshawks reduces numbers of wild or released game available for humans in autumn, it is important to realise that the impact may neither be ameliorated through being focussed on weak prey individuals nor reduced by territorial exclusion of juvenile predators. Perhaps Goshawk predation on corvids may sometimes benefit humans, by reducing predation on eggs and young game by corvids (Milonoff 1994), but not where there is heavy dependence on hunting released game.

Goshawks also kill poultry and other domestic species. More than 3% of prey remains collected round Saarbrücken were domestic hens, which were taken throughout the year only slightly less often than Magpies (Dietrich & Ellenberg 1981), rising to 23% of winter diet records in Belarus (Ivanovsky 1998). Free-range poultry kept on smallholdings to recycle household scraps are tempting for Goshawks. Goshawks may mainly take racing pigeons that are lost (Bijlsma 1993, 2005a), but the loss of high-value birds round training lofts is liable to cause anguish even to the most tolerant pigeon fancier (Kutscher 1981).

Predation by Goshawks can therefore evince strong opinions from hunters, scientists and other conservationists (e.g. Kalchreuter 1981a, Brüll *et al.* 1981, Bengtsson 1997, Ryttman 1999, Lester 1999). Wildlife biologists may then be asked to estimate predation impact on particular species at minimal cost, and thus without measuring kill rates year-round by radio-tracking. It is then important to remember that diet estimated when emergent young prey are available should not be used for the whole year. The diet during courtship and incubation may be representative for the winter and spring periods, but the diet in the nestling period at best represents only the summer. Estimates of diet based on prey remains from unspecified periods, and especially from pellets, should be viewed with strong reservations and not used in arguments about impacts of hawks on species of socio-economic significance. Moreover, great care is needed when converting from prey biomass to kill-rates of individual predators and from kill-rates to predation impact.

To try to obtain a balanced assessment of the importance of raptor predation for different interest groups, a survey of government conservation authorities and main national hunting organisations was conducted as part of a European Union project on 'Reconciling Game-bird Hunting and Biodiversity' (Viñuela 2002). Each department or organisation was asked in 2001 to give a score, on a scale of 0–5, for how serious a

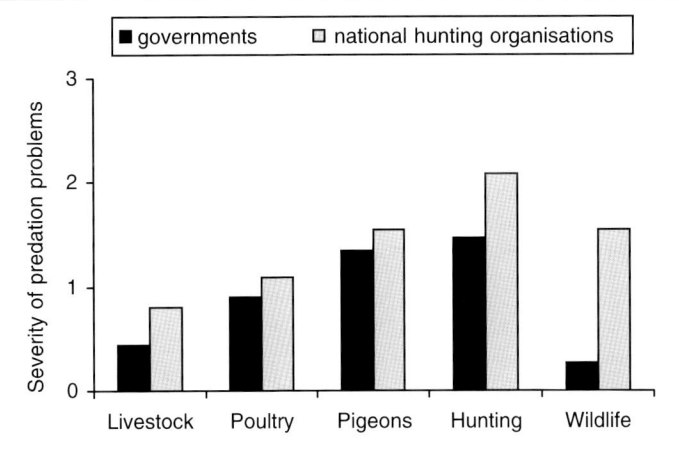

Figure 101. The severity of problems from raptor predation for different human interests, as perceived by 20 European Union governments and 12 national hunting organisations, as average ratings on a scale of 0–5. Data from Kenward 2002a.

problem they considered raptor predation to be in their country for different interest groups (Figure 101).

The hunting organisations and the governments viewed problems from raptor predation as being worst for hunters. There was general agreement between governments and hunters on the relative extent of problems for other interests, except in the case of conserving non-game wildlife. Hunters were more pessimistic than governments about raptor problems for all interests, and especially for conserving rare wildlife species. The representatives of government and hunting were also asked to name the species of raptors that they considered most problematic in their countries (Figure 102). The Northern Goshawk stood out as the species considered most problematic by governments, although the Common Buzzard ran it a close second among hunting organisations.

Buzzards do not easily catch full-grown poultry and pigeons, and are therefore unlikely to create the same problems as Goshawks for small-holders and pigeon fanciers. However, it is worth briefly considering studies that have compared their predation on game. Jens Dahl Mikkelsen (1984) recorded raptors at two nearby pheasant release sites in Denmark. Goshawks were seen by him less than Buzzards, but were more often close to release sites and attacking pheasants (Figure 103). The Goshawk attacks were five times more likely at a site with adjacent woodland than in one >200m from nearest woodland. In a questionnaire, raptors were considered to kill more than 10% of pheasants at only 11 of 47 estates, with less (29%) killed by Goshawks than by Buzzards (55%, and 10% by Sparrowhawks). The single factor that best explained the level of problems with Goshawks was the ease with which hawks could approach a pen unseen. The best predictors of problems were the proximity of a pen to woodland and the age of pheasants when released. The pens with least predation were far from trees and contained very well-grown poults.

214   *The Goshawk*

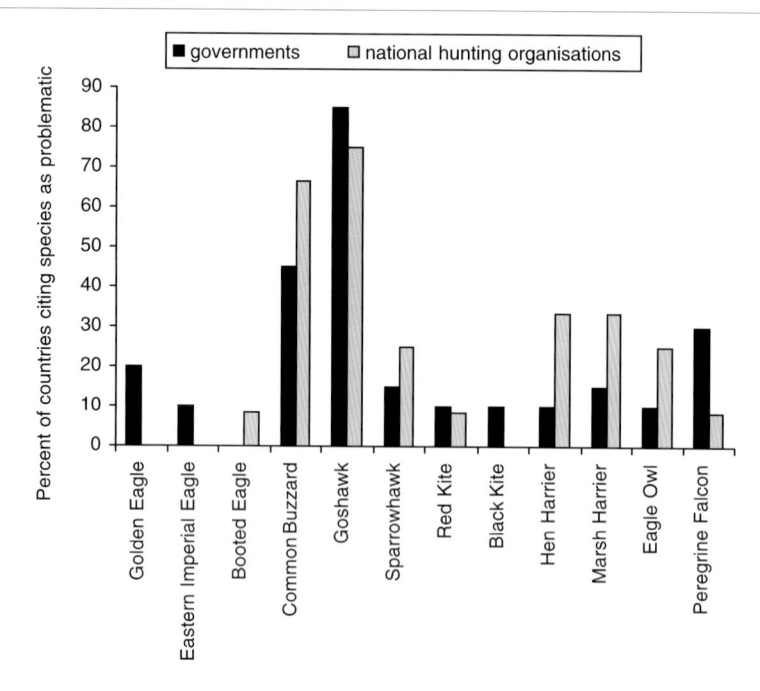

Figure 102. The frequency that different raptor species were cited as problematic by 20 European Union governments and 12 national hunting organisations. Data from Kenward 2002a.

In a British questionnaire survey, Harradine *et al.* (1997) found that Goshawks accounted for only 8% of avian predation incident-reports in 1995, with 2.5 times as many reports for Buzzards and 4.5 times as many for Sparrowhawks. However, there are fewer than 1,000 breeding pairs of Goshawks in Britain, compared with more than 30,000 pairs of Buzzards and at least 40,000 of Sparrowhawks (RSPB 2005), so the impact of Goshawks is disproportionate. Goshawk foraging areas are ten times as large as for Buzzards, so Goshawks are more likely than Buzzards to find any pheasant pens nearby. In a part of Dorset with one pheasant release pen per 4km^2, two of three released Goshawks found pheasant pens within a week (the third found free-range poultry), and a juvenile female hawk killed 17 pheasants in 22 days. In contrast, less than half of 136 radio-tagged Buzzards encountered pheasant pens early in life, only 8% associated strongly with pens and only 15% of Buzzard nests had pheasant kills (Kenward *et al.* 2001d). Compared with Goshawk kill rates of 0.16–0.43 pheasants/day in areas with released pheasants, the rate for breeding Buzzards was 0.025 pheasants/day. Therefore, a Goshawk is more likely than a Buzzard in the same area to encounter pheasant pens and kill pheasants.

I was once told by a Swedish raptor enthusiast that no more work on Goshawk predation was needed, because it had been recognised that there were no substantial impacts and my work on pheasants was disturbing the *status quo*. There is now

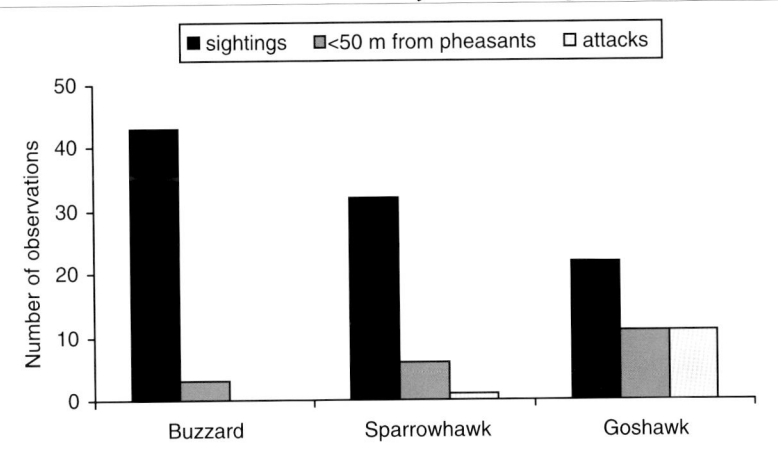

Figure 103. At a Danish pheasant release site, Goshawks were seen less often than Buzzards and Sparrowhawks, but most often in attacks. Data from Mikkelsen 1984.

evidence of strong predation impacts by raptors other than Goshawks (Valkama *et al.* 2005), so the time is ripe to move on from such a state of denial. There is also growing scope for moving on from conflicts about raptor predation, by adopting innovative solutions that involve cooperation between different interests.

For example, falconers were involved in much early work on raptor predation, especially by Goshawks (Craighead & Craighead 1956, Brüll 1956, Pielowski 1961, Meng 1971, Kenward 1978b, Fox 1981), and could still contribute substantially to understanding the factors that reduce attack success. Just as trained raptors were used to test power lines (Nelson 1978), so there is scope for using trained Goshawks to test the value of different pen designs and cover configurations for released and wild game. Falconers could also accumulate valuable information for designing low-predation landscapes by routinely recording prey species, habitats, attack type, distance and success in each attack. Another role for falconers is to find ways of influencing what raptors choose to attack, for instance by further developing the pioneering trials with taste deterrents (Nicholls *et al.* 2000).

CONCLUSIONS

1. Goshawk diet reflects the vulnerability (the risk of detection and capture) of individual prey species as well as their abundance. Capture risk can in principle be measured as attack success, but is liable to bias except when hawks hunt in flight.

2. Variation in success during attacks arranged with trained hawks showed that hares and Blackbirds have relatively low risk of capture, as do healthy Woodpigeons in flocks because of increased ability to detect the hawk. However, risk increased when pigeons filled their crops and if they were in poor condition.
3. Selection that could mitigate the population impact of predation was most likely when prey sought to outfly the attacking hawk but were too weak, and least likely when prey like pheasants were likely to escape in cover unless taken by surprise.
4. For assessing impact of predation on prey populations, kill-rates can be measured by direct observation or estimated from diet. However, diet estimated when young prey emerge in summer should not be used to estimate predation impact throughout the year. Care is also needed when converting from prey biomass to kill-rates of individual predators and from kill-rates to predation impact.
5. Selection of vulnerable prey could also be a result of active choice by an experienced predator, leading to increased attack success with age and to specialisation in prey that had low risk of capture by inexperienced hawks.
6. Daily biomass requirements of hawks hunting in Sweden were 133g for males and 169g for females. A higher proportion of large prey was wasted, due to having more inedible material and losses from scavenging before hawks had taken several days to consume the kill.
7. In Oxfordshire, free-living hawks had high attack success on Woodpigeons, which dominated their diet, but relatively low success attacking pheasants. Goshawks killed pheasants more extensively in Sweden, potentially killing 19% at an estate where 4,000 were released and exerting an unsustainable pressure on wild populations in two areas.
8. In Finland, predation pressure on woodland grouse species was also sometimes unsustainable and may have been strong enough to be responsible to drive population cycles. Breeding of corvids was reduced near Goshawk nests in Germany and the Kestrel population was apparently being depressed by them in Kielder Forest in England.
9. A survey showed that governments across the European Union recognised problems from avian predation for hunters, pigeon-fanciers and poultry-keepers, with the Goshawk seen as the most problematic species. Perhaps the previous denial of predation impacts should be replaced by cooperation of different interests to find innovative solutions.

CHAPTER 8

Death and demography

It is only five miles from where 'Foot' was flying at the pigeons on Brussels sprouts, but it is six years later, almost to the day. Despite the cheering sun of this winter morning, Bridget and I are worried about the signal from a young female Goshawk that I released three days ago. She had been trained while on loan to a falconer and was now to remain in the wild if she could, to see how well trained hawks would survive if lost unintentionally.

This bird had not killed in the last two days. She had not been hunting effectively, and could easily have been called to the fist at any time. That was the intention this morning, to rescue her as a failure. However, the signal was unexpectedly weak and coming from within a blackthorn thicket.

I abandoned the cumbersome antenna and crawled under the sharp twigs, with a glove on my left hand in case she would come to me, and the receiver in my right. A paper-clip in the antenna socket gave enough reception to show when I was approaching the source of the signals. There lay a fox, as if asleep, but this half-starved young vixen was cold. Next to her, tail feathers protruded from disturbed ground. I pulled out the legs and back of the Goshawk. The journey back to Oxford was sombre and smelly.

A *post mortem* revealed the rest of the hawk in the stomach of the fox. There was also a single puncture wound in the fox which went from the edge of the ribs through the liver capsule and diaphragm into the lungs, with a fatal internal haemorrhage from the liver injury. Perhaps the vixen found the hawk on the ground with a kill. At any event, the hawk had footed her once in the chest before succumbing. The fox would have become drowsy from loss of blood as it finished its meal, and lay down for a sleep from which it never woke. Other species of predator can kill Goshawks, but not always with impunity.

The death of this young hawk was a special disappointment. Her trainer had encouraged her to hunt rabbits and other pest species. She had not killed pheasants. We wanted to know whether early hunting experience might be used to promote a choice of prey that minimised socio-economic conflicts. However, a more fundamental research aim was to test in a rigorous way whether unplanned loss into the wild of Goshawks imported for falconry was helping to establish the British Goshawk population. We had therefore stacked the odds against her and the other released hawks, through simulated loss in an unfamiliar area in mid-winter. Of 11 hawks released without prior experience in the wild, she was one of four that died. Although radio tagging was an elegant way to find dead hawks, each was a bitter experience.

POPULATION DYNAMICS

In order to conserve a species, it is important to understand how and why its numbers change. Is a population stable? Is it declining, and if so is that towards a new stable state or towards extinction? Is it increasing, and if so may that have an impact on other species? What factors affect stability, decline or increase? All of these questions are hard to answer.

In principle, change in size of a population can be assessed in two ways. One is by counts at regular intervals, as is done in censuses of human populations. Just as a census of humans is done by households, so populations of raptors have traditionally been assessed by counting nests in an area, to estimate the number of breeding pairs.

Counting nests has been convenient as a population indicator for elusive species like raptors. Moreover, visiting nests enables the measurement of reproductive rates and monitoring of diets. Reproductive rates are necessary for the second method of estimating population change. It depends on knowing the size of the 'cohort' of young raised each year. Mortality rates can then be used to estimate the number that will remain alive in each year in future, and hence to estimate the total population size from the surviving cohorts of successive years.

This is a powerful approach because it does not merely measure the *status quo*, but can also use factors that influence birth and death rates to predict future populations. Excluding immigration and emigration, the size of a population remains the same if the sum of births for each year will on average equal the sum of deaths. In this case the rate of population change, denoted as λ, is equal to 1. A wildlife population with λ = 1 is considered stable. If λ = 1.071, it will double in 10 years.

Another important aspect of this approach is that, given a census of breeding sites, it can also estimate the size of the population that is not breeding. When a Buzzard population was estimated from mortality rates and reproductive rates in a Dorset study area, the model predicted that only one in four birds present in spring would be breeding. Independent estimation of total Buzzard numbers showed this to be the case (Kenward *et al.* 2000). This aspect is important for understanding how raptors relate to other species, because a predation impact estimated from the density of pairs with nests might be much too low.

The value of using vital rates in conservation was demonstrated during the period in which pesticides reduced raptor populations. The organochlorine pesticide DDT was accumulating in the fat of predators and interfering with breeding, most noticeably by reducing thickness of egg-shells and thus destroying clutches (Newton 1979, 1998). By estimating mortality rates from ringing data, and making an assumption that all adults would breed in a depressed population, it was possible to estimate the minimum annual 'replacement rate' productivity that would balance the annual deaths.

However, we must not forget that mortality rates can vary. In the 'pesticide era' it is now clear that cyclodiene organochlorines such as Dieldrin and Aldrin may have affected some raptor populations more through mortality than through productivity (Newton 1986, 1998). It is also important not to overlook the breeding rate, estimated as the proportion of potential breeders that start a brood. The breeding rate may be a crucial regulator in healthy raptor populations, yet is easily forgotten when focussing on replacement rates.

The estimation of population change from mortality and reproductive rates at any scale smaller than a global scale is complicated by emigration and immigration. These dispersal movements may not only maintain genetic links, but may also be important for the dynamics of fragmented populations, for example for Goshawks in the south-west USA (Wiens 2004, de Volo *et al.* 2005). Moreover, irruptions from the north may lead to temporary populations in areas south of those normally occupied by Goshawks. Nevertheless, most studies assume that change in size and structure of Goshawk populations depends on increase by breeding and decrease by death.

This chapter considers how Goshawk populations are influenced by the factors that affect the vital rates, including mortality rates, breeding rates and reproduction at active nests. It reviews how these factors vary in space and time, and how this affects the rates, and hence the populations, at different levels of scale in space (local to continental) and time (annual to evolutionary). The possibility of deriving and modelling population processes from individual performance will also be noted, because this is likely to be a focus of future models to predict populations. Productivity will be considered first, because the relative ease of measurement at nests means that this rate has been the most studied.

EGG-LAYING AND CLUTCH SIZE

The vital rates, including mortality rates, breeding rates and productivity, are typically further subdivided. For example, mortality is estimated separately for different years of life (e.g. juvenile, adult), and productivity can be separated into nest success (the proportion of breeding attempts that fledge broods) and brood size (for broods that fledge). Productivity conventions have been usefully defined by Ian Newton (1979), Karen Steenhof and Mike Kochert (1982).

The boundaries between productivity, mortality and breeding rates also need careful attention. The majority of raptor biologists end the measurement of productivity at fledging. The estimation of juvenile mortality therefore starts at that point. As noted in Chapter 3, egg-laying is a relatively unambiguous start-point for measuring productivity (Newton 1979), because shell fragments provide evidence even in nests that fail early. On that basis, the breeding rate is the proportion of hawks in each age group that produce a clutch. Site 'occupancy' may also best be defined by the egg-laying that makes a nest 'active', and not by sightings (because radio-tagging shows that birds visit neighbouring sites) or the presence of fresh greenery on nests (because it can persist from autumn).

At local level, Goshawk laying dates and clutch sizes vary annually, with food supply and spring temperature (Figures 35, 36), and with the age of breeding females. The number of eggs laid sets a limit on the number of young that can be raised, but there are several factors which act to reduce brood size below this level. We can learn about these factors by looking at variation in success between clutches of different size, between different areas and between years in the same area.

TOTAL AND PARTIAL BROOD FAILURES

In a particularly thorough study of nest success, Santi Mañosa (1991) traced the fate of 220 eggs in 104 clutches laid by Spanish Goshawks. He found that all the four clutches with a single egg were abandoned (Figure 104). As clutch size increased, the proportion of deserted nests declined, but death of individual eggs and of chicks increased for clutches of four and five. As a result, the highest proportion of young fledged from clutches of 3–4 eggs.

However, whereas failure of nesting attempts due to early abandonment decreased with brood size in Spain, later nest failures due to extrinsic factors such as predation or human disturbance tended to increase (Mañosa 1991). Predators caused loss of 8% of eggs and chicks, and humans 12%. The result was that total losses averaged 36% across clutch sizes of 2–5. Partial losses due to infertility also increased for large clutches, because 25% of the eggs laid fourth in order were infertile, compared with 4% of earlier eggs.

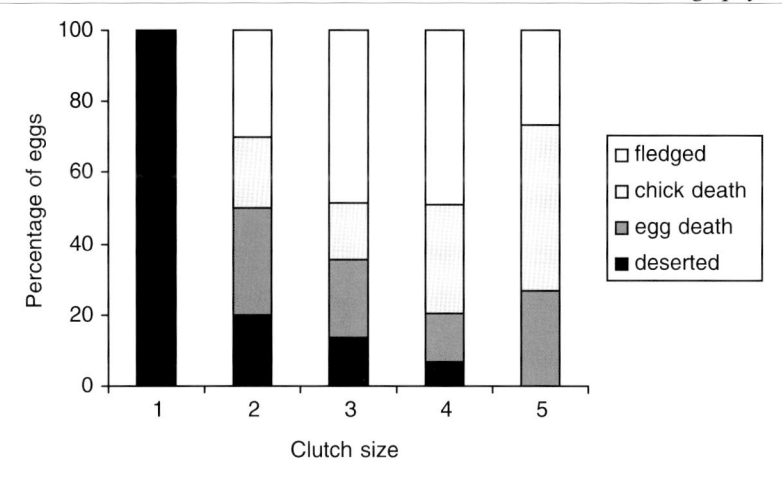

Figure 104. As clutch size increased at 104 Goshawk nests in Spain, the tendency to desert decreased but partial brood losses increased. Data from Mañosa 1991.

Similar data come from two sites to the north, in Schleswig Holstein (Looft & Biesterfeld 1981) and Finland (Wikman & Lindén 1981). Again, one-egg clutches were unsuccessful and there was a tendency for compensation between different causes of loss as clutch size increased, especially due to an increase in partial losses (Figure 105). As a result, the number of young fledged from clutches of 5–6 was no greater than from clutches of four. Although there were only three clutches of five eggs in Finland and three in Spain, the effect was clear in Schleswig-Holstein for 29 clutches of five eggs and one of six (from which only one chick fledged). Indeed, only 3% of those 30 German pairs raised a chick from all 5–6 eggs, compared with 16% for four eggs, a peak of 30% for three eggs and again only 13% with two eggs. It seems that an optimal clutch size for Goshawks in Europe is 3–4 eggs.

The decline in total brood loss with increase in clutch size probably reflects the breeding ability of the parents. Hawks that lay large clutches are presumably the most competent to rear at least one of their chicks. Indeed, total loss of large clutches tends to reflect extrinsic factors, such as nest predation, rather than abandonment (Mañosa 1991). It may therefore be best, when comparing nest loss between areas, to focus first on the two largest categories of clutch size. These 'competent-parent' losses were 7% for 56 nests with 4–5 eggs in Finland, 17% for 30 nests with 5–6 eggs in Schleswig-Holstein and 31% for clutches of 4–5 eggs in Spain. Subtracting these total losses from those recorded for the smaller clutches indicated that residual 'low-competence' losses were 8–11% of nests in all three areas.

The extrinsic causes of complete brood failure included, in order of increasing importance, collapse of the nest, bad weather and nest destruction by humans and other predators. Nest collapse does not always cause loss of the brood. In Finland, two of four young were reared on the ground after a nest collapsed (Huhtala & Sulkava 1976).

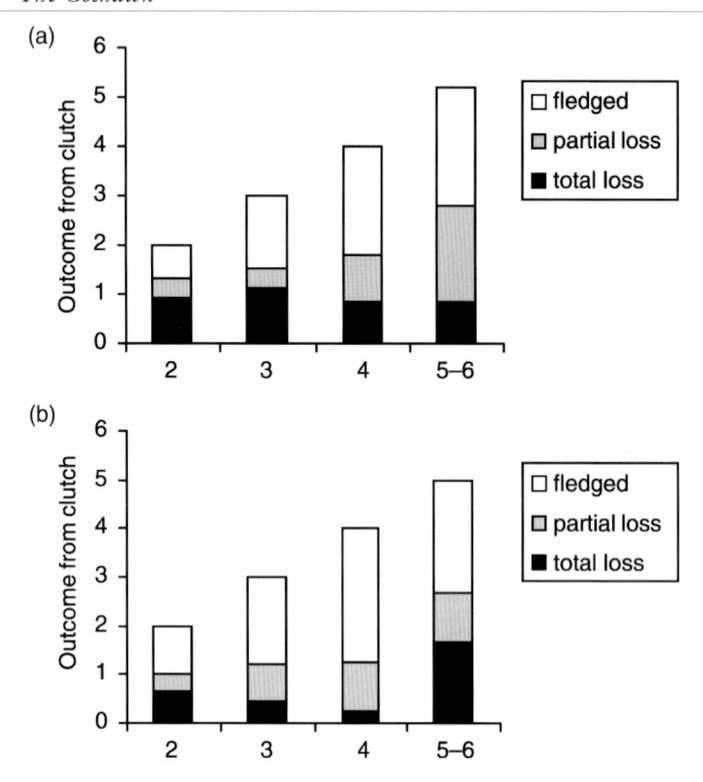

Figure 105. The success of different clutch sizes from 335 nests in Schleswig Holstein (a) and 140 in Finland (b). Data from Looft & Biesterfeld 1981, Wikman & Lindén 1981.

Poor weather can affect all stages of the breeding cycle. Laying is delayed in cold springs (Bijlsma 1993, Sulkava *et al.* 1994, Drachman & Nielsen 2002). Few pairs may lay when there is high rainfall during courtship, with subsequent cold weather and high rainfall linked both to nest failure and small brood sizes in a German study (Kostrzewa & Kostrzewa 1990). Likewise, Fairhurst & Bechard (2005) found that low April temperatures and high April-May precipitation increased nest failures and hence reduced productivity in Nevada. Poor weather is liable to affect the ability of hawks to forage and the breeding success of their prey.

Nest destruction is normally the most important cause of total failure. Link (1977) attributed seven of 97 Bavarian brood failures to martens, which were sometimes later found using the nest to rest on. Martens (Fishers) were also considered the main predator of broods in Wisconsin (Erdman *et al.* 1998). In Alaska, four of 12 nests failed due to mammal predators in one study year, possibly because of martens, although McGowan (1975) felt that Lynx and Porcupine should not be overlooked as possible culprits. Doyle (1995) found evidence of Wolverine predation on a nest in the Yukon and Raccoons raid Goshawk nests in other parts of North America. Eagle

Owls are on record taking young Goshawks in Finland, Germany, Sweden and Spain (Curry-Lindahl 1950, Üttendorfer 1952, Huhtala & Sulkava 1976, Hennessy 1978, Tella & Mañosa 1993), as are Great Horned Owls in the USA (Squires & Kennedy 2006), and Golden Eagles could also eat whole Goshawk broods. Even female Goshawks have raided other Goshawk nests (Bergström, in Höglund 1964b). Corvids occasionally eat Goshawk eggs (Joubert 1987), but maybe not unless the adults have already deserted or have been scared away (Link 1977).

Predation may sometimes be independent of brood size, but may also interact with brood size or factors that affect brood size. If food is abundant, in which cases broods are likely to be large, female parents can spend more time in nest stands ready to defend nests (Dewey and Kennedy 2001). On the other hand, if there is inadequate food during rearing, greater activity (e.g. calling) by larger broods than smaller ones may attract predators (Mañosa 1991). Large broods may also be most likely to weaken the nest so that it falls or is blown down (Höglund 1964a, Marquiss & Newton 1982b).

Similar interactive effects may encourage disease. Large broods will be most at risk of contacting prey with infectious disease if risk increases with number of prey deliveries. Although disease rarely kills whole broods, individual nestlings in Alaska, Germany and Wales have been killed by trichomoniasis (McGowan 1975, Looft & Biesterfeld 1981, Link 1986, Cooper & Petty 1988), which is caused by a flagellate protozoon often found in pigeons. More recent surveys (Wieliczko *et al.* 2003, Krone *et al.* 2005) indicate a high prevalence of nestling infection in some Goshawk populations with low pathogenicity, which indicates that Goshawks can adapt to *Trichomonas* infection. Nestling Goshawks may also harbour *Leucocytozoon* blood parasites, but these seem not to have adverse effects (Toyne & Ashford 1997, Krone *et al.* 2001). Likewise they have *Streptococcus aureus* on their feet without this becoming pathogenic (Needham *et al.* 1979). Possible impacts on Goshawks of West Nile Virus and avian influenza have yet to be described.

Partial brood failure due to nestling deaths may sometimes result purely from starvation or cainism due to food-shortage (Chapter 4), and sometimes from disease induced by malnutrition. Chicks also occasionally die as a result of developmental defects of uncertain origin. These can include plumage defects (Bijlsma 1997, Rutz *et al.* 2004), and also skeletal malformations such as a double tarsus (polydactyly) or beak deformation (Dekker & Hut 2004). Repeated problems at the same nest suggest genetic origins, but other possibilities include polyoma virus infections (Vedder 2000) or accumulation of teratogens such as PCBs.

In some areas, human activities become the most important cause of total brood failure. This may be deliberate, as is the killing of hawks by gamekeepers, farmers and pigeon-fanciers, or the collection of eggs or young birds. It may also be the incidental result of other activities, as when nest trees are felled or nests abandoned because of disturbance by forest workers, or if clutches fail due to contamination with pollutants or pesticides.

The effect of deliberate interference is easier to demonstrate than that of contaminants. Link (1977) reported that 32% of 97 Bavarian Goshawk broods failed because of humans, with only 9–13% failing for other reasons. Losses reached 59% of 111 broods in areas of Britain where deliberate human interference occurred, compared with 17% elsewhere (Marquiss & Newton 1982b, Marquiss *et al.* 2003). In six other areas of central Europe, total brood losses averaged 30–47% during the 1960s (Fischer 1980). The relative importance of persecution and pesticides will be considered further in Chapter 10.

Studies have now started to seek relationships between habitat and Goshawk breeding performance. Bijlsma (1993) showed that nests in larch, despite a preference of hawks for this tree, were less successful than in pines, spruce and firs. Krüger (2002b) found that nests were most successful in stands with *large* trees (of high diameters at breast height, DBH) and in areas with most woodland. Kostrzewa (1996) found nests to be most successful where there was 40–50% woodland.

SPATIAL AND TEMPORAL VARIATION IN PRODUCTIVITY

Both Europe and North America experience a similar wide variation in the success rate of Goshawk nests (= number of nests fledging at least one young / number of clutches). The median values, of 77% in Europe, and 82% in North America (Figure 106) do not differ significantly between the two continents. Although two of the European studies have data from the 1960s through the 1990s, from Finland since 1963 (Sulkava *et al.* 1994) and Schleswig-Holstein since 1968 (Looft 2000), the majority of data from these and other sources were collected after 1975, and hence after the introduction of protective legislation and bans on extensive agricultural use of organochlorine pesticides. Nest success now rarely falls below 60% (i.e. nest failures rarely exceed 40%).

In contrast to the transatlantic similarity in nest success rates, the median Goshawk brood in North America has half a chick fewer than in Europe. Only four of 25 American sites have values greater than the median value of 2.55 young per fledged brood from the 64 sites in Europe. Two of these four relatively productive American sites had 31–33 nests in the far north (McGowan 1975, Doyle 2000), where breeding is enhanced during Snowshoe Hare abundance but otherwise tends to fail completely. At a third site, an early estimate from 22 nests on the Kaibab Plateau (Boal and Mannan 1994) is superseded by lower values from 282 nests (Reynolds & Joy 2006). However, the 2.64 young from successful broods in 211 nests in the Humbold-Toiyade forest of Nevada (Bechard *et al.* 2006) truly represents an area of North America with unusually good breeding conditions.

The number of young per active nest is a product of nest success and brood size, i.e. *productivity = nest success rate × size of fledging broods*. The relative contribution of total failures and brood size to productivity differed between Europe and North America. In Europe, variation in productivity stemmed almost equally from nest failures and brood reduction, whereas in North America productivity depended

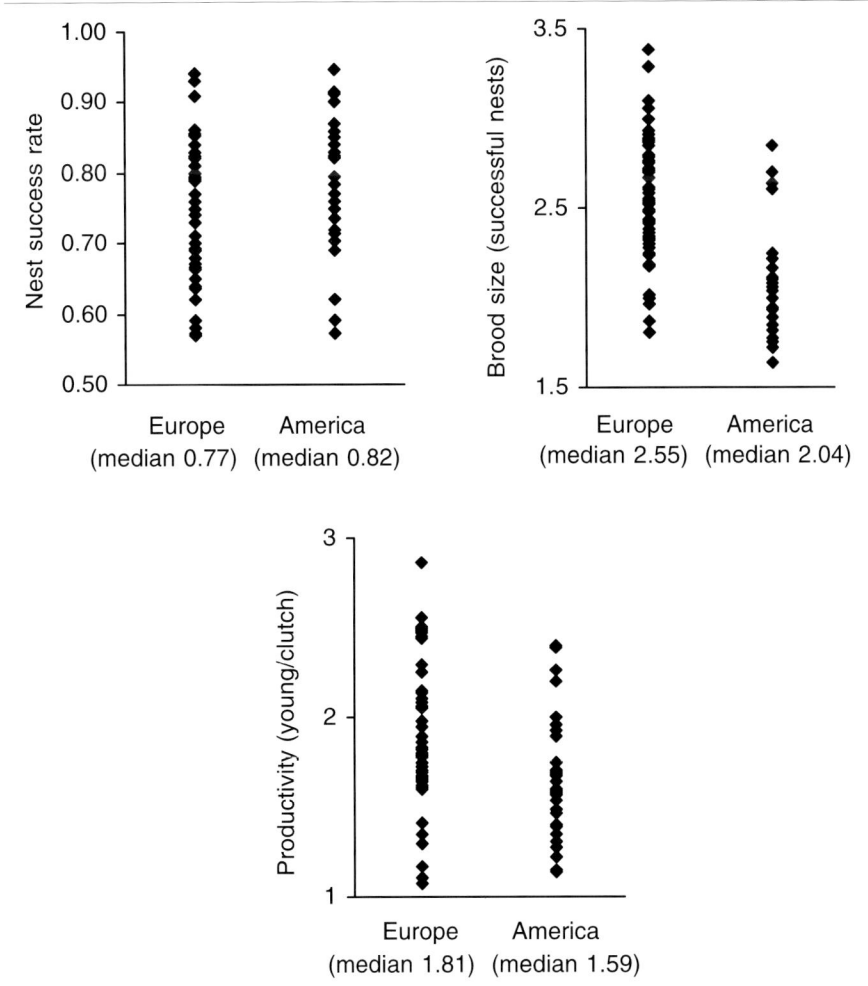

Figure 106. Whereas goshawk nest success was similar at 68 nests in Europe and 27 in America, the fledged brood sizes were much lower ($p < 0.001$) and hence there was also lower productivity ($p = 0.003$) in America. Sources from Rutz et al. 2006, as listed in Appendix 2, for studies with at least 20 nests or 5 years of data.

most strongly on brood size. None of 12 American studies with more than 50 nest records had less than 70% success, compared with seven of 42 in Europe, where high levels of destruction by humans were recorded in several cases. Nest success consistently below 70% seems to indicate unusual pressure on a population.

In most cases the variation in productivity between areas and years probably reflects differing influences of food, weather and other natural factors. Unfortunately, few studies have undertaken the daunting task of measuring food supplies. A recent exception used a large number of transect surveys to estimate abundance of seven bird and three mammal prey species during four years on the Kaibab Plateau (Salafsky et al. 2005, 2006). Variation in young per territory (which includes site occupancy and clutch-based productivity) related most strongly in this southerly study to abundance of the three squirrel species. However, the longest studies of productivity in relation to food supplies concern Goshawk populations exposed to cyclical changes in abundance of hares and grouse in northern study areas.

Especially comprehensive local data come from the studies started in the 1960s by Seppo Sulkava in Finland (e.g. Sulkava et al. 1994). Clutch and brood sizes can be related to annual counts by hunters of woodland grouse, which are such important Goshawk prey (Chapter 7). There were strong correlations between clutch size, laying date and prey density, possibly mediated in part by effects of weather on both predators and prey (Chapter 3). There was also a strong linkage between brood size and clutch size, although brood size tended to diverge from a 1 : 1 relationship (Figure 107). A weaker correlation between brood size and prey was

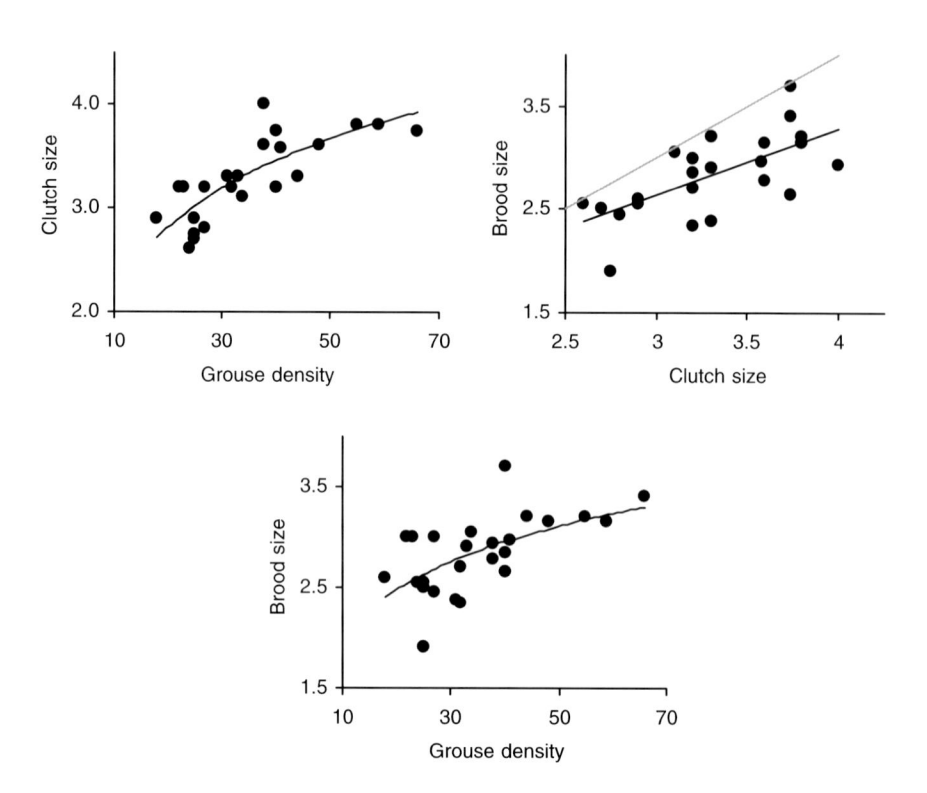

Figure 107. In Southern Ostrobothnia during 1967-1991 clutch size correlated strongly with woodland grouse density (top left, $p < 0.001$). Brood size also correlated with clutch size (top right, $p < 0.001$), but diverged from the grey line that would result if 1 young was reared from each egg, so that there was a weaker link (bottom, $p < 0.005$) between brood size and prey density. Data from Sulkava et al. 1994.

only marginally significant after the linkage between clutch size and brood size had been taken into account.

Similarly, although early laying associated with good weather tends to produce large clutches (Drachmann & Nielsen 2002, data from Sulkava *et al.* 1994 and Bijlsma 2004), correlations between pre-laying weather and brood size at fledging tend to be weak if detectable at all. Poor weather before and during laying seems to influence productivity mainly by causing brood failures (Kostrzewa & Kostrzewa 1990, Penteriani 1997, Fairhurst & Bechard 2005).

Thus, in a situation where grouse were important food during laying and rearing, much of the influence of food supply on productivity occurred early in the breeding season. There was a similar effect in a study through two peaks of the Snowshoe Hare cycle at Kluane Lake in the Yukon (Doyle 2000). There were correlations with hare density in spring for the number of active Goshawk nests and laying dates, but not for brood size or productivity.

In Finland, there was strong synchrony between Goshawk brood size and grouse numbers across nearby regions (Ranta *et al.* 2003). However, Goshawk productivity can also vary considerably between nearby areas. Mick Marquiss and Ian Newton (1982b) were able to demonstrate that clutch size, hatching success and the size of successful broods all decreased with increasing altitude of the nest site in Britain. These changes may have reflected poorer weather or diet differences at higher altitude (Chapter 6).

Goshawks are unusual because their breeding has been studied in detail for most 5° quadrants of latitude and longitude across Europe (Rutz *et al.* 2005). Europe is a small part of the Northern Goshawk's global distribution. However, it covers all latitudes of the Palaearctic distribution, from upland woods of Spain to the Fennoscandian taiga, and from maritime western countries to the continental climates of Eastern Europe. This is almost the entire regional distribution of the nominate race *Accipiter gentilis gentilis*.

At this largest scale below an inter-continental comparison, it would be reasonable to expect the geographic trends in Goshawk morphology, foraging and phenology (timing) to be reflected in trends of reproductive success. Quite remarkably, there is no consistent variation in clutch size, nest success and brood size (hence productivity) at this scale across Europe. General rates for European Goshawks are 3.3 eggs per clutch, a nest success of 76% and 2.5 nestlings per successful brood, with implied productivity (nest success × brood size) of 1.9 young per clutch.

PRODUCTIVITY OF INDIVIDUALS AND AT SINGLE SITES

Laying dates have been relatively late wherever females have been studied in their first year, with small clutches and reduced productivity (Chapter 3). Breeding tends to be earliest in the season when Goshawks are 6–9 years old, the age-range in which highest productivity was recorded (Bijlsma 2003a, Nielsen & Drachmann 2003, Risch *et al.* 2004, Krüger 2005). Productivity tended to decline among older hawks, surprisingly abruptly in the last three studies. There was no sign of senescence until females were more than 12 years old in Schleswig-Holstein and one female reared three young 17 years after her feather marks were first registered. Further data are needed to confirm the timing of senescence in female Goshawks, and it may not exist at all in males (see Chapter 9).

Although too few feathers are collected at breeding sites to investigate male senescence, there are enough data to show that the productivity of males is especially low in their first year and increases after their second year (Figure 108). Although productivity is reduced for young breeders of both sexes, the process differs. Nests of young males are very likely to fail, but successful broods are as large as for older males, whereas nests of young females tend to have smaller broods as well as greater risk of failure (Kenward *et al.* 1999, Nielsen & Drachmann 2003).

The brood reduction of young females, especially those in their first year, may depend partly on physiological factors, because females that breed in their first year lay late and small clutches irrespective of the age of their mate and hence his competence. However, it is not clear whether they lay late because males are reluctant to provision hawks in juvenile plumage or purely because of their physiology. Their reduced nest success may also reflect inexperience at incubation and rearing chicks, including nest defence.

The increase of success in older males probably mainly reflects competence of provisioning. Indirect evidence for the benefits of increased experience with age comes from predation on pigeons. In almost all studies of Goshawk diet at nests in Germany, Woodpigeons or Feral Pigeons are the dominant prey. Pigeons are

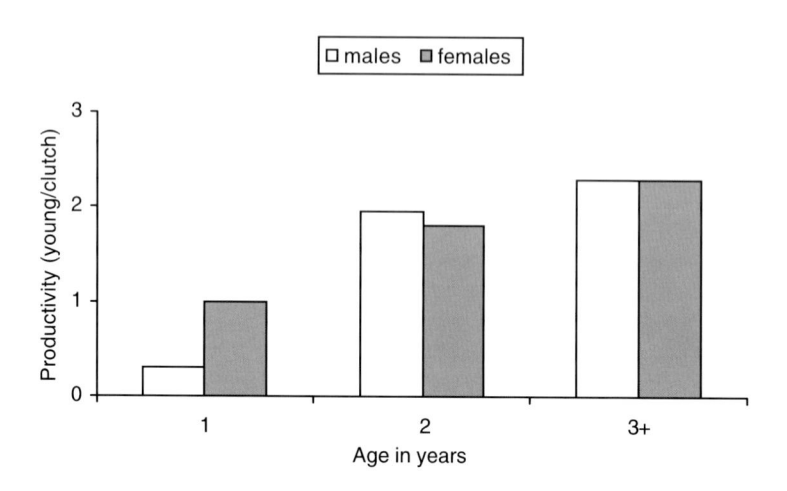

Figure 108. Productivity increases with age in male and female Goshawks. Data from Nielsen & Drachmann 2003.

high-biomass prey that are convenient for males to carry to nests, they tend to be attacked successfully (Chapter 7) and are abundant. Nests where the diet contains a high proportion of pigeons have high productivity, high nestling mass and high survival of male breeders between years (Krüger & Stefener 1996, Rutz 2005a). Not only does the proportion of pigeons in the diet increase with age (Rutz *et al.* 2006b), but so does the proportion of pale-coloured Feral Pigeons, which are taken disproportionately by the best breeders (Rutz 2005b).

In contrast to the regional scale, where there is little variation in productivity, areas used by individual pairs differ greatly in their output (e.g. McClaren *et al.* 2002). Occupancy can be highly divergent across such areas (Kostrzewa 1996), such that some fledge young almost every year, whereas others may rarely have a clutch of eggs, let alone a successful brood. With a relatively long-lived species like the Goshawk, which has strong spatial fidelity when breeding (Chapter 5), this raises the question of whether some areas are highly productive because they have highly competent hawks or because they have the best foraging opportunities.

Oliver Krüger and Jan Lindström (2001) used 25 years of data from a study in Westfalia, in Germany, to show that the sites that are least often occupied also produce the smallest broods when they are occupied. With a study that was several times the life-span of most hawks, the result indicates strongly that some sites are of poorer quality whichever hawks occupy them. A result of poor sites being occupied only at high density, in a distribution of individuals that is know as Ideal Pre-emptive (because the best sites are held first, Pulliam & Danielson 1991), is that the average productivity is reduced at high density. This is because poor breeding in the low quality sites dilutes the effect of good breeding elsewhere. As a further result, population growth slows when occupation of the poor sites increases breeding density. Such an effect can contribute to the regulation of density.

OCCUPANCY AND BREEDING RATES

At Kluane Lake, only one Goshawk nest was found in the year with lowest Snowshoe Hare density, and the nest failed. However, in the year with most nests (nine) the productivity was 3.3 young per occupied nesting area (Doyle 2000). In studies of similar or greater length in Wisconsin, Nevada and Idaho, productivity of nests varied about three-fold between years, from 0.8 to 3.4 per active nest (Erdman *et al.* 1998, Bechard *et al.* 2006). However, as at Kluane, there was a tendency for low occupancy to associate with nest failure (Figure 109), so productivity per known nesting area (and hence for areas as a whole) tends to fluctuate more than is implied by measures at active nests. Thus, with site occupancy varying from 23–83%, production of young per site varied seven-fold (0.32–2.26) for the Nevada Goshawks (Figure 109). In a comparable study in Idaho, variation was nine-fold (Bechard *et al.* 2006), and eight-fold during just four years (0.14–1.23) on the Kaibab Plateau of Arizona (Salafsky *et al.* 2005, 2006). A tendency for productivity to be good

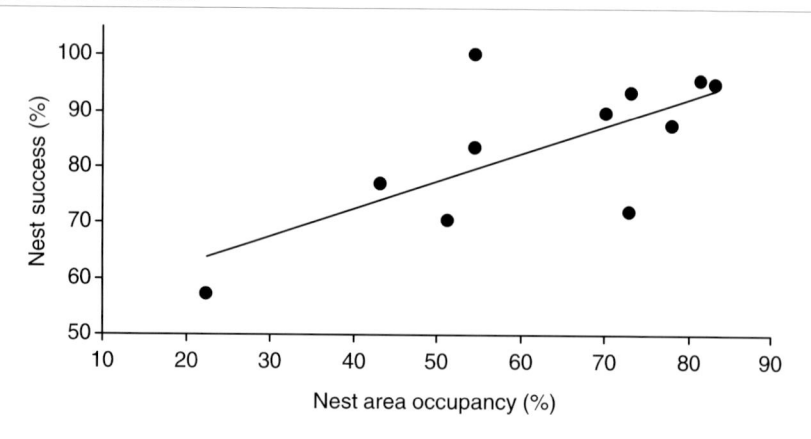

Figure 109. For Goshawks in Nevada, nest success tended to be low when occupancy of nest areas was also low. Trend line $p < 0.02$. Data from Bechard et al. 2006.

in 2000 and poor in 2002 was common to all three areas, which may suggest a climatic influence.

Occupancy of nest areas is hard to define, partly because areas that are used infrequently may take several years to discover at the start of a study, and partly because change in land-use may make some unsuitable while also creating new ones during the course of a study. In prolonged studies, a convenient index of total occupancy is the maximum number of nests occupied in any one year. Relative occupancy can then be defined against this maximum. There may be trends of population size during the course of a study, and even changes in the area searched. However, short-term variation in occupancy can be expressed by the change in number of active sites between years, as a percentage of the maximum. Changes in occupancy between consecutive years were greatest in the three longest-running North American studies, despite the European studies having a much longer duration (Figure 110). Although the European populations fluctuated quite strongly, change from year to year was more gradual. Variable occupancy in North America makes it hard to obtain reliable estimates of density and population trends without prolonged study of many nests (Reynolds et al. 2005).

Individual hawks that are recognisable from feathers or other markers sometimes occur near nests without breeding, either before starting to breed or in years when they do not manage to lay eggs. On Gotland, when adult hawks were caught away from nests in winter and radio-tagged, a few males and a larger proportion of females subsequently associated with nest sites where no eggs were laid (Figure 111). Average distances travelled from the nest during the courtship period were greater among males and females that failed to raise young than for successful breeders, reflecting poor breeding by males that had to forage far at this time and by females that were wandering between sites.

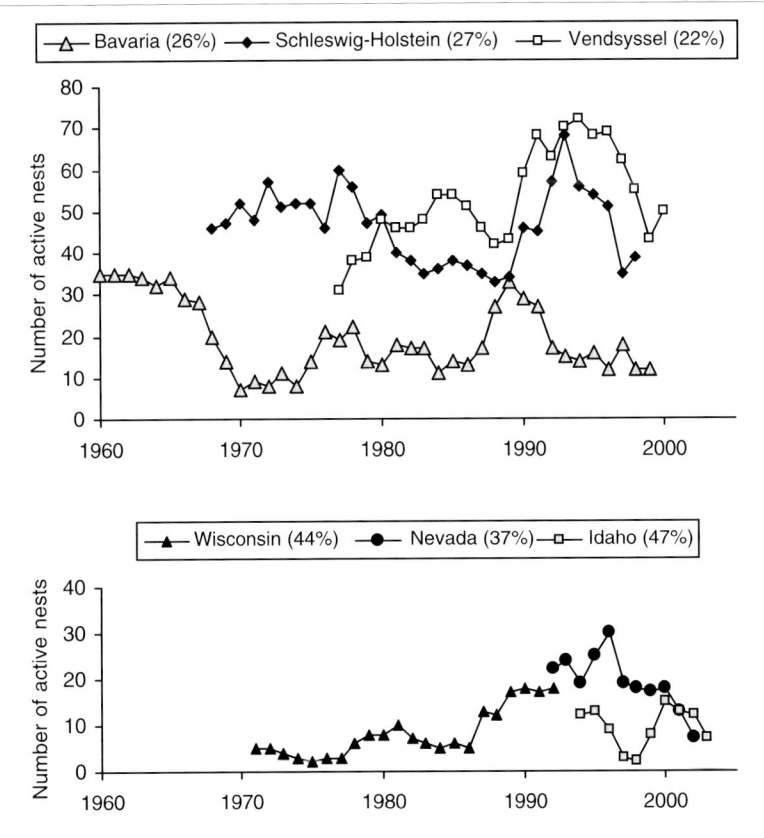

Figure 110. Active nest areas for three long-term studies in Europe and three in America, with maximum occupancy change between years as a percentage of the greatest number of nests in each area. Data from Looft 2000, Drachmann & Nielsen 2002, Rust & Mischler 2001, Erdman *et al.* 1988, Bechard *et al.* 2006.

There were other radio-tagged adults of both sexes that did not associate with nest sites at all, and no males or females attempted to breed in their first year on Gotland (Figure 112).

The figure shows that no juveniles were recorded associating with nests. The proportion of older hawks recorded at nests without laying was small and largest for second-year females, but not statistically different between age and sex categories of adults. However, even with radio-tags it is unlikely that all nest visits were recorded by checking birds at intervals of several days. It was much easier to be sure about the proportion of hawks associated with eggs (unshaded) and hence to record the breeding rates.

Breeding rates on Gotland did not differ between second-year males and older males, but were lowest for the younger adults among females (Figure 112). A high

232   *The Goshawk*

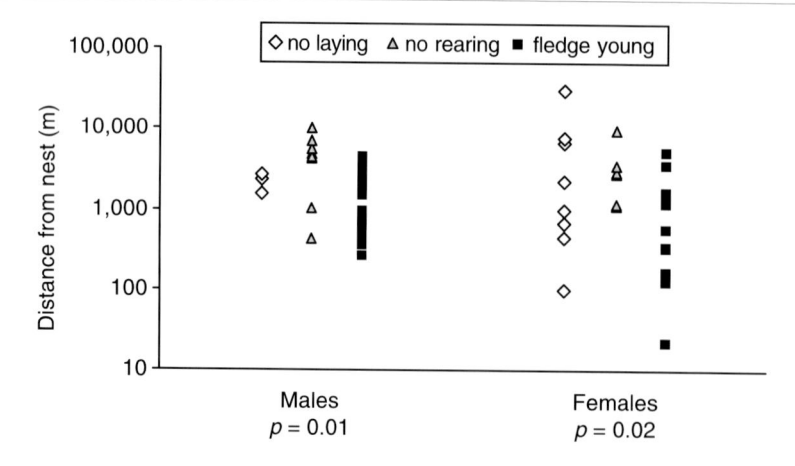

Figure 111. Radio-tagged adult Goshawks that fledged young on Gotland were on average located closer to nests during courtship (March) than birds that had no clutch or reared no young. Data from Kenward *et al.* unpublished.

failure rate for male first-breeders was also as expected from other studies. Indeed, a high turnover of adult hawks tends to result in low productivity because there are so many one-time breeders (Rust & Mischler 2001).

However, the higher adult breeding rates for males than females were surprising, as was the total lack of breeding by first-year hawks. Fritjof Ziesemer (1983) recorded that the proportion of nests in Schleswig-Holstein with hawks in first-year plumage varied between 4.4% and 10.5%, and estimated that 22% of hawks bred first as

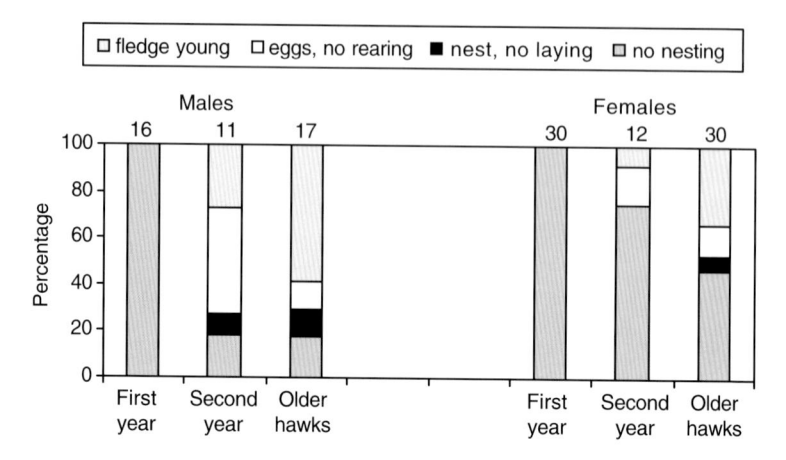

Figure 112. The breeding rates of radio-tagged Goshawks that were alive in spring on Gotland. A value for *n* is given at the top of each column. Data from Kenward *et al.* 1999 and unpublished.

juveniles. Other studies that collected moulted feather found less than 10% of juvenile breeders in stable populations, but rising to 20–25% in growing populations (Thissen et al. 1981, Nielsen 1986) or even beyond 30% when populations were low (Link 1986) or increasing strongly (McGowan 1975). In developing urban populations, 36% of 56 early breeding attempts and ten of 17 new breeders were in their first year (Würfels 1994, Rutz 2005b). In North America south of inland Alaska, breeding by first-year Goshawks is rare (F. Doyle, pers. comm., S. Lewis, pers. comm., G. Fairhurst, pers. comm.) or unrecorded despite prolonged study (Wiens & Reynolds 2005).

MORTALITY AND SURVIVAL RATES

When assessing causes of death or their impact on populations, it is convenient to think in terms of mortality rates. However, when modelling populations from the numbers in each age category that are alive and producing young, it is more convenient to think in terms of survival rates. The mortality and survival rates, if given as proportions of a population, must sum to 1 (or to 100 as percentages). So, for example, 30% mortality is 70% survival.

Most data on mortality of birds comes from ringing. In a very simple estimation, when all the ringed birds are likely to have died, if 60% were recovered from dead birds in their first year, first-year mortality was 60%. Modern models can accommodate different rates in all age categories, making allowance (after Haldane 1955) for the impossibility of recovering old birds among nestling cohorts ringed recently, and for differences in reporting rate between age- and sex-classes. The Achilles heel of all ringing methods is lack of data on reporting rates, which may differ in undetected ways between sex- and age-classes. Assumption of a uniform rate may be reasonable for estimating rates of older adults from several adult age-classes. However, juvenile mortality is then overestimated if inexperienced birds are most likely to be recovered, for example because they are most likely to hit human artefacts or to be killed at game farms when they disperse and find an easy source of food.

Another approach is to estimate the turnover of breeding adults from feathers or markers at nests. This method is unbiased if feathers can always be correctly assigned, if birds do not change breeding areas and if they always breed. It overestimates mortality if birds that have bred before are not always recognised or if they move outside the study area. It is therefore most likely to be reliable for obtaining estimates of mortality from stable populations of adult breeders. It is likely to overestimate mortality when there are many juveniles breeding, because their feathers change most between years (Chapter 5) and also if an abundance of first-time breeders lowers the site-fidelity. Although site-fidelity of 96–98% has been estimated from feather markings in Europe, studies with colour bands in North America have recorded only 90% fidelity for females (Chapter 5).

Even the banding approach will over-estimate mortality unless areas are large or self-contained, so that emigration of breeders is rare. Over-estimates will also occur unless there is a concerted effort to identify every breeder by recapture or resighting. The use of feathers for DNA analysis (Rudnick *et al.* in press) may be more cost-effective, but all nest-based records have the drawback of not registering birds until they start to breed.

Radio-tagging can provide very detailed information on performance of individuals in relation to habitat and social factors, in principle from fledging onwards. Records from long-life tags can inform on dispersal, further range changes before breeding, breeding performance, breeding-site fidelity and cause of death, as the necessary background for individual-based models that predict population development. However, a potential for bias due to tag life, reliability and possible impact on wearers must be avoided by cross-checking with other markers and training for tag attachment. With well-funded projects, radio-tagging can be more cost-effective than the alternatives (Kenward 1993). However, the expense of equipment and training has been a deterrent for volunteer work.

CAUSES OF DEATH

During 1880–1930, when a bounty was being paid for Goshawks killed in Norway, 3,000–5,500 were being killed annually (Munthe-Kaas Lund 1950), although the recorded bag had fallen to 654 for the period 1965–70 (Bergo 1996). Some 4,000–8,000 were still being killed annually during 1964–1975 in Finland, where Goshawks had no protection (Moilanen 1976). It is not surprising that 83% of rings recovered before 1975 in Finland (Saurola 1976), and 53% for Norway (Sollien 1978), were from hawks killed by humans.

Now that Goshawks have varying levels of protection, the proportion of rings reported from killed hawks has fallen below 50% in recent analyses of ring recoveries, although high proportions of death from 'unknown causes' may represent some hawks killed illegally. Generally speaking, the proportion of hawks killed deliberately by humans, or unwittingly by their artefacts (cars, trains, wires, buildings, pollutants), has tended to decline as human populations become sparser to the north (Figure 113).

Pollution was at one time another unwitting cause of death, as discussed in Chapter 10. However, levels of organochlorine and heavy metal pollutants are now generally far below those during the 1960s and 1970s, with occasional exceptions. Thus, Kenntner *et al.* (2003) found higher DDT concentrations in Goshawks in the former East Germany, where this pesticide was used until 1988, than in the former West Germany, and unusually high levels in hawks from Berlin, where DDT was manufactured and remained in food chains. These authors also found three hawks with toxic levels of lead, probably from ingested shot.

In countries with low human population density, more than 50% of ring recoveries may have deaths attributed to natural causes, including accidents,

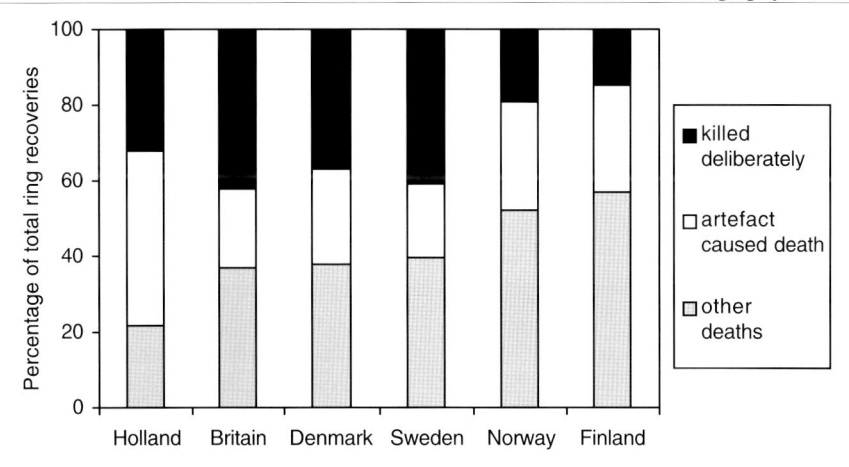

Figure 113. Ringing indicates that the proportion of Goshawks killed by humans or human artefacts (collisions, pollutants) decreases as human population density declines to the north. Data from Niedeman & Schönbeck 1990, Bijlsma 1993, Halley 1996, Tornberg & Virtanen 1997, Nielsen & Drachmann 1999a, Petty 2002.

predation, disease and starvation. Bizarre accidents are described in the early literature. One hawk was found hanging up a tree, with its foot trapped in a hole from which it had presumably been trying to extract a squirrel or bird (Böhnsack 1971). Others have apparently been torn apart when they grasped a hare in one foot and attempted to halt its progress by seizing vegetation in the other foot (Fischer 1980). Another hawk was attacked by a fox as it ate, but escaped with a chewed wing into a nearby stream, where it drowned (Klawes 1956). Two others appear to have choked on parts of mammalian prey (Bloxton *et al.* 2002).

Predators that routinely eat diseased prey might be expected to evolve strong resistance to common pathogens, which may then merely act in concert with other debilitating factors. Thus, experimental feeding of breeding Kestrels showed that stressors such as food-shortage can pre-dispose hawks to infection (Wiehn & Korpimäki 1998). This would explain why 53% of 49 hawks trapped at Hawk Ridge in Minnesota had detectable levels of *Aspergillus* fungi (which can cause fatal Aspergillosis) during an irruption year (1972), compared with only 7% of 45 a year after the main irruption (Redig *et al.* 1980). Beyond this early quantitative survey, recent reviews by Squires & Kennedy (2006) and Rutz *et al.* (2006a) provide access to an extensive literature on disease in wild Goshawks.

Surveys for presence of pathogens do not reveal how many infected birds will die and may therefore over-rate the importance of disease, as may submissions to veterinary centres of hawks found by the public with debilitating conditions. On the other hand, disease may be under-recorded in collections made by the public, because rings are often reported from birds long dead, and other dead hawks submitted for analysis by the public are most likely to be fresh when they are trauma victims. The same problems apply for assessing the importance of deaths from starvation. There can be many records of starved hawks in winters that are unusually severe (Linkola 1956), but the percentage of hawks reported as starved is maximally 3–6% in ring recoveries (Saurola 1976, Sollien 1978).

Nevertheless, Peter Sunde (2002) gained interesting results from hawks handed in by members of the public, by examining changes in the relative frequency of starvation along the latitudinal extent of Norway, which extends 1,300 km from north to south. Just 9% of the hawks were registered as starved, but this percentage increased to the north (Figure 114). Moreover, starvation deaths were significantly more frequent among juvenile males than in other age and sex groups. Males were also a smaller proportion of the hawks found dead in the north, presumably because they starved without being found or because they emigrated.

The least-biased data on causes of death is likely from radio-tagging. If hawks are checked frequently, carcasses can be collected fresh enough for detailed *post mortem* examination. If long dead, the finding location and X-ray examination for lead, which may leave smears on bones even if pellets are missing (John Cooper, pers. comm.) can reveal whether hawks were killed deliberately or died other traumatic deaths, to produce the three categories used for ring recoveries in Figure 113. Comparisons with contemporary ringing data have confirmed that ringing recoveries substantially over-represent the proportion of deaths of Goshawks and Buzzards for which humans were responsible (Kenward *et al.* 1999, 2000).

On Gotland, humans killed 35% of 63 Goshawks recovered by radio tracking, mostly at farms where hawks could legally be shot while attacking poultry, compared with 46% of hawks reported as killed in 79 ring recoveries from Gotland just before and after the study. Most of the other ringing records had no cause of death given, with no records of starvation or disease. However, 28% of the radio-tagged birds died of starvation or disease and another 24% may have done so but were too decomposed

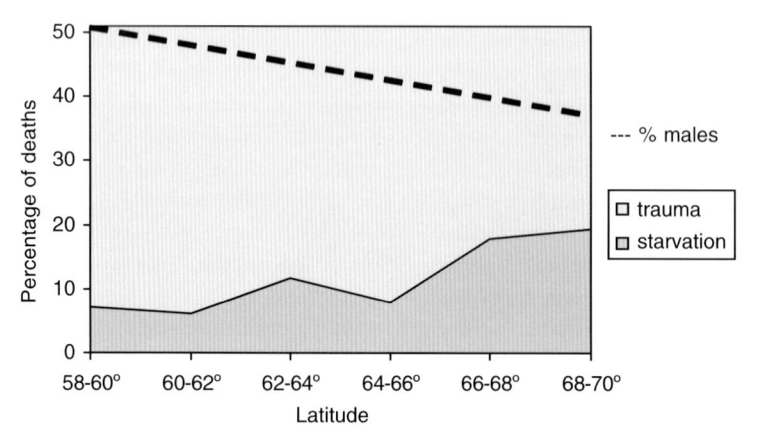

Figure 114. Among Goshawks found dead in Norway, the proportion that are starved increases to the north and the proportion of males declines. Data from Sunde 2002.

for study, with only 10% succumbing to accidents and 3% to predation (by other Goshawks). Only two hawks definitely died from disease; the proventricular ulcers in three hawks and abundant gut parasites in another three birds were probably secondary results of their poor nutritional state (Kenward *et al.* 1999).

Predation appears to be more of a problem in North America, where predators accounted for six of 12 deaths of radio-tagged young hawks in one study, albeit only one of them after fledging (Ward and Kennedy 1996), and five of nine adults in another (Boal *et al.* 2006). Great Horned Owls, which kill adult hawks as well as nestlings (Rohner & Doyle 1992, Erdman *et al.* 1998, Doyle 2000, Boal *et al.* 2006), may well be the worst predator problem for Goshawks in North America. The ecological equivalent in Europe is the Eagle Owl, the world's largest owl, which also kills Goshawks. In a review of limiting factors for Goshawks breeding in Europe, Rutz *et al.* (2006a) concluded that although there is currently no widespread impact from Eagle Owls, re-establishment of these owls across their historic range may cause problems for Goshawks in future.

Adult Goshawks are probably killed seldom by mammals, although killing of females on nests by Fishers (martens) became a serious problem in the Wisconsin study of Tom Erdman, Dave Brinker and colleagues (1998). Martens take young at nests in Europe (Link 1986) but not adults, perhaps because Goshawks defend nests in Europe less aggressively than in North America against natural predators as well as from humans.

JUVENILE AND ADULT MORTALITY RATES

Early estimates of first-year mortality for Goshawks were high. The proportion of rings recovered within a year ranged from 73–80% in Sweden and Finland, with 40–49% by the end of the hawks' first October (Höglund 1964a, Haukioja & Haukioja 1970). This implied that 40–80% of the Goshawks died within their first year, However, up to 83% of recoveries were from hawks killed (legally) by humans, and such hawks can be 84% juveniles (Chapter 5), which could greatly over-represent juveniles in the ring recoveries. Pertti Saurola (1976) estimated a correction for this bias by partitioning the mortality. He found that 25.3% of all rings were returned from hawks killed by humans. Therefore, only 74.7% of ringed hawks were available to die of 'natural causes'. For these, 58% of ring recoveries were for juveniles, representing 43.3% (=58% of 74.7%) dying naturally in their first year. To the 43.3% was added 18.5% (for all the rings from killed juveniles) to give a 'best estimate' of 61.8% for first-year mortality. Humans were responsible for 30% (=18.5/61.8) of the deaths. This first-year mortality only increased to 64% if gamekeepers failed to report 40% of the ringed birds they killed.

There was a comparable estimate of 58% juvenile mortality from ringing in Germany (Kramer 1973) and more recent estimates for Denmark were 59% (Noer & Secher 1990, Drachmann & Nielsen 2002). Fridtjof Ziesemer (1983) obtained an estimate of 47% from ringing in Schleswig-Holstein, but noted that his estimate of 38% for adult mortality was higher than the 33% turnover of adults revealed by feather markings. He estimated that juvenile mortality could be 40% in a balanced population with most adults breeding.

The estimates by Saurola (1976) gave a mortality of 31–35% for second-year Goshawks and 15–18% for older hawks. Other turnover rate estimates from feathers at nests in the early 1970s were lower than in Ziesemer's analysis of data from 14 years in Schleswig-Holstein. Helmut Link (1977) estimated turnover of 21% in Bavaria between 1975 and 1976 and study populations that increased during 1972–79 in the Netherlands had 21–26% (Thissen *et al.* 1981). However, the turnover for a longer period in Bavaria rose to 33%, comparable with Schleswig-Holstein and an estimate of 34% from Denmark (Drachmann & Nielsen 2002). Even higher adult turnover rates have been recorded subsequently from feathers in Bavaria (Link 1986, Bezzel *et al.* 1997a), but the estimate was only 20% from a study in Switzerland where humans killed few hawks (Bühler *et al.* 1987).

In North America, turnover of breeders has been recorded mainly by trapping and marking adults. Three studies with more than 100 records during periods of 6–20 years have provided estimates of 17–21% turnover (Erdman *et al.* 1998, Bechard *et al.* 2006, Reynolds & Joy 2006). Another four smaller studies have given estimates ranging from 14% (Kennedy 1997) to 30% (DeStefano *et al.* 1994a, Detrich & Woodbridge 1994, Bechard *et al.* 2006). The seven American studies have a median of 21%.

Two studies in North America have estimated survival of radio-tagged adults, at 28% for 39 birds in Alaska (Iverson *et al.* 1996) and 26% for 32 in Minnesota (Boal *et al.* 2005). In northern Finland, a higher estimate of 37% was obtained from 28 hawks (Tornberg *et al.* 2006). On Gotland, the mortality rate for hawks aged at least three years was 17–21% for 36 males (depending on the estimation method) and the same for 42 females. Mortality of 29% for 20 females in their second year was significantly higher than this. In combination with 41% mortality from 16 males, the overall rate was 32% for second-year hawks. Similarly, data from 104 radio-tagged juvenile males and 101 females could be combined to give an overall first-year mortality of 42%.

The mortality estimates from three age categories on Gotland, where ringing was used to check for absence of bias (Kenward *et al.* 1999), help to interpret the results from other studies. Firstly, adults have a higher mortality in their second year than when older. An estimate of 32% for Gotland agrees rather well with the estimates of 31–35% from ringing in Finland. Likewise, 17–21% for older hawks on Gotland agrees with 15–18% from Finland and with the median of 21% of more southerly studies in North America.

Juvenile estimates agree too; with 42% on Gotland, 40% in Schleswig Holstein and 41% from Dutch ringing estimates in Bijlsma (1993). They also agree with the most extensive study of juveniles in North America. Of 109 nestlings radio-tagged in Utah and New Mexico, 94 survived for at least three months (Ward & Kennedy

1996, Dewey & Kennedy 2001). The 86% survival for three months is the same as for 205 juveniles on Gotland.

It would be reasonable to expect higher mortality for juveniles and adults during short radio-tracking studies, especially as winter conditions can become harsh in northern Finland, Alaska and Minnesota. However, it is less easy to reconcile the estimates of turnover from feather records. These estimates are high even if the populations contained many second-year birds, unless there is a residual need to correct for site-infidelity. Or perhaps adult mortality really was exceptional in Bavaria. The long-term studies there (Bezzel *et al.* 1997a, Rust & Mischler 2001) include two populations, of which only one is depicted in Figure 110. After the 1960s they both declined, showing pulses of failing recruitment reminiscent of cycles and with virtual extinction of the second population.

TEMPORAL AND SEX-LINKED VARIATION IN MORTALITY

Mortality tends to be estimated by pooling small amounts of data from several years to derive one overall value. However, mortality rates are likely to vary from year to year. This effect is probably less than variation in reproductive rates, but was great enough for the proportion of hawks recorded after their first winter to be reduced when temperatures were low during December-March (Marcström & Kenward 1981a).

On Gotland, the deaths of all four starved adults with radio tags were in late winter, whereas juveniles also starved in autumn. About 15% of the young males and females died of starvation, trauma or unknown causes within three months of fledging (Figure 115), but mortality of males was higher than for females in every month between September and March, so that only 49% of males survived to April compared with 69% of females. However, no males died in the next three months, whereas female survival dropped to 64% for the whole of the first year.

As a result of relatively low survival of males in their second year, by the breeding season of their third year there were 56% more females than males in the population. At this point, most males were associated with active nests. However, there was no evidence of polygamy, so a much smaller proportion of females were able to breed (Figure 112). Polygamy is extremely rare in Goshawks, with two observations of males with two females at 503 broods in The Netherlands (Bijlsma 1993) and two among urban Goshawks in Germany (Rutz 2005a), none of which reared young.

It is not clear whether older hawks show senescence in terms of reduced survival. Nielsen & Drachmann (2003) found reduced survival after the sixth year in Denmark. However, the decline in known-age breeders from year five in Schleswig-Holstein (Risch *et al.* 2004) fits constant survival of 0.73, which is an improvement on the 0.67 estimate for an earlier period by Ziesemer (1983).

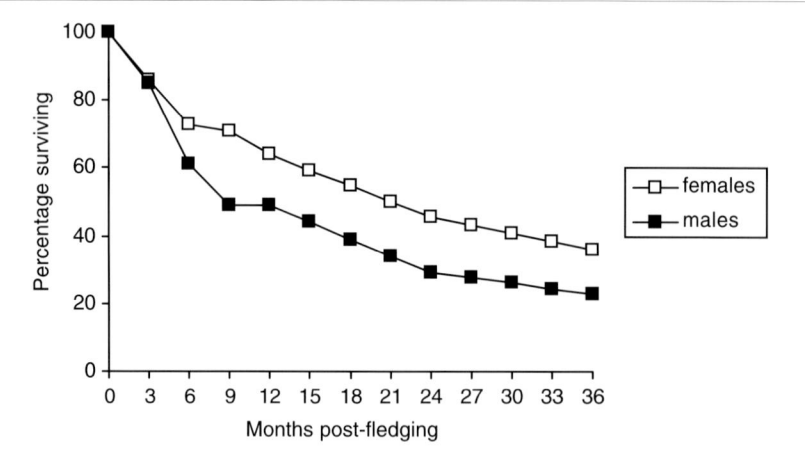

Figure 115. The survival of 104 male and 101 female Goshawks with radio tags on Gotland diverged substantially in their first year (shown at three-month intervals) but their relative numbers retained a similar ratio thereafter. Data from Kenward *et al.* 1999.

Therefore, a population reduction since 1992 in the Danish study may relate to reduced survival of older hawks for reasons other than senescence.

LIFE-TIME REPRODUCTIVE SUCCESS

As a result of death before first attempting to breed, or failing the first attempt, a majority of Goshawks have no offspring. Two studies lasting 25–30 years in Europe have estimated productivity of individual females that nested at least once (and hence had moulted feathers collected) in a Danish population (Drachmann & Nielsen 2002) and in Upper Bavaria (Bezzel *et al.* 1997a). Remarkably, each study estimated that 15.5% of nesting females produced 50% of the young (Figure 116), which was the proportion of young from 20% of females in 12 years on the Kaibab Plateau (Wiens & Reynolds 2005). With fewer than 50% of hawks living long enough to breed, even where many nest in their first year, in each case fewer than 8% of females from one generation produced half the young in the next. In such a situation, a trait that improves evolutionary fitness can spread rapidly.

Turnover of nesting hawks was lower in the Danish population than in Bavaria, so the average output by each nesting female was 5.7 young, compared with 3.3 in Bavaria. Although there was a strong tendency for females that survived longest to rear most young, this was not always the case. One of the Danish hawks did not attempt to breed until aged seven and was then unsuccessful.

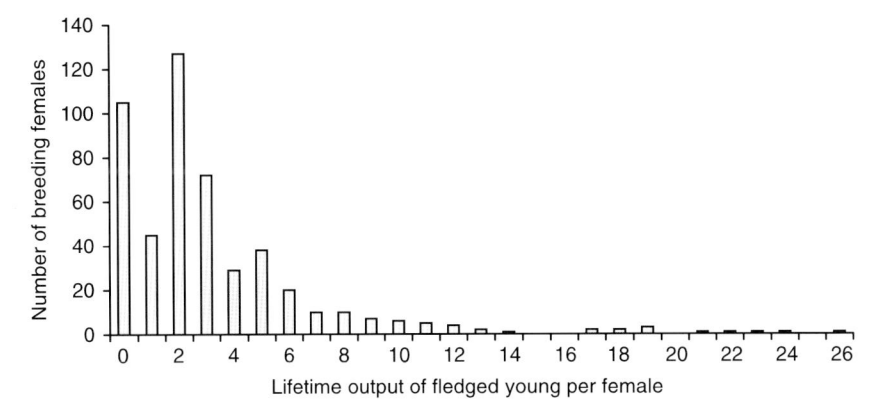

Figure 116. The lifetime production by 489 female Goshawks identified by moulted feathers during a 30-year study in Bavaria. Data from Bezzel *et al.* 1997a.

POPULATION MODELS

Many studies have used models to predict the development of Goshawk populations. A simple example is to examine whether a population is likely to persist with the mortality rates that have been estimated (Höglund 1964a, Ziesemer 1983). This is a question of whether the observed productivity will replace the estimated deaths.

A more sophisticated approach was to predict how Goshawks might recover after the extreme emigration of an irruption in North America. From the age of hawks passing Cedar Grove Observation Station, Mueller *et al.* (1977) estimated that the population had declined by 70% in 1972–3. Based on ringing estimates of 65% juvenile mortality and 30% adult mortality, and the assumption that 75% of all hawks (including first-year birds) would breed, they found that a productivity of three young per pair would take 12 years to make good the 1972–3 decline. This gave the prediction 'that the Goshawks will require several 10-year cycles to regain the population levels existing in 1971.'

Tom Erdman and Dave Brinker (pers. comm.) challenged the prediction with a bet that there would be another irruption in the early 1980s. Their mortality rate estimates of 50%, 30% and 25% for first-year, second-year and older hawks predicted an increase of 21% per annum with 2.5 young per breeding attempt, and hence the recovery of a 70% loss by 1980. In the autumn of 1982 there was a Goshawk irruption to rival that of 1973, and a crate of beer was won.

Models of this type are relatively easy to program on a computer. They can now be created as a 'Leslie matrix' in a spreadsheet to allow different breeding rates and productivity, as well as mortality, for each age and sex class. When these data were entered for the Gotland study, the model was balanced ($\lambda = 1$) for males but gave a 2% decline for females (Kenward *et al.* 1999). The female population balanced in the model if mortality for the oldest hawks was set just 0.5% lower than the estimate from radio-tagging (Figure 117).

The strengths of the Gotland model are that (i) it included age-specific measurements of breeding rate as well as mortality and productivity, and (ii) it fitted the observed stability of the population. Moreover, (iii) if mortality and productivity alone were used to predict a breeding rate for adults of each sex in a balanced population, the rates actually observed fitted the model's predictions of breeding by 73% of males but just 40% of females: among hawks trapped away from nests in winter, 71% of males and 42% of females were at nests with eggs next spring. The model is also robust to reduced productivity through senescence, because it used observed values without assuming how they distribute within older hawks. Despite the relatively low productivity of 1.7 young per active nest (compare with Figures 106 and 108), the high juvenile survival resulted in large numbers of non-breeders (see Figure 115).

Predictive modelling is a growing area of ecology, and at least three other types have been developed for Goshawks. Lensink (1997) used the distance distribution of ring recoveries to predict spread rates of recolonising raptors, including Goshawks. Johansson *et al.* (1994) and Kimmel & Yahner (1994) used habitat and altitude measurements to predict where Goshawk nest areas were most likely to be found, as did Carroll *et al.* (2006) with maps obtained by remote sensing (by satellite). Kenward & Marcström (1988) built a model based on functional and numerical responses to show how individual differences in competence could enable Goshawks to stably suppress numbers of a wild game-bird species or generate cycles.

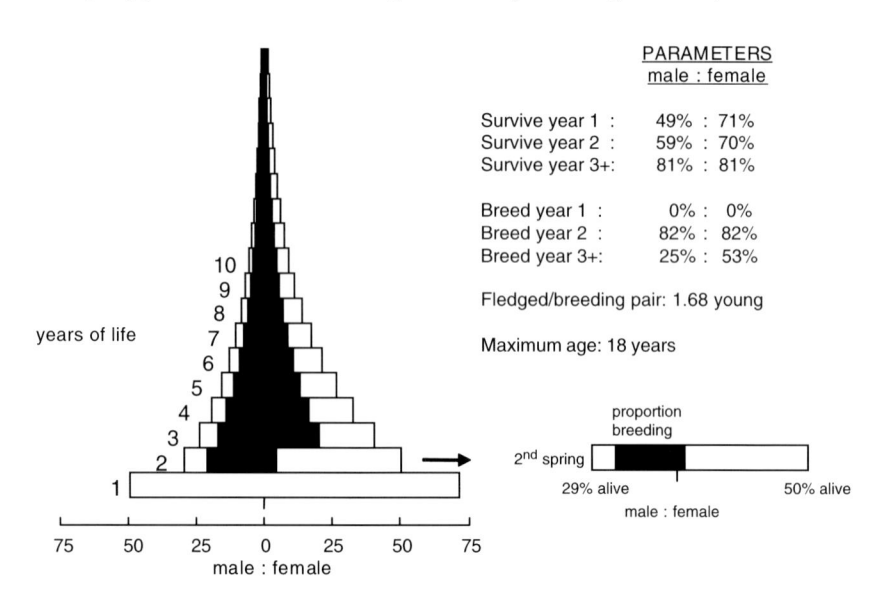

Figure 117. The Goshawk population on Gotland in spring. Bars represent the number surviving in each of 20 years from 100 fledged of each sex. Breeders (black) are a higher proportion for males than females. Data from Kenward *et al.* 1999.

DYNAMICS AND DENSITY

Although Goshawk clutch size, nest success and productivity do not vary at a regional scale across Europe, there is 30-fold variation in breeding density (Rutz *et al.* 2005). Although median densities in European and North American study areas are similar, Goshawks reach much greater densities in Europe (Figure 118).

Goshawk densities across Europe have been linked to densities of woodland and to frequencies of pigeons in the diet at Goshawk nests (as a proxy for the abundance of this staple prey). In five regions of Europe (Bavaria, Saxony, Denmark, Holland and Switzerland) Goshawk density and woodland have been assessed in 6–9 study areas. Within these regions as a group, and across all studies in Europe as a whole, there is a weak but statistically significant tendency for hawk density to increase as the proportion of woodland increases (Rutz *et al.* 2005, 2006a).

If the whole of Europe is then divided into 17 quadrats of 5° × 5°, there are two separate 'hotspots' of maximum Goshawk density. One is in east-central Europe, focussed on Poland, and the other is a west-central area focussed on The Netherlands and western Germany. Both areas have high pigeon densities, but there is on average much less woodland for the ten study populations in the western quadrat (27%) than for the five populations in the east (61%). If the relationship with woodland is also taken into account, there is a strong relationship between Goshawk density and the frequency of pigeons in the diet (Rutz *et al.* 2005). Goshawk densities across Europe thus depend on both food supply and landscape structure, as do the sizes of their home-ranges in Sweden (Kenward 1982, see Chapter 5).

Although these relationships are informative about factors that limit the maximum density of Goshawks on a large geographic scale, they are not informative

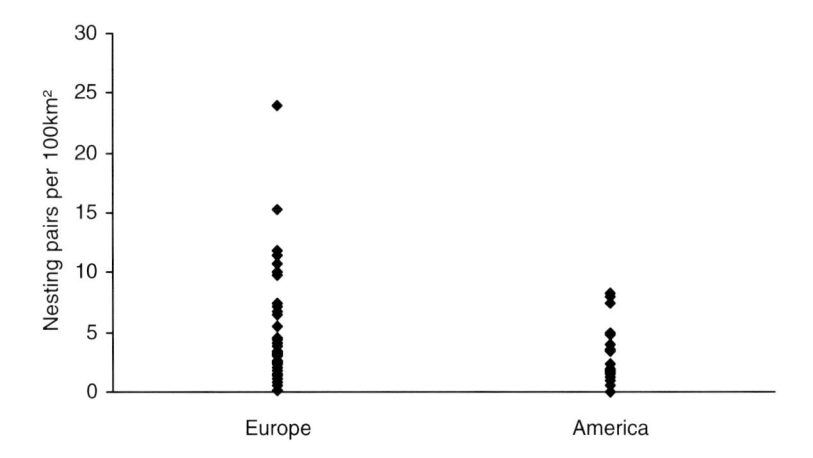

Figure 118. The mean annual density of active Goshawk nests in 46 study areas in Europe and 17 in North America. The median value for Europe is 3.38; the median for America is 3.39. Sources in Rutz *et al.* 2006a and Appendix 2.

about the regulation of Goshawk density. The lack of variation in nest-based productivity indicated the absence of a relationship between density and productivity at that scale (Rutz *et al.* 2005), for which there are at least two possible explanations: (i) the large scale may tend to smooth out the local and temporal variations involved in regulation of Goshawk density, or (ii) nest-based productivity may be a relatively small part of density regulation compared with mortality and breeding rates. These explanations are not mutually exclusive, but we have information to examine only the second at present.

POPULATION REGULATION

In a stable environment, numbers of active raptor nests in an area may fluctuate somewhat but, except where cycles occur, they tend to maintain a relatively constant level, as noted in Ian Newton's classic book on *Raptor Population Ecology* (1979). When Goshawks are new to an area, numbers of nests at first rise rapidly, and then stabilise, in a characteristic 'S'-shaped curve, as shown by Christian Rutz and colleagues (2006a) for the colonisation of Cologne (with data from Würfels 1999) and Hamburg (Figure 119, and see the Dutch population recovery in Chapter 10). If future Goshawk populations are to be predicted, it is important to understand the processes which result in population expansion at low density and stability at high density.

To regulate a population about a limiting level, at high density either mortality rates could increase or reproductive rates could decline. There is most information on the

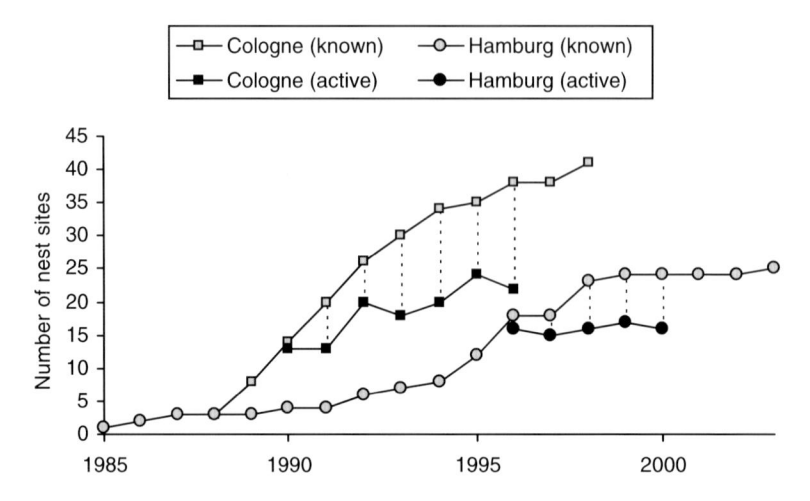

Figure 119. The growth in number of known breeding territories, and active nests when these were studied systematically, in Cologne and Hamburg. Dates of first settlement are unknown. Data from Würfels 1999, Rutz 2005c.

productivity of active nests. Huhtala & Sulkava (1981) noted that the highest average clutch size (3.8) was found in the Finnish area with the lowest Goshawk density. The only equal or larger clutch sizes were averages of 3.80 and 3.96 from expanding populations in Britain (Marquiss & Newton 1982b, Anonymous 1990). Link (1986) found that annual productivity was high when density was low in Bavarian study areas, and also that four areas with varying density had high productivity where density was low. In Schleswig-Holstein, Looft & Biesterfeld (1981) noted that 44% of 111 broods failed completely in nests less than two kilometres apart, compared with only 30% of 385 broods separated by more than two kilometres, which suggests interference when adjacent pairs are very close. Moreover, a spacing for most nests in Drenthe of 600—1,600 m was associated with a 41% failure rate (Bijlsma 1993). However, Krüger & Lindström (2001) did not find a strong link between population growth and density. Other studies have found high productivity at high densities (Bezzel *et al.* 1997a, Olech 1998), but possibly as an artefact of reduced non-breeder competition through killing by humans (Rutz *et al.* 2006a).

There probably is some reduction in nest-productivity at high density, but the ambiguous evidence for it suggests that it is not strong enough to be a major regulating factor. It is also important to remember that because food and woodland are major limiting resources, as indicated by the relationships of maximum density with food supply and woodland across Europe, the limits on density are likely to differ between different areas. Regulating factors may therefore best be sought during population fluctuations within one area at a time, or using 'resource-share' variables (such as prey biomass per hawk or area of hunting habitat per hawk) rather than density itself.

If nest-productivity is unlikely to regulate Goshawk populations, what about mortality? This has proved much harder to study than nest-productivity, either within or between study areas. However, for the five years of intensive radio-tagging on Gotland, variation in first-year mortality did not clearly relate to numbers of rabbits or pheasants, or to the weather (Kenward, Marcström & Karlbom, unpublished).

There is more information on the final vital rate that could be regulatory, the breeding rate. That rate turns out to be extremely variable, for Goshawks in their first and second years at least, between populations at different stages of development. Although the breeding rate in each age class can only be measured by radio-tracking or estimated when mortality rates are known accurately, an index of breeding rate is given by the proportion of new breeders recruited from each age class, which is available for five studies, in one case (Drachmann & Nielsen 2002) separately during and after the growth phase of a Danish population (Figure 120). The Gotland population (Kenward *et al.* 1999) differs from others in that there was no recorded killing of hawks at nests, whereas this occurred during the stable phase of populations in Schleswig-Holstein (Ziesemer 1983, Looft 2000) and Denmark, but probably less during growth of the Danish population and not d ing the colonisation of Cologne (Würfels 1994) and Hamburg (Rutz *et al.* 2006a).

The strength of these changes makes it reasonable to propose that change in breeding rates is the main process regulating Goshawk populations in relation to

246   *The Goshawk*

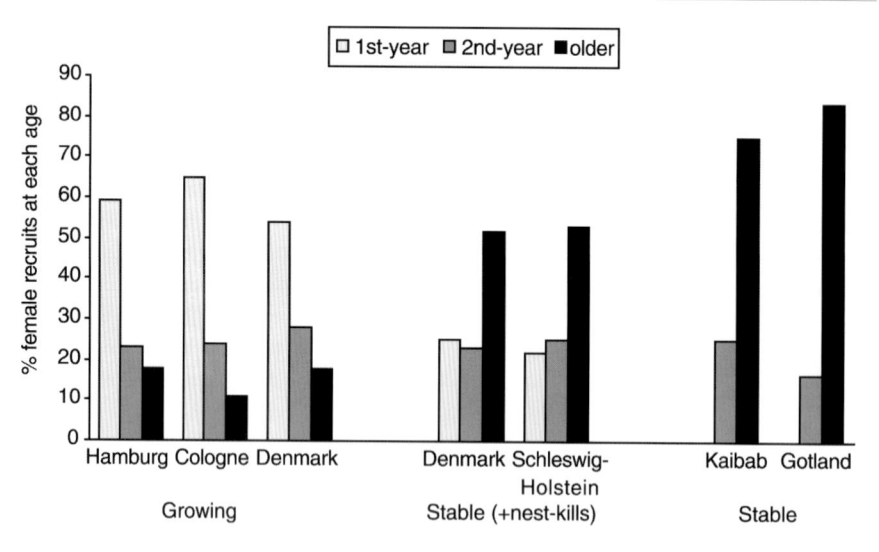

Figure 120. The proportion of female Goshawks recruited to breed before their third year was much greater in growing populations than where nest numbers were stable, with least young breeders where there was no killing at nests. Data from Ziesemer 1983, Würfels 1994, Kenward *et al.* 1999, Drachmann & Nielsen 2002, Rutz 2005c, Wiens & Reynolds 2005.

limits set by food, landscape structure and perhaps aspects of Goshawk behaviour. The challenge is then to explain how breeding rates of young hawks become so low in stable populations.

The nests of Goshawks tend to be regularly spaced (Rutz *et al.* 2006a), as for many raptors (Newton 1979) and especially those with evasive prey (Nilsson *et al.* 1982). Richard Reynolds and Suzanne Joy (2006) showed also that the regularity is greatest at small scale whereas over large areas the nests tend to clump, presumably in the best habitat. Regular spacing suggests that social behaviour, in the form of territoriality, limits the number of hawks that breed in an area, with non-breeders constrained to areas unsuitable for breeding. The use of the term 'floaters' implies that the non-breeders are a mobile part of the population.

However, continuous tracking of Buzzards for two to four years from fledging indicates that they usually settle within their first 18 months, with no long prenuptial movements (Kenward *et al.* 2001b, Walls 2005). Male Goshawks too do not move far to breed (Chapter 5). The majority probably have home-ranges between the breeding pairs, rather than 'floating' about (Rutz & Bijlsma 2005). Why then don't all these hawks breed?

The courtship period is one of minimum resources. It is also a time when males must more than double their daily kill-rate in order to bring a female into breeding condition through courtship feeding (Chapter 3), i.e. to cross the provisioning threshold. There is spacing behaviour at this time and home-ranges become smallest, indicating territoriality. However, the males that bred least well on Gotland were those with largest ranges (Chapter 7), indicating that they had to travel furthest to forage rather than that they were most constrained by territoriality. It therefore seems conceivable that breeding rates (i) are determined by male competence to reach a provisioning threshold and (ii) are the main resource-share dependent factor in the regulation of nest density in Goshawks and similar raptors. Territoriality may well influence provisioning, but not directly determine breeding rates. Total density (of breeders and non-breeders) could then be set by average mortality across years, depending on local resources (or humans).

The main problem with the 'provisioning threshold' hypothesis is in explaining why the less competent hawks do not breed much later in the season, when the appearance of young prey enhances food supplies. In fact, some pairs are delayed and then lay fewer eggs (another way in which male provisioning affects productivity). However, late breeding has implications for adult moult and perhaps also juvenile survival, and is probably precluded by gonad regression. This regression would not be expected until after the end of the normal laying period and may be later for males than females. Thus, early-laying females that lose their mates can still replace them from the non-breeding pool (Newton 1979, Bijlsma 1991).

Neither the 'territoriality' nor the 'provisioning' hypotheses fully explain why nest spacing in an area remains similar from year to year, rather than increasing in years when fewer birds breed. The probable answer is that the established birds have favoured nest sites, which creates a tradition of use to dampen any sudden change in spacing. Evidence for this is that when several nearby breeders were killed during a period of increasing food supply in Norway, an increased number of new pairs managed to take their place (Selås 1997b). In a pattern of established birds, sites where it is not easy for a male to bring a female to laying will be used only occasionally. Although maps often show all the nests in study areas, in any year there may be many unoccupied sites (see Figure 119).

Under the provisioning hypothesis, the ability of each surviving settled male to cross the provisioning threshold in any year will depend on individual experience, which should increase with time. Hassle from neighbours might also decline as pairs become familiar with each other. This would predict that numbers of active nests might tend to increase slightly with time in a stable population, especially if females in established pairs are more inclined to self-feed during courtship.

The provisioning hypothesis accommodates considerable variation in breeding density, in the medium term as well as annual change in 'occupancy', with relatively little change in the total density of hawks. Vidar Selås (1998a) showed that a recent temporary decrease in Norwegian fox numbers due to sarcoptic mange was associated with an increase in grouse numbers, as has also been demonstrated by experimental removal of foxes from islands (Marcström et al. 1988). Goshawk breeding density responded like the grouse and both increased towards levels observed in the 1950s, when foxes were hunted more intensively for skins. There was an equivalent decrease in grouse and Goshawk numbers as the fox population recovered Selås (1998a).

Goshawks also bred best when vole numbers were high, not because Goshawks ate many voles (Selås & Steel 1998) but probably because there was then least predation on grouse by foxes and martens (Widén *et al.* 1987, Marcström *et al.* 1988).

DYNAMICS AND EVOLUTION

In a situation where 8% of one generation produce half the offspring, natural selection could act rapidly. This seems to be the case for Goshawks. Risto Tornberg and colleagues (1999) noted that the density of woodland grouse and their proportion in the diet of Goshawks had declined during the period 1960–2000 in northern Finland. As a result, the average weight of prey brought to nests had also declined. They predicted that the need to catch and transport more small prey would favour small males, leading to a decline in size. This decline in male size was shown by the collection of Goshawk skins and bones in Oulu museum.

They also predicted that with shortage of grouse, female Goshawks would become more dependent on hares as alternative prey in winter, making it an advantage for female size to increase. This size increase was present for females in the museum collection too.

Such rapid change in populations is sometimes called 'contemporary evolution'. It led to an increase in size dimorphism, as females got larger and males smaller. The dimorphism index of adult females increased from 11.8 to 12.7 for wing measurements (see Figure 11 in Chapter 2), and from 7.4 to 11.8 for sternum size. This shows just how rapidly a change in dimorphism could evolve, for example to converge the size of males and females, as found in North America where Snowshoe Hares are an important winter prey for both sexes (F.I.B. Doyle pers. comm.). My suggestion (Kenward 1996), that size of female Goshawks across their global distribution relates to the size of the largest common lagomorph, has yet to be tested.

Even on Gotland there were signs of contemporary evolution in progress (Penteriani & Kenward in review). The smallest males were most likely to pair with a female that laid eggs and to raise the heaviest young. The largest females tended to raise the most young. Although these trends in reproductive fitness could have produced the same effects as at Oulu in Finland, they may also merely have been part of the process that maintains sexual dimorphism, for example if balanced by opposite trends in survival at other times of year.

IMPLICATIONS FOR CONSERVATION AND MANAGEMENT

The flexibility in age of first-breeding is very important for Goshawk conservation, at least in Europe. Variation between populations in the proportion of hawks that can breed nine to ten months after fledging might have a genetic component, but this seems unlikely because the widespread nature of the trait suggests that it would not be in an individual's interest to refrain if an opportunity arises. Goshawks would benefit from flexibility, both where prey numbers fluctuate at high latitudes and after severe winters at high or low latitudes (or altitudes). Nevertheless, breeding of first-year hawks seems to be rare in North America. If there is a genetic predisposition to breeding in first-year plumage, the trait is presumably more prevalent in Europe.

Possible causes of apparent high mortality of Goshawks in Denmark and Bavaria are discussed in Chapter 10, as is the past impact of pesticides. However, the association of high mortality with a low age of first breeding provides a very convenient method of monitoring impacts on Goshawk populations, merely by collecting feathers at nests of breeders (Kenward & Marcström 1981, Grünhagen 1983). This approach can also be used to check for any increase in turnover of breeders, although in this case there is a clear need for work on possible bias in turnover estimates. Checking could involve use of DNA analyses to confirm that the subjective assessments are accurate, or with radios that could also indicate dispersal beyond the study area. The development of small leg-mounted radios with long life would enable safe attachment by volunteers.

Such checking would not be necessary if monitoring were intended only to record the proportion of juveniles in the breeding population. There might be some under-estimation of first-year breeders if such monitoring were biased towards records at the sites most frequently used, as would be the case without systematic survey. However, a constant level of bias should not prevent detection of increased juvenile presence as an early-warning of increased mortality. Together with productivity records, feather collection seems well suited for low-cost monitoring of dynamics in Goshawk populations. Low cost monitoring of density is considered in Chapter 9.

The interactions between weather, food, laying dates and clutch size are an interesting area for further study. Do young and old females lay the smallest clutches mainly because of age-linked physiology (e.g. ripening of fewer follicles), or because they tend to breed with the least competent males, initially through poor competition with established females and later through needing to replace experienced long-term mates? Does poor spring weather mainly act to hinder male hunting, thus reducing female food supply during laying, or do females start the breeding season with low reserves in cold springs or do females tend to ripen fewer follicles when they are cold or breeding is delayed? As usual in science, any increase in knowledge, in this case about factors which influence breeding, raises a host of new questions, in this case relevant to an understanding of how Goshawks might respond to rapidly changing climate.

Mortality, emigration, immigration and reproduction are not only predictors of change in populations, but also measures of performance for individuals. Does each individual die young or survive to dispersal, emigrate or remain in the natal population, settle or die if it emigrates, produce a clutch or not, rear a brood or not, survive another year? The answers to these questions, for the individuals in a population, sum to the vital rates for the population. However, they are measured for individuals, and this is why it is crucial to identify individual hawks reliably. Individual-based population models are superior to those that use pooled samples for 'unknown' birds for predicting how populations will respond to future landscapes (Goss-Custard 1996, Sutherland 1996).

For effective modelling, it is also important to understand whether breeding density is regulated primarily by territoriality or provisioning. As the provisioning hypothesis is based on variation in individual competence, it is testable by feeding an experimental group of radio-tagged males that are predicted to be non-breeders and comparing their performance with controls.

Good access routes and terrain through much of Europe have facilitated the development of Goshawk study areas for recording nest densities, productivity, and feather-based turnover estimates that can be checked through a high return rate for rings from the dense human population. These conditions do not exist through much of the global range of Goshawks and other species. However, provided hawks can be caught in winter in areas where populations are sedentary, radio-tagging can permit individual-based recording of breeding rates, nest productivity and survival for adults and juveniles. With reliable long-life radios, this is an alternative to an area-based approach in terrain that is hard to search for nests, although it does not measure densities. Future funding authorities may find it cost-effective to support this approach in less accessible areas, perhaps to increase coverage of the Goshawk in North America and east across Eurasia.

The opportunity to examine a major component of a species' population dynamics within and between continents is rare. Much is due to dedicated volunteers, as well as professional ornithologists. Volunteer effort already extends to the mapping of breeding presence for birds across Europe (e.g. Hagemeijer & Blair 1997), which brings a challenge to increase coverage and fill gaps. The measurement of nest productivity over large areas requires agreement on standardisation of methods. Development of standards and new methods is an important area in which professionals can take a lead, and then transfer their knowledge to volunteers via ornithological organisations. The 'Goshawk challenge' for systematic recording of vital rates is now set for other species too.

CONCLUSIONS

1. Productivity per active nest (with eggs) is reduced by total and partial loss. Causes of complete brood failure include abandonment or collapse of the nest, bad weather and nest destruction. Humans, martens and large owls are the most important predators.
2. Productivity is lowest for clutches of one and two eggs, which most often fail completely. However, partial losses increase with brood size, especially with 5–6 eggs, such that clutches of 3–4 eggs often rear most young. Brood size is linked to clutch size, and can therefore be affected by pre-laying food supplies and weather.

3. Study of female residency by recognising moulted feathers shows that productivity increases after first breeding, but peaks at age 6—9 years and then tends to decline. Males that breed in their first year are susceptible to total brood loss, and females to partial loss.
4. Although clutch size, nest success and productivity vary between individuals, sites and years within a study area, they vary less between areas and do not vary regionally across Europe. However, productivity is higher in Europe than North America, due to median brood sizes of 2.5 and 2.0 respectively.
5. In North America, populations showed greater annual changes in site occupancy than in Europe, where populations showed most variability of breeding in first-year plumage. Up to 65% of European females first bred as juveniles in expanding populations and where breeder-turnover was high.
6. Ring recoveries and collection of dead hawks showed that starvation increased to the north while death due to humans and their artefacts declined. Hawks were rarely killed by predation or disease in Europe, but perhaps more often in North America.
7. Ringing and radio-tagging show that an annual adult mortality of 15–25% is typical, but it can be more than 30% at high latitudes and with killing by humans at nests (if turnover records from feathers are unbiased). Ringing tends to overestimate juvenile mortality, which can be as low as 40%. Mortality is closer to 30% for hawks in their second year.
8. Taking mortality into account, about 8% of females produced 50% of the offspring in two populations, indicating that there is scope for rapid evolutionary change in Goshawks. In fact, change in food availability in northern Finland had selected for small males and large females, resulting in significantly increased sexual dimorphism in just 40 years.
9. Goshawk density reaches higher levels in Europe, where it varies regionally with abundance of pigeons and woodland, than in North America. Due to high variability in breeding rates, these, rather than nest-productivity, are considered the key regulator of Goshawk populations, in a process based more on individual competence and provisioning than territoriality.
10. Collection of moulted feathers as well as productivity records are recommended for monitoring dynamics of Goshawk populations, with increased interaction of professionals and volunteers to gather extensive data on all vital rates as well as presence and density.

CHAPTER 9

Falconry and management methods

18 February 1983

Dear Dr. Y,

I understand that in the areas of Schleswig-Holstein and Lower Saxony it is permissible to take Goshawks (probably under licence) from areas where they are doing damage to game stocks and re-release them elsewhere. In the last few years a number of Goshawks have been finding their way to this country from West Germany for falconry purposes. I wonder what the attitude of the West German authorities would be if they were aware that these birds are coming here for falconry and not release to the wild?

I look forward to hearing from you,

Yours etc.

Z

Your feelings about a letter like that will depend on how you view falconry. Previous chapters have indicated the considerable involvement of falconers in past research on Goshawks. Chapter 10 will consider how falconry might benefit Goshawk conservation in future, but it is important first to explain how falconry developed, how it relates to the biology of Goshawks and how it gave rise to domestic breeding, release techniques and other management methods. It is a matter of discovering the opportunities in falconry and not merely viewing it with concern.

Falconry is as rich in stories, methods and ideas as any other human interaction with other animals, such as dogs or horses, so this chapter can only skim the subject. For more explanation of falconry's biological background and involvement in conservation, a good start would be recent books by Nick Fox (1995) and Tom Cade with Bill Burnham (2003). It is worth visiting the web-site of the International Association for Falconry and Conservation of Birds of Prey (www.i-a-f.org). A large collection of falconry books and journals is available, courtesy of the British Falconers' Club, in the library of the Edward Grey Institute of Field Ornithology at Oxford University, and in the Archives of American Falconry at Boise, Idaho. The annual journal of the Deutscher Falkenorden (German Falconers' Club, www.falkenorden.de) contains much on raptor biology and conservation politics.

HISTORY OF FALCONRY

The origin of falconry is unclear. The problem is that raptors may have been kept to be admired, or as aids when netting or liming birds, long before trained raptors first caught wild prey for their owners. In his *Bibliotheca Accipitraria*, Harting (1891) noted that a bas-relief thought to depict a falconer had been found in the ruins of Khosabad and dated to around 1700 BC. However, depiction of a man with a raptor is not good evidence that falconry existed in Mesopotamia nearly 4,000 years ago. From an examination of many mosaics and early writings, Kurt Lindner (1973) concluded that tame raptors and owls were used in Greek and early Roman times for fowling, probably to attract mobbing birds down to nets or to twigs covered with bird-lime, but not for falconry. The Thracians also used flying raptors to frighten birds down into nets at ground level.

A Japanese writer, Akizato Rito (1808), reported that falcons were given as presents to Chinese princes of the Hiu dynasty around 2200 BC. These birds may merely have been exclusive pets. However, the verbal tradition in the ancient falconry of the Naxi people in China is that falconry originated in Northern Manchuria and passed from there to the Mongolian sources of westward emigrants from central Asia (Gott 2000). Moreover, the earliest indisputable evidence of falconry comes from the Far East, possibly with a Chinese description as early as 700 BC (Ye 2005). According to Japanese records, trained Goshawks were introduced from China in 244 AD, although falconry did not become popular in Japan until more than a century had passed (Jameson 1962).

It seems that falconry reached Europe from Central Asia with the Vandals (Lindner 1973). An origin in East Asia, where genetics also indicates the origin of dogs from wolves, is therefore very plausible. There are opposing claims in the Shahnamei epic of Ferdowsi, written after the Muslim conquests, that falconry originated with the Pishdadid kings in ancient Iran, before the time of Zoroaster and thus perhaps more than 5000 years BC. However, if falconry had persisted since that time, there should have been signs in the relics of Persepolis (Yazdani 2005) and in the writings from the Egyptian, Greek and early Roman civilisations. The challenge is for Chinese and Arab archaeologists to produce scripts more than 2,000 years old on hunting with trained hawks.

The evidence that falconry came to Europe with the Germanic tribes is its first appearance in Roman culture shortly after Vandal immigration. It was practised in England in Saxon times, with Aethelbald, King of Mercia, being sent a hawk (presumably a Goshawk) and two falcons by the Archbishop of Mayence between 733 and 750 AD (Glasier 1978). Saxon falconers tended to liberate their hawks into woods in the spring, taking young from the nests in the summer (Cox & Lascelles 1892). The Bayeux tapestry shows that King Harold took a jessed raptor and hounds on his visit to William of Normandy in 1064; the raptor's size and large tail depict a Goshawk.

The earliest record of Goshawk density comes from Britain in the eleventh century AD, thanks to an unusually complete record of eyries in the Domesday Book for the county of Cheshire. In a county of about 2,700km^2 which was 27–36% wooded, Derek Yalden (1987) noted that 24 nesting areas were recorded, or 0.9 pairs per 100km^2. The nests were recorded because they had considerable levy value. At that time, tax could be paid as a hawk or £8–10 paid in lieu.

Falconry was thriving during this period in Muslim culture. The first treatise in Arabic appeared in the eighth or ninth century AD and gave particular attention to the Goshawk (Allen 1980). During the last millennium at least, trained accipiters have been widely used by the peoples of Asia, from Turkey, Iran, Iraq, Georgia, Mongolia and China in the north, to Arabia and the Indian subcontinent in the south, with extension into North Africa as far as Morocco. Sparrowhawk species are still trapped on passage in the eastern parts of this range. They are used to take quail and other small game before release in spring. Goshawks are trained less often than the smaller accipiters in some areas, partly because there is less prey for them and partly because they need more food. These are working hawks and they are expected to earn their keep. However, a saying persists that 'There is no finer sight in the world than a man on a horse with a Goshawk on his fist' (Majid al Mansouri, pers. comm.). European falconers apparently gained much falconry knowledge during the crusades, including use of the hood. Hooded raptors are protected from alarming sights during training and from being diverted by other hawks or unsuitable prey at inopportune moments on hunting trips.

Falconry was also richly documented in Japan from the ninth century onwards (Morimoto 2005), with the first surviving falconry treatise 'Shinshuu Youkyou' dated 818 AD. In the subsequent millennium there was a strong cultural association with military power and hence the ruling classes, with repeated bans on practise of falconry depending on who was in power. Goshawks were the most favoured species in both Japan and China, where prized hawks in the eighth to tenth centuries came from the Ganjun Shan (Ganjun Mountains), and the occasional vagrant *albidus* obtained almost mythical status. A white Goshawk caught in winter was so important that its journey as gift to the Uyghurs in 884 was recorded in detail (Mayo 2002).

The first surviving 'western' falconry treatise was written around 1247 by Emperor Frederich II of Hohenstaufen. As a result of his book, *De Arte Venandi cum Avibus*, Frederich II has been called the father of ornithology. His principle of testing hypotheses, for instance by sending a trusted servant to the north to see whether barnacles really metamorphosed into geese, was an important step in the development of modern science.

Falconry flourished in Europe during the subsequent half-millennium, and one of the best English books on the subject, *An Approved Treatise on Hawks and Hawking* (Bert 1619), reveals how sophisticated the veterinary treatment of trained raptors had become by the 17th century (Cooper 1979). Bert's book deals at length with the training and flying of Goshawks, which in the strict sense is hawking rather than falconry. Likewise, those who train Goshawks are austringers, from the French 'autour' for Goshawk and 'autoursier' for its trainer (originally Latin *astur*).

The Goshawk was clearly a popular raptor for training in Saxon times, and remained so throughout the Middle Ages, when general interest in falconry may even have exceeded the present enthusiasm for football. Chaucer liked to go out 'with grey Goshawk on fist' and Shakespeare used many falconry metaphors. As *The Boke of St Albans* indicates (Berners 1468), falcons were probably flown mainly by the aristocracy, whereas a Goshawk was 'for a yeoman'. While the falcons could provide spectacular flights for noble falconry excursions, but required a stable of reliable horses, a well-trained Goshawk could be used on foot to keep the larder stocked with common small-game. Thus the doings of common austringers would largely have gone unrecorded, compared with the falcons in the paintings and writings of the upper classes.

In Britain, falconry lost its popularity following the Civil War and with development of effective sporting guns, Land Enclosure Acts (which restricted access to good hawking land) and rural de-population through the Industrial Revolution. Falconry remained popular only among gentry by the late 17th century, as indicated by the title of Richard Blome's *The Gentleman's Recreation* (Blome 1686), which contains interesting observations: 'There are divers sorts and sizes of Goshawks, which are different in goodness, force and hardiness according to the different countries where they are bred; but no place affords so good as those of Muscovy [Russia], Norway and the North of Ireland, especially in the county of Tyrone.'

By the late 18th century, the practice of falconry was restricted to a few landowners in Britain and Ireland, including James Campbell (1780), who recorded the Goshawk still present in the north of Scotland and Ireland, and Colonel Thomas Thornton who obtained a nestling from Rothiemurchus forest in 1786

(Upton 1980). These few enthusiasts formed a series of clubs and kept the sport alive until the present British Falconers' Club (BFC) was founded in 1927 (Upton 1980).

With the loss of interest in falconry came an upsurge of persecution of raptors in the name of game conservation. Falconry had been responsible for some of the earliest legislation protecting raptors, as when Henry VII protected the Goshawk's eggs 'in pain of a year and a day's imprisonment, and to incur a fine' (Cooper 1981). In the 19th century, falconers were virtually alone in championing these predators in Britain, and the efforts of those like Gerald Lascelles (Cox & Lascelles 1892), who sought to preserve New Forest raptors, were not always appreciated by the shooting fraternity. Morant (1875) wrote in scorn of the 1873 Select Committee on Bird Preservation: 'No doubt, beside certain naturalists, it is our falconers who are anxious to make birds of prey more numerous'. Falconry's role in protecting raptors continued in the 20th century, with a partnership of the British Falconers' Club and the Royal Society for the Protection of Birds in a bounty scheme for landowners who preserved raptor nests and especially with the role of Lord Tweedsmuir, falconer and protectionist, in the passage of the United Kingdom's Protection of Birds Acts.

Unfortunately the relationship between falconers and wildlife protectionists has been less happy during the last three decades. Falconry became re-established in many European countries and in North America between the two World Wars, but was restricted to a score or so participants in each national club. However, during the 1960s there was an explosion of interest in the sport, coincident with increasing general interest in wildlife and in recreational keeping of animals (e.g. pigeons). This renaissance in Western falconry was stimulated in Britain by several books featuring modern falconry, such as *The Goshawk* (White 1951) and the film *Kes* (Hines 1968). Falconry displays at country shows and in small zoos added to publicity and journalists eagerly rediscovered a 'lost art'. Membership of the British Falconers' Club (BFC) grew from 51 in 1937 to 140 in 1957 and was relatively constant within a range of 948–1,111 during 1988–97 (J. R. Fairclough, pers. comm.).

The problem with falconry's expansion was that it coincided with steep declines in raptor populations, especially the Peregrine and Sparrowhawk, that we now know were due primarily to agricultural use of organochlorine pesticides (Hickey 1969, Ratcliffe 1980, Newton 1986). Once aware of the pesticide problem, the majority of falconers were content to train species which remained relatively common, and members of the BFC voluntarily stopped seeking licences to obtain British Peregrines. Sadly, not all falconers were so responsible. Young were taken from the last active Peregrine eyries in Denmark and Schleswig-Holstein. Some protection interests sought to ban falconry, and achieved this aim in Denmark, Sweden and (for recreation) in Australia.

The majority of countries recognised the work that falconers were doing to breed and restore raptors that had been extirpated by pesticides (e.g. Cade *et al.* 1988, Cade & Burnham 2003) and were content to tighten controls on ownership of raptors. In fact, bans and re-instatement have affected falconry in many countries throughout their history (Macdonald 2005, Morimoto 2005, Yazdani 2005). Bans typically followed revolutions (in China, France, Iran, Portugal, Japan), mainly as a reaction to dynasties for which the sport had became an important recreation.

Modern falconry in Europe and North America is controlled either by licensing the individual falconer or by requiring birds to be marked to denote legal origin. In those countries that licensed individuals, often with organised mentoring and examination of competence, active falconers remained relatively few (less than one in 100,000 of the human population) and often retained legal access to wild raptors. Until recently at least, this was the case for the few hundred active falconers in countries with good populations of wild Goshawks, including the USA, Austria, France and Germany. Falconers in the USA have also recently regained access to the Peregrine populations that they helped to restore.

In Britain, where few falconers have access to the large open spaces needed to fly falcons at their best, hawks for hunting in wooded country were more appropriate. A survey in 1970 showed the Goshawk to be the most commonly held species (Kenward 1974). With very few wild Goshawks in Britain, these hawks were all imported. The United Kingdom then adopted a system of certifying legal ownership of each raptor, but not licensing individual falconers. This system encouraged domestic breeding. The British development of 'DNA-fingerprinting' (Jeffreys *et al.* 1985) was soon applied to test that domestic parentage was genuine (Parkin 1987). Despite early evidence that wild Goshawks had been 'laundered' through domestic pairs, a random survey of ten families in 1995–6 found no cases (Williams & Evans 2000). DNA tests can also be used to prevent transfer of certification marks by storing a biological sample at the time of marking. Such a 'mark-and-bank' scheme is a powerful tool for ensuring that falconers adhere to controls (Kenward 2004).

Although recreational falconry remains banned in a small minority of countries, the techniques that it has developed are nevertheless used with tame raptors in most of them for wildlife management. These techniques include taming and training, domestic breeding, the restoration of wild populations and the management of predation and other wildlife nuisance issues.

TRAINING GOSHAWKS

The methods used by falconers to tame and train Goshawks and other raptors are based on aspects of their behaviour and physiology that we have partly considered in earlier chapters. Falconers gain 'green fingers' (Ripley 1975) in raptor psychology that require hands-on experience far beyond book-learning. The increasing importance for raptors of hands-on management makes it worth considering in more detail what is learned by these 'applied psychologists' of Goshawks. As with all training, there are many 'right' ways. Some of the following is merely personal experience.

Raptors have traditionally been obtained for falconry as 'eyasses' (taken from the nest), 'passagers' (trapped during their first autumn or winter) or 'haggards' (trapped as adults). Haggard Goshawks are seldom trained in modern falconry, on principle (and hence in falconry codes of conduct) because they represent wild breeding stock, but also because they have had longer in the wild to accumulate latent diseases and to learn behaviour that may complicate hunting with them.

Important behaviour is still developing in eyasses. Learning is involved in recognition by young raptors of their own species, of potential predators and of their nest environment. Thus, young Peregrines 'hacked' (fed free through the post-fledging dependence period) from platforms on buildings readily adopt such sites for breeding (Tordoff *et al.* 1998, also Rosenfield *et al.* 2000 for Cooper's Hawks in plantations). The same probably applies to the types of site used by Goshawks that nest in forests or the very different situations in towns. Learned recognition, often called 'imprinting', starts around hatching and has separate but overlapping sensitive phases for three main social aspects as well as for enemies and the nest (Fox 1995). The timing of each phase has been found by trial and error, and then used for good management. Recognition of the nest-parent as brooder and feeder occurs very early. Chicks that are introduced to a nest after their downy stages may avoid the foster parent. The time of maximum feather growth is the peak period both for learning to recognise siblings and the fear response, a phase of 'friend or foe' recognition.

An extremely important type of learned recognition is that of the provisioning parent, which occurs late during growth of feathers and can extend through the post-fledging period, and is later used to recognise potential mates. If hawks see a person feeding them prior to imprinting on adults, they start using the 'heee-yah' call as they would to adult hawks. They may continue this 'scream', especially when hungry, for as long as they remain in the hands of falconers, sometimes repeating it every few seconds until fed, with each feed reinforcing the habit. Such birds are undesirable for hunting, for domestic breeding with natural mating and for release back into the wild.

Satisfactory 'provisioner-imprinting' of eyass accipiters is achieved in three ways. One is to start handling them as soon as they can pull at food for themselves, as their feathers start to grow at two to three weeks old. These early-contact eyasses remain in the world of the trainer and become completely tame (McElroy 1977), maybe through treating humans as siblings and not as provisioners because they are never fed by hand, or even on the fist until well after feathers harden. A second approach applies to wild hawks that must make a long journey from nest to trainer. These must only be moved close to fledging, and then reared as a group until after feather growth is complete, with food put in their enclosure at night to avoid its association with humans. Lastly, domestic-bred young can be left with parents as late as possible, which produces especially tame young if they are fostered by birds that are themselves tame and they have regular sight and sound of human activities (Fox 1995).

My last Goshawk, Miss Piggy, was treated in the second way. Although she was reared to fledging in the wild, she was then in quarantine quarters with another young Goshawk until dispersal age. She saw me for a few minutes each day, but not in association with food, which was given during the night. However, she was fed several times a day at the start of her first annual moult, because her appetite declined so much that I thought her ill. Although she had accepted me as a 'courtship

Figure 121. Anklets designed by the British Falconers' Club to identify Goshawks in the wild if lost by their trainers.

provisioner' (Chapter 3), she might have treated a male hawk in the same way if given the chance.

Before a hawk can be trained, it must have a jess attached to each leg. Each jess is typically a thin strip of supple leather, about one centimetre wide and extending some 20cm from a Goshawk's leg. Jesses are held to prevent a bird flying off if it is alarmed by something unexpected during handling or prematurely during hunting. A modern jess system has an anklet on the leg, which can carry an identify tag (Figure 121), with an eyelet from which slit-ended jesses can attach to a swivel and leash, for added safety when handling, or be replaced by hunting jesses that cannot catch on vegetation when a hawk is released. Some falconers fly hawks with anklets alone, secured on the glove by a cord looped through each eyelet and released to slip through the eyelet as the hawk leaves the fist. In most countries, Goshawks must now wear a registration marker, either a close-ring slipped onto a domestic-bred bird before the foot reached full size, or a tamper-proof leg band placed on a bird obtained from the wild under licence.

There are three stages in the training process, all of which overlap to some extent. The first stage, known as 'manning', is one of habituating the hawk to the trainer and to all the strange sights and sounds associated with humans. Late-contact hawks must first become familiar with the trainer. The process is typically started by feel, in darkness. By pressing a gloved hand against the back of the hawk's legs on its perch, it can be persuaded to step back onto the fist. The hawk is nervous, and will grip quite hard, possibly even clenching its feet convulsively as it feels the hand move like a living prey. To protect hand and wrist, the trainer therefore wears a gauntlet of tough but pliable leather, such as buckskin. This glove must withstand the talons yet be supple enough for the hawk to grasp comfortably and for the falconer to sense the bird's mood as it tenses or relaxes its grip.

The trainer's next aim is to get the hawk to feed on the fist. After 'prey-killing' clutching of the fist, which can also be elicited by a rabbit-like squeak from the trainer, a hawk which has not yet fed that day tends to peck at the glove. The trainer arranges for the beak to meet some tender meat so that feeding starts. It can take five minutes or five hours to reach this point, talking softly to the hawk, maybe arranging very dim lighting on the food but not the human and perhaps squeezing its toes to draw its attention down to the food.

The early-contact eyass has long been familiar with the trainer by the time other eyasses take a first meal on the fist, at around the time of independence from parents. Passage and haggard hawks are also often placid or hungry enough to feed quickly on the fist. From this point on, the training for all depends on feeding. The hawk is encouraged to eat good meals of low-fat meat, such as beef heart or rabbit, to manage the mass of an eyass at around the mean of wild juveniles in September (see Figure 18). If the hawk is fed one to two hours later each day, it remains as keen as a dog in training.

Each feed is prolonged, by starting with tough wings or necks of birds as 'tirings' (from the French *tirer*, to pull) on which there is little meat. When the hawk is at ease eating in dim light, with the trainer being careful to avoid intimidation by eye-contact, the light can be increased and the trainer will start to walk with the hawk, perhaps give it glimpses of the outside world through a door, and spend time talking quietly to the hawk with food out of sight. In a week, a skilful trainer will be walking outside with a late-contact hawk, giving food before introduction to anything new. By noting if the bird changes from relaxed to tightened feathers, twitters or starts looking for an alternative perch, the trainer will draw back and encourage the hawk to eat before moving on. Otherwise the hawk will try to fly from the fist, to be checked by its jesses and helped back by the trainer. Perfection is to train a hawk without one such 'bate' (From the French *battre*, to beat), but very few have ever achieved it.

The second stage of training is 'calling off'. After learning to look to the fist for food, a hawk can be encouraged to step onto the glove for its meal. With a bit more persuasion it will jump onto the fist, and then progress quite rapidly to flying a short distance for its reward of food. For early flights, the trainer will attach a light line or 'creance', but will be careful that there are no sudden frights from which the hawk might fly and be checked by the creance. The trainer will have accustomed the hawk to a call or whistle during earlier meals, and now uses this to encourage prompt response to the outstretched fist. There should be no more flights than achieve a prompt response, lest appetite is lost and the hawk overflies the fist.

For the skilled austringer, training an eyass Goshawk to come 50m can take no more than two to three hours a day during two to three weeks. However, Goshawks (especially males) are relatively nervous compared with species more suitable for a novice, and first training takes longer. A Goshawk is not a hawk for a raw beginner to train without expert supervision. It is important to move rapidly to increased flying, so that muscle mass can be built up by exercise, if a hawk is to be strong and keen to hunt.

When not being handled, an unmanned hawk remains in the mews, with movements usually restricted to a single chest-high perch so that it can be approached without the trainer towering overhead. However, once the hawk sees its trainer as a source of rewards, it can be kept loose, free to chose from a variety of perches. There are close vertical bars at windows, where perching might cause feather damage or otherwise be unsafe. A manned Goshawk can also be put outside to 'weather', ideally with its leash attached to a running line on the ground between perches and always in a place safe from predators. One perch should be low and have a bath nearby, which some hawks use daily. Healthy hawks seldom drink, except for a few mouthfuls at the start of a bath. They can be considered at ease when they 'rouse', by slowly erecting all their contour feathers and then shaking, before perching upright on one leg with the other drawn up into slightly fluffed plumage.

Once a hawk is well manned and coming promptly to the fist from a reasonable distance, it is ready for the third training stage, of 'entering' at quarry. This is when the trainer needs an even deeper understanding of wildlife than in other forms of hunting, because a hawk's hunting confidence comes from early kills with few misses. The trainer must know where and how to approach prey which will have difficulty escaping. This requires familiarity with the likely haunts of young rabbits, hares, squirrels, pigeons or gamebirds, and when their behaviour makes them vulnerable. It is good to get a hawk trained in late summer, before young prey become strong and experienced. Later in the year it may be possible to find prey further from cover, or weak from food shortage in winter, for example when emaciated pigeons can be found at *Brassica* fields. However, trained hawks seem to hunt best if they are not kept too long before learning to catch their own prey.

Goshawks are normally entered from the fist, with the outcome of first attacks often decided within 30m of the trainer. A typical flight would be at a half-grown rabbit surprised 20m from the nearest bramble patch, or at the last to fly in a family of young game-birds. If the hawk succeeds it is allowed to feed well on the prey, which will have been despatched by the trainer if the hawk's initial grip was not fatal. Accipiters lack the notched beak and killing bite of falcons, which have less strong feet to immobilise prey while feeding, and wild hawks sometimes remove flesh and entrails from prey for some time before it dies (e.g. Bijlsma 2004). To ensure that the hawk does not feel threatened by the trainer and tries to fly off with its prey, the hawk on its first kill is given tit-bits to make the trainer's presence pleasurable while it takes a good meal. A hawk which has been allowed to 'feed up' cannot be flown again, but on subsequent outings it may only be given a snack the first time it kills, and then given further opportunity to take quarry.

A Goshawk that is really keen to hunt will rouse frequently and sit on the fist with especially the head feathers slightly fluffed, looking eagerly about. This eager state is known by its Arabic word '*yarak*' (N.B. also Greek ιεραξ for hawk), and has been considered to indicate strength in muscle but little fat. However, as wild Goshawks do not have low fat without also having low muscle protein levels (Chapter 2, Figure 16), a 'no-fat' explanation is probably inadequate. Yarak differs from the listless apathy and sunken eyes of a hawk in poor condition, which falconers avoid, and may resemble a point beyond maximal flight hunting activity in wild hawks but without loss of muscle.

A hawk that is to be flown at a difficult quarry may initially be flown at little else. Thus a large female might be entered at leverets for subsequent flying at full-grown hares. Although some wild females in adult plumage regularly take full-grown hares, trained hawks reject this quarry unless they learn quickly to grasp the head. Small females should probably not be flown at hares. Small male Goshawks have similar difficulties with rabbits and are usually flown at feathered prey. A really fit Goshawk can overtake a partridge in level flight or follow a pheasant to 'bind' to it as it slows to put-in to cover. Tony Jack (1970, 1971) noted how an attacking hawk may use wind-sheer, flying low near the ground to follow a pheasant upwind or gaining height to make the best use of a following wind. Walter Bednarek (1998) suggests that the relatively low wing-loading of the northern Goshawk races *buteoides* and *albidus* makes these more inclined to gain height when chasing prey and during hunting in general.

Hawks learn a variety of tactics to deal with each prey. For instance a hawk which has followed a rabbit to a bramble-patch may use its flight speed to 'throw-up' a few metres overhead, hanging for several seconds in a clumsy hover. If the rabbit's escape route takes it through a gap in the cover, the hawk can dive and seize it (Fox 1981). Next time the hawk misses prey, it will more readily repeat this, and may learn to circle overhead if there is no tree nearby. My hawk 'Foot' learned to 'wait-on' like this over *Brassica* fields in which he had spotted pheasants, because I learned to reward him by running to flush the prey. If the prey was a large cock pheasant, Foot would feint at it but not bind, because he had learned that the cocks were less easy to hold than the smaller hens.

If a hawk misses its quarry, it will 'take stand' on a suitable vantage point and can be called back to the fist for a titbit. If the bird is reluctant to fly down, it may be called to the 'lure' instead. For a Goshawk, this is often a rabbit skin, stuffed and weighted to be just too heavy for the hawk to carry and with a morsel of meat tied to one end. Hawks are not trained to stoop at the lure for exercise, as are falcons. A Goshawk's lure is used mainly as an ultimate temptation, twitched on a line through the grass under a bird which has perched in a tree and disdained the fist.

As a falconer becomes more confident, a Goshawk may be encouraged to fly at quarry from trees rather than solely from the fist. This 'Freie-folge' technique was developed in Germany and takes advantage of a tendency for wild hawks to follow predators such as foxes to exploit the prey they disturb (Hodder 1993). It can produce more spectacular flights than from the fist, with the hawk having a height advantage and also the chance to spot prey which would otherwise be missed. With rewards of an occasional call to the fist for a titbit, as well as quarry disturbed by the falconer, a hawk learns to follow from tree to tree along hedgerows and through woods, becoming a strong flier and experienced hunter. The main problem is that the hawk may spot suitable quarry some distance away, and fly off in an unnoticed attack. If the falconer is busy crashing through a bramble patch to try and flush rabbits, it is all too easy to miss the bird's departure until it is out of sight. I lost three Goshawks in that way before starting to use radio-tags.

LOCATION AIDS

The traditional aid on hawks flown free is a 1–2cm-long closed bell of beaten brass or monel. A pair of bells with a semi-tone between them may be mounted on fine leather 'bewits' round the legs above the jesses, but a bell may also be attached to the base of the tail feathers. A Goshawk moves its tail far more than its legs when feeding on a kill and a perched hawk will occasionally wag its tail from side to side. As a visible cue at a distance, the Naxi people in China attach a long white tail feather from a cockerel to the tail of their Goshawks (Gott 2000).

The start of radio-tracking wildlife (Le Munyan *et al.* 1959, Cochran & Lord 1963) included use on eagles (Southern 1964). Application for falconry followed rapidly and my first research tags were designed for that purpose, together with an RB–4 receiver named after the falconer Robert Berry who first bred Goshawks by artificial insemination (Berry 1972). That design by Bill Cochran was the first reliable receiver that became widely used for tracking wildlife, as the LA–12. Falconry is probably a larger market than research for radio-tags on raptors and will drive further developments that aid conservation, such as small tags with GPS to give effortless accuracy that would greatly benefit research on habitat use.

Radio-tags are a huge benefit in falconry, making it much safer than in the past to fly raptors, initially at hack (see below) and later for Freie-folge with Goshawks, to get them more fit and skilled than when flown mainly from the fist. Nevertheless, bells can be useful for checking the position of nearby hawks, hearing when they fly and locating them on kills or perches in cover, so that they can be recovered to the fist without anxiety. Goshawk behaviour can change state from relaxation to angst like the flick of a switch. It is not hard to upset a hawk, after which it may not come to food or tolerate approach for some time.

Moreover, radios can fail, especially when too little attention is paid to changing batteries in falconry tags or receivers. At these times, as when seeking quarry, it helps to understand behaviour of hawks and prey. Goshawks change track readily when hunting (Chapter 5) and rarely go far very quickly. It is often wise to wait a while where the bird was last seen, listening carefully for bells or for the sound of mobbing birds, perhaps climbing quickly to a better vantage point from which to look and listen and check the radio again, in case reception was temporarily poor. Having failed to kill, the hawk may wait a few minutes and then return to seek the trainer. If all else fails, it is a good idea to be at the same spot at dusk, in case the hawk was sitting still on a nearby kill. There is a particular moment at dusk, when the Blackbirds cease scolding and even the wind seems to still, when a lost hawk's bell may be heard in the distance as it leaves a hidden kill to fly to roost.

If the hawk is lost overnight, the best cue is invariably the cawing or sight of a dipping or circling crow or Rook. On a clear frosty dawn, a bell's faint tinkle may carry from more than 500m as the hawk rouses for the first time in the new day. However, a passage or haggard hawk may disdain both fist and lure by that time, especially if it killed the previous day. Such birds can become too wild to approach within 24 hours of loss, although eyasses have come to the lure or been picked up from kills more than a week later.

WHAT HAPPENED TO TRAINED GOSHAWKS?

A survey of British Falconers' Club members in 1970 showed that Goshawks had been one of the two species most frequently trained (Figure 122). Kestrels were the other commonly flown raptor because this species was recommended for beginners at a time of growth in numbers of falconers and was still available from the wild under licence in the late 1960s. Most Kestrels were released back into the wild within a year after training. The Peregrines, Merlins and Sparrowhawks had been obtained earlier under licence, and these too were more likely to have been lost or released in their first year than Goshawks, for which falconers faced the complication and expense of import. The mortality rate of trained eyass Goshawks was 26% in the first year and similar in subsequent years, comparable with the rates for Peregrines and lower than for Sparrowhawks. Ultimately, about 52% of Goshawks were lost or released, compared with 86% of Kestrels, 80% of Peregrines and 60% of Sparrowhawks (Kenward 1974).

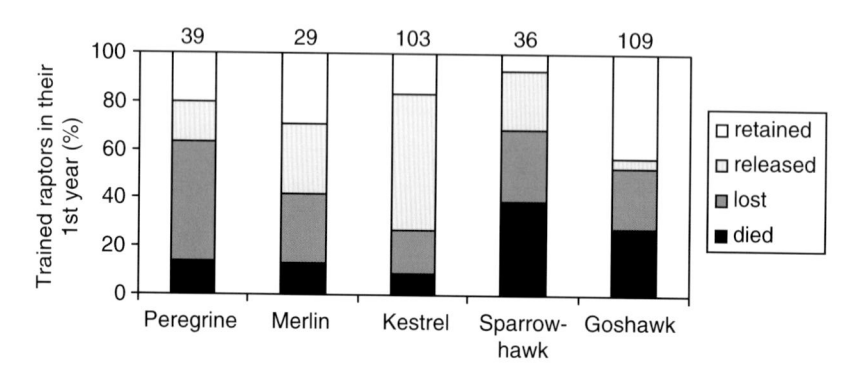

Figure 122. Goshawks were frequently trained by members of the British Falconers' Club before 1970, and the proportion lost or released in their first year of life was lower than for other species. Data from Kenward 1974.

The 1970 survey was a voluntary response from 113 of 400 BFC members, and therefore perhaps biased towards results from the most responsible falconers. A second survey was conducted in 1978 of all 143 members who had received 216 Goshawks from the club after the introduction of import licensing in 1970. Two hawks (1%) had died in transit and data were obtained for 90% in all. The proportions of hawks eventually lost was 30% and 3% were unofficially released. These rates were lower than the 37% lost and 13% released in the earlier survey, with first-year mortality also down from 26% to 22%.

Causes of death in the trained Goshawks could be compared with wild hawks on Gotland that were fit for detailed post mortem examination and had not been killed deliberately by humans (Figure 123). A third of deaths in trained and wild hawks were from trauma while hunting. However, wild hawks tended to starve whereas trained hawks were more vulnerable to pathogens and other clinical conditions.

Trained hawks that are incompetent hunters do not face starvation as in the wild, and parasites are controlled well by falconers. These results may under-estimate deaths from malnutrition in wild hawks, for which six cases with proventricular ulcers or high loads of parasites may have already been starving (Chapter 8), but also in the trained hawks because 12 died from 'fits', which can have multiple origins including hypoglycaemia (Cooper 1981). Such conditions have been a strong stimulus for veterinary work on raptors, as have natural pathogens. Aspergillosis accounted for six of ten deaths from pathogens in trained hawks, which may already have been infected in the wild.

The BFC's Goshawk imports in the 1970s, which came mostly from Finland, also funded experimental releases to investigate the role of lost and released Goshawks in re-establishing the British population. There were 22 hawks released, 14 with

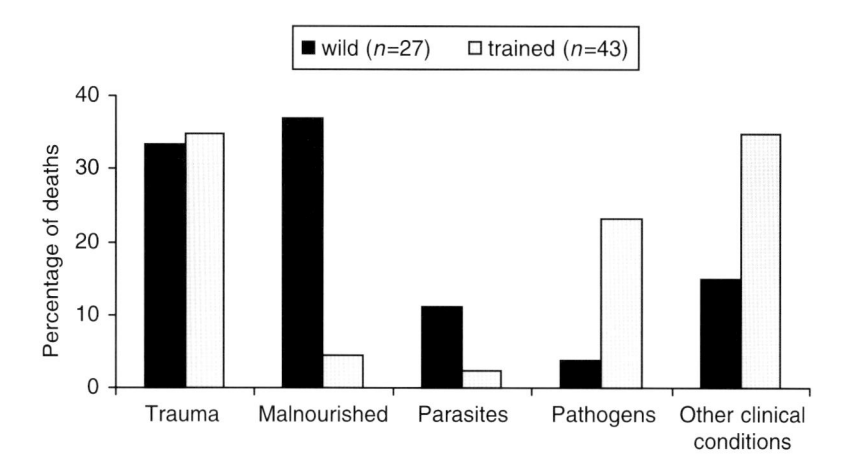

Figure 123. Deaths of trained Goshawks from trauma while hunting were as frequent as in the wild on Gotland, but less frequently associated with malnourishment and parasites. Data from Kenward 1981a, Kenward *et al.* 1999.

radio-tags. Three of these with prior experience in the wild survived, but one of the 11 eyasses was poisoned on a pheasant kill shortly after release. Among the other 10 radio-tagged eyasses, four survived without help until the first radio stopped after 17 days, as did three others, but only after being given food when they stopped active hunting; three died (two of them killed by foxes). Among 33 eyasses that were accidentally lost, 13 were known from subsequent sightings, recapture or death to have survived at least 20 days in the wild, again giving a minimum survival rate of 40% after loss of eyasses (Kenward *et al.* 1981b).

Radio-tagged hawks that made at least two kills required no further help, and it was relatively easy to rescue those that stopped hunting. Wild Goshawks increased the frequency of their flights as time passed after a kill (Chapter 5), but also eventually ceased moving if they remained unsuccessful. They sat with fluffed plumage conserving their energy (Plate 23), with a last chance of ambushing any prey that moved close enough. At this point they could be rescued by providing food below them, as we also did experimentally under cover of darkness for one hawk on Gotland. Thus, by using radio tags and knowledge of the expected hunting behaviour, a 'hard release' could be turned into a 'soft release'.

DOMESTIC BREEDING

There is a single record in the late 16th century, from a Buddhist monk in Japan, for a pair of trained Goshawks that produced and reared a male and a female chick (Hokiiche 1822). This seems to be the first record for domestic breeding of raptors. Goshawks were next bred in captivity by artificial insemination in North America (Berry 1972). Trained female Goshawks had laid eggs on a number of occasions (e.g. Mavrogordato 1937 and Chapter 3), and it was found that semen could be forced from male hawks in the same way as from poultry. Male Goshawks that are used to restraint for abdominal massage will give semen with very little pressure, and can even be persuaded to 'mate' voluntarily with a hand or waterproof clothing from which semen can be collected. After checking for sperm viability, a syringe and soft, flexible tube are used to insert semen into the oviduct. As the oviduct opens inside the cloaca, the female is usually restrained for artificial insemination, although that is not necessary with females that are very tame (Berry 1972, Schulz 1981).

The early 1970s were a time of huge interest in rescuing raptor populations from feared extinction through pesticide contamination. Domestic breeding was being developed rapidly by falconers, especially for Peregrines. The number of large falcons bred annually rose from fewer than 20 in 1972 to more than 200 in 1975 (Kenward 1976b). After four or five unsuccessful attempts at breeding Goshawks by natural mating, which frequently resulted in males being killed by females, young were produced in Britain, Poland and Germany in the 1970s (Bresinski *et al.* 1978, Fentzloff 1980, Görze 1981). The conditions conducive to production of young gradually became clear enough to be summarised in reviews by the Hawk Board (1988), Capp (1993), Döttlinger (1993), Kimson (1993) and Pöppelmann (1997).

Apart from the importance of factors general to all domestic breeding of raptors (excellent instructions are in Fox 1995), the key to success is the psychological condition of the male. He is easily intimidated by the female before he comes into full physiological condition for breeding, and then at best will not fertilise her and at worst will be killed. One solution has been to have a sufficiently large enclosure, with internal shrubs or other visual obstructions, for the male to be able to rest, and especially to pluck prey, out of her sight. The female can be wing-clipped by cutting five or six of her primaries quite short, so that she can still fly but is hindered from chasing the male (Haddon 1981, Pöppelmann 1997).

A safer approach than a single enclosure is to use two or three linked compartments, each typically of 2–3m across with a 5–6m side in common (Görze 1981). A height of 2.5–3m is adequate, with a space of 1m above perches for copulation (Henckell 1997). The compartments are linked by hatches through which hawks can see each other and display, and which are opened for them to meet only when a female has passed through her aggressive begging phase and the male is trying to give food to her. There is some debate about whether flight display with the male carrying food is important (Kimson 1993), and large interlocking ring aviaries with 5m height have been proposed as a safety measure (Grünhagen *et al.* 1999). However, Pöppelmann (2000) considers that modest linked compartments are adequate if the birds are selected and introduced correctly. Circular flights remain possible if the central partition has hatches at each end and the enclosure is large enough.

Timing the access of the hawks to each other seems to be critical. The male's breeding condition may be set back if he is with the female too early. On the other hand, if he is introduced late, the female may lay before copulation is properly synchronised. The female's cloaca appears to pulse and thus draw in semen in rhythm with her wailing during effective copulations (Bednarek 1997). He can be considered ready when he comes straight away for food and plucks it slowly and deliberately while 'chupping' and with tail coverts spread (Kimson 1993).

For added safety, the male may be separated again when laying is complete and her attitude to him becomes more aggressive again. Walter Bednarek (1997) describes increased aggression after laying but renewed tolerance around hatching, perhaps linked to changes in her hormonal or nutritional status. One male was separated after fertilising a female and then given access to another female, with which he copulated 20 times on the first day and again produced young (Henckell 1997). Both females were passage hawks. The male was 19 years old, as another proof that early senescence is no safe assumption in this species.

Goshawks tend to be especially sensitive to disturbance, so enclosures normally have solid sides and an open roof, perhaps covered for a 1–1.5m above the nest in one corner (Plate 12). In these 'skylight-and-seclusion' enclosures (Hurrell 1970, 1977), noises outside can cause a panic in which the male is at risk of attack. However, the sounds of visiting to feed or observe can be covered by continuous playing of radio broadcasts.

The open roof provides sunlight to help sterilise the area at a time when no visits are possible. Safe feeding is also important, with food that improves in quality towards laying. Professional raptor breeders tend to have their own colonies of rats and quail to ensure absence of disease and other contaminants. When the male is together with the female, she can be fed a small amount separately to engage her just long enough for the male to pluck food for her without risk from an impatient mate (Pöppelmann 2000).

Successful breeding seems to have occurred most often when the female is placid and unaggressive while the male is bold. This seems to be most likely if she had minimal contact with humans until around the age of dispersal, and thus beyond the age at which learned recognition ('imprinting') of the provisioner occurs, and was then put in an enclosure without being trained for hunting (Pöppelmann 1997). In contrast, it may be no disadvantage if the male has gained confidence and flight skills through being flown at quarry, or is even wild-caught (Döttlinger 1993).

However, female Goshawks trained for falconry can successfully mate naturally. One female kept weathering on a tether-line even laid fertile eggs after being mated by a wild male (Aufderheide 1997). An approach practical in a single enclosure, without the need for partitions, is to tether the female in this way, so that the male can avoid her, until she is prepared to copulate (Domingo Garcia Llano, pers. comm.)

Domestic breeding has permitted observations of Goshawk reproduction in more detail than possible in the wild. Bednarek (1997) provides drawings of how the down is first lost in a growing area round the cloaca, which must reduce risk of bacterial build-up in semen spilled during the frequent copulations. The bare area then extends forward as the brood patch, which is completely formed with the laying of the penultimate egg, which marks the start of full incubation. A most delicate interplay of hormonal secretion and physiological responses must be at work.

Breeding experiments have also indicated the great flexibility of learned recognition behaviour in Goshawks. A German hawk successfully reared two chicks of a domestic hen, although a third disappeared after getting blood from food on its head. The young hens roosted on a perch next to the Goshawk when old enough (Pöppelmann 1994). With such flexibility, the failure to recognise offspring from other broods after fledging (Chapter 5) becomes entirely understandable.

It is easy to overlook the difficulties of breeding, which is routine for relatively few species. In his Chairman's report to the Hawk Trust (now the Hawk and Owl Trust) in 1973, founder Philip Glasier wrote 'I realise there are many who have said and still maintain that the breeding of birds of prey that are suitable for falconry will never take place.' However, he could also include an Addendum: 'We now know that what might well be called the 'Big Break-Through' has happened. Cornell University have bred 20 Peregrines, some Prairie Falcons and some Lanners this year.'

Despite some success in developing domestic breeding of Goshawks, this species has not been bred for falconry to the same extent as many others. As the mass

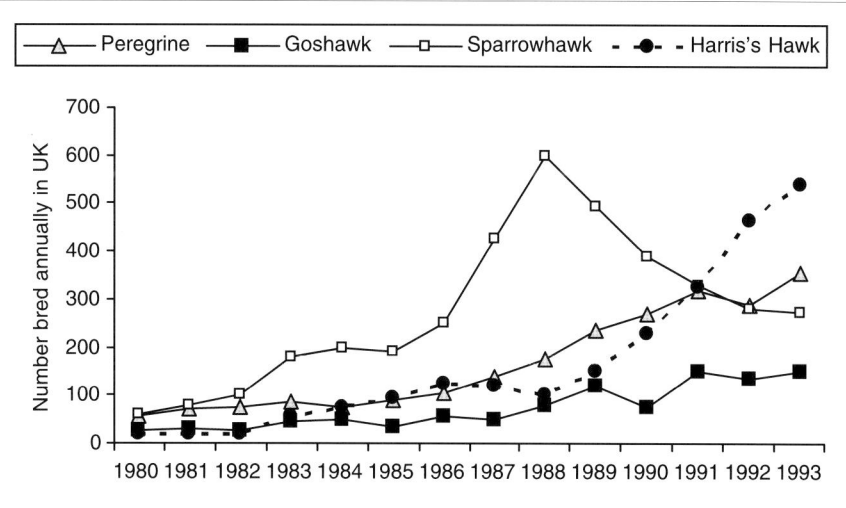

Figure 124. During the period of government registration of all domestic breeding of raptors in the United Kingdom, production of Goshawks developed more slowly than for other species favoured by falconers. Data from Hawk Board 1992 and Fox 1995.

breeding developed in the United Kingdom during the 1980s, government data on registration of domestic progeny were kept on all raptors until lack of evidence of laundering showed this to be unnecessary. Although about 150 Goshawks were being bred annually in the early 1990s (Figure 124), and the maintained registration of this species shows continuing production, domestic breeding of Peregrine Falcons and especially Harris's Hawks has increased to a much greater extent. In contrast, the production of Sparrowhawks peaked at around 600 in 1988, as did that of Kestrels at more than 1,000 in 1987 (not shown). Subsequently, fewer Sparrowhawks and Kestrels were bred because supply far exceeded demand and wild populations needed no re-stocking.

The placid temperament and unusually social habits of the Harris's Hawk, which occurs from southern North America into South America, make it popular for training. Prices reached more than £1,000 before breeding developed strongly. Without the need for multiple compartments and carefully timed separation, it is also much easier than Goshawks to breed. It is not surprising that breeders turned to this species and, with a ready supply, the Harris's Hawk has extensively replaced the Goshawk in Britain for flying in wooded country. Other countries are poised for the same situation. The German falconry club increased its annual production to 18 Goshawks in 1980, but was still producing only 18–36 in 1993–95 and 29–32 in 1996–98, compared with 98–121 Peregrines and a small but growing number of Harris's Hawks (Bednarek 1996, 1997, 1998, 1999). After breeding 117 Goshawks between 1989 and 1998, the BFC discontinued a 20-year national breeding scheme in 2002 (Fairclough 2003).

HACKING

Hacking is a process developed by falconers to mimic the natural post-fledging period, with provision of food for eyasses at an artificial nest site. It also sometimes involves the feeding of older birds at a site from which they are 'hacked back' to the wild. Young hacked raptors are free to develop their flying skills, and are then either recovered for training before they disperse or are left to disperse naturally in re-establishment projects. Hacking has become a highly efficient technique of choice for soft-release of young to re-stock or re-introduce many raptor species, including falcons, eagles, kites and condors (reviews by Sherrod *et al.* 1981, Cade 2000).

Before the advent of radio tags, falconers used hacking more for falcons than for hawks, because it is more difficult to keep an eye on young birds in woodland than in open 'falcon country'. Hacking is now used for Goshawks, although primarily as 'tame-hack' for early-contact eyasses, which can be given a great deal of freedom to develop fitness and agility, with reduced risk of loss because they are oriented towards humans (e.g. Jones 2003).

Goshawks which were obtained as passagers or haggards already have flying skills, and can be released without hack. They may go wild very quickly if nervous of humans. Wild Goshawks can also be unafraid of humans. A falconer in the Colorado Foothills set up a loft with 100 pigeons to enjoy the skilled attacks of adult male hawks, which were also trapped and banded. One even fed free on the fist without any training before flying off (Moran 1995).

Incapacitated wild Goshawks can therefore be released after treatment without hack if they are reasonably fit and in sound condition, which can be achieved by a training course. Holz & Naisbitt (2000) found that rehabilitated Australasian Brown Goshawks maintained weight better in the wild if flown with falconry techniques before release.

PREDATION MANAGEMENT

The release of trapped Goshawks at a distance is also an effective way to manage their predation on game-birds. Few Goshawks returned to the original capture site in central Sweden after being ringed and released more than 30km away (Figure 125). The large home-range of Goshawks makes it likely that release within 10km of a capture site is within sight of an original home-range for a hawk that soars. Some hawks released up to 30km away may either have recognised their location or have been guided back by geographic features such as lakes or the coast. However, those hawks transported beyond 30km appeared to survive well, as a high proportion of them were eventually recorded again, through recoveries by the general public and at other sites (Marcström & Kenward 1981b). There was a similar pattern, with the

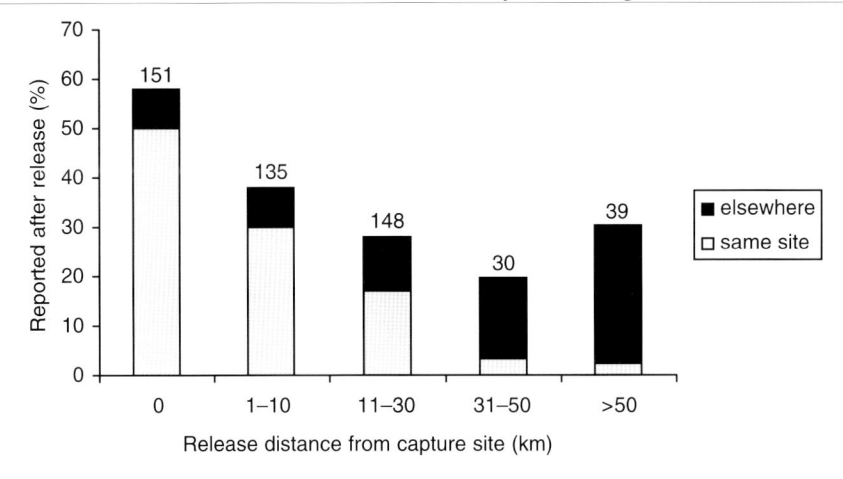

Figure 125. When Goshawks were released more than 30km from pheasant feeding sites in central Sweden during 1970–77, few returned. Release distance of 0km = on site. Values of *n* are shown above each bar. Data from Marcström & Kenward 1981b.

same sharp fall in 'same-site' recoveries at 30km, for 324 hawks relocated in southern Sweden, but with fewer recovered overall than further north.

Among 894 hawks trapped at pheasant sites, 82% were juveniles. Similarly, among 693 Goshawks caught at four estates in southern Sweden during the 1980s, 91% were juveniles (Neideman & Schönbeck 1990). In both studies, there was a slight tendency to catch more females among the adults, but 57–58% of the juveniles were male. With trapping starting in August, the median capture date for juveniles was in October in central Sweden, but in November in the south.

The preponderance of juveniles, especially males, and the timing of captures suggested that the hawks at pheasant feeding sites were mostly dispersers and may in many cases have been making a second movement after leaving the nest area. When hawks were trapped for radio-tagging in central Sweden at Frötuna, most were caught in September and October and several continued dispersing after being tagged. Three of four hawks that continued dispersing within a week of capture had been caught in compartment traps baited with pigeons (Figure 127), whereas all of nine hawks that stayed were caught at some point in a spring net on a killed pheasant (Kenward *et al.* 1983).

Thus, if hawks are to be relocated to reduce predation on pheasants, it is a good idea to trap them in spring nets and to move juveniles at least 30km. Adults should be moved further, because a small experiment with radio-tagging on Gotland showed that such birds could rapidly return from 30–40km away. Trapping in spring nets is not only selective, but easy to implement when presence of kills indicates predation. This occasional use is less work than maintaining live pigeons in traps for long periods. Compared with shooting, selection of target species is secure and injury is very rare (see Chapter 5 for trap details).

Trapping can also be useful for understanding and conserving Goshawks. It can provide data on physiology, movements and survival (Marcström & Kenward 1981b, Neideman & Schönbeck 1990). Perhaps most importantly for the future, if the trapping is done with no economic expense for conservation beyond licensing, it provides a cost-effective way to monitor the size of Goshawk populations by mark-recapture techniques. If hawks are being marked at nests, the size of the total population of juveniles is, in principle, the total number marked divided by the proportion of marked hawks among recaptures. The number of breeding pairs is then the juvenile population divided by the productivity per pair. Various corrections may be needed to this simple estimation. However, this approach gave a similar estimate to atlas mapping for the size of the total Goshawk population in Sweden (Svensson 2002), and to density-extrapolation for the population on Gotland (Kenward *et al.* 1999).

Nevertheless, trapping should not be a first resort for managing predation by Goshawks. Problems can also be reduced in the first place by diversion of hawk attacks, pre-emption and perhaps by deterrence. Based on work in Denmark, Germany and Sweden, feeding areas for game should have (i) dense ground cover 1–2m high, (ii) no trees suitable for perching Goshawks within 100–200m and (iii) a network of hedges and dykes to give game cover as it approaches the site (Kenward & Marcström 1981, Mikkelsen 1984, F. Ziesemer pers. comm.). Any trees giving additional cover and roost sites at pheasant release pens should be dense conifer thicket, for example of young spruce, and released poults should be as old as is practical (at least eight weeks).

Such habitat should completely deter predation by Buzzards, which mainly affects young pheasants (Kenward *et al.* 2001d), but Goshawks are bolder and more skilful and will kill pheasants right through the winter, especially during snow-cover (Chapter 7). To pre-empt their competition with hunters, it is a good idea to shoot game in the autumn rather than in mid-winter. To pre-empt reduction of the spring population of pheasants, it is also a good idea to trap surviving hen pheasants soon after the shooting season and to keep them through the hardest months in large enclosures (Göransson 1982), at least until they have laid first clutches and spring vegetation is well developed.

Deterrence with scaring devices at game release pens has been tried in Denmark, Germany, Sweden and Britain, but without clear success (Kenward & Marcström 1981, Mikkelsen 1984, Harradine *et al.* 1997). Nevertheless, there is scope for further innovatory work on acoustic and taste-aversive techniques (Musgrove 1996, Kenward 2000, Nicholls *et al.* 2000), perhaps using radio-tagged and trained hawks, as suggested by Mikkelsen (1984), to monitor the response of individuals to specific measures. Another suggestion is that reduction of medium-sized predator (meso-predator) populations by top predators may reduce predation on game stocks (Tapper 1999).

Intensively managed poultry is enclosed and therefore not at risk from Goshawks. However, owners of poultry that is truly 'free-range' are liable to suffer losses (Figure 126), especially to the small breeds (e.g. bantams) that are the most efficient

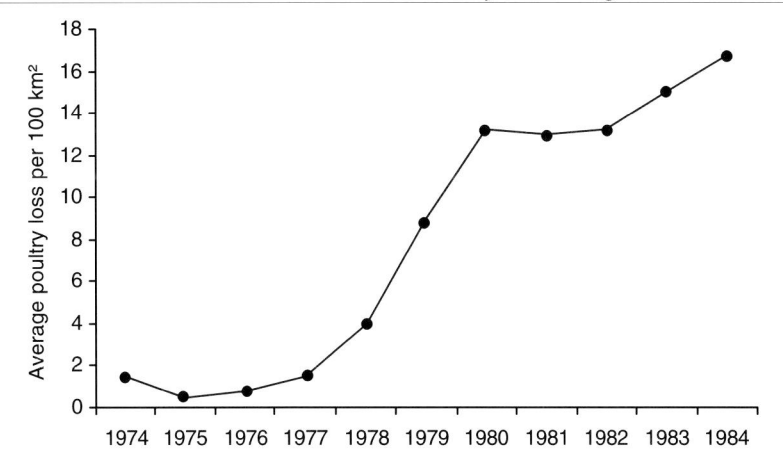

Figure 126. Compensation paid for loss of poultry to goshawks in 2 Swiss cantons increased after hawks were protected and pesticides restricted in the early 1970s. Data from Bühler & Oggier 1987.

converters of household scraps into eggs. Holding large breeds of chicken and improving cover may reduce problems in areas where risk from Goshawks is high. If a few individual hawks become poultry killers despite such measures, the use of enclosures, compensation or relocation will depend on social and economic priorities.

It is worth remembering that the predatory abilities of Goshawks can be a benefit as well as a disadvantage, at least if the hawks are trained. Trained Goshawks have been used to deter gulls from airfields (Arendonk 1980), where other scaring methods are most effective if reinforced by raptor attacks, although most such work is done with falcons. Goshawks have also proved useful for clearing rabbits that were causing damage estimated at £130,000 annually to an oil refinery in Germany, in a situation unsuitable for guns. A team of volunteers used three hawks to remove 650 rabbits in two years, by which time most burrows were unoccupied (Saar *et al.* 1999).

IMPLICATIONS FOR CONSERVATION

Goshawks have been removed from areas to try to save remnant populations of Black Grouse in central Europe. Goshawks are an important predator for Black Grouse (Chapter 7) and may be critical for survival of small grouse populations where high Goshawk density is enabled by other prey species (Chapter 8). However, removing Goshawks from an area with enough habitat for a viable grouse population is a more demanding task than managing predation in small areas where game are released or fed. The expense, as well as the social acceptability, may not be sustainable in the long term. At a site in Baden-Württemberg, the removal of 98 hawks was criticised because they were thought to have been killed instead of relocated (Dobler & Siedle 1993). In

Schleswig-Holstein, 742 Goshawks were removed in 1980–93 from one area (with 10% given to falconers and the rest killed) without saving a grouse population (Busche & Looft 2003). On socio-economic grounds, remnant populations of very vulnerable prey can probably only be secured in the long term if an area can be managed to reduce predator success with minimal need of predator removal.

The socio-economic issues of managing predation by Goshawks will be addressed again in Chapter 10. However, it is worth noting here that relocated Goshawks were important for re-establishing a Goshawk population in Britain in the late 1960s and 1970s. The most successful releases were of birds trapped to conserve game in Sweden, which were arranged by falconers David Kent and Russell Coope in areas bordering Wales in the late 1960s. Earlier in the 1960s, hawks from central Europe had been released in the north of England by Sergeant Bill Ruddock, who provided some for falconry and released the older ones. With the advent of import licensing in 1970, there were 75–140 hawks entering the United Kingdom annually for falconry until 1978 (Marquiss 1981). It was estimated, using rates of loss and minimum survival after loss, that 15–33 of these were entering the wild in Great Britain each year (Kenward *et al.* 1981b). There was also an official (licensed) release from British Falconers' Club imports, which established small breeding populations in two areas.

Mick Marquiss and Ian Newton (Marquiss 1981, Marquiss & Newton 1982b) documented the resulting increase in the British Goshawk population from nine to 59 pairs between 1970 and 1980. The number of areas containing one or more pairs also rose during this period, from four in 1970 to a peak of 18 in 1977. The most rapid expansion in numbers and areas was in 1975–77, correlating with peaks of imports two years earlier (Marquiss 1981). The location and timing of the re-establishment also indicated that imports rather than natural immigration were the main source of colonists. The hawks settled mainly in the north and west of Great Britain, where many were lost by falconers or released, rather than in the south and east, where the sea crossing from mainland Europe is shortest. Moreover, nine of the established breeding birds were seen to carry the remains of falconry equipment. The size of second primary feathers (P2) from the earliest settlers resembled those of central European hawks, as imported by Bill Ruddock, whereas settlers in the 1970s had larger feathers typical of Goshawks from Scandinavia. It is therefore fairly certain that the re-established British Goshawks crossed the sea in aircraft rather than under their own power. It is hard to imagine more cost-effective conservation than the schemes that used the finance and skills of falconers, working with other biologists, to re-establish the Goshawk in Britain.

A final lesson from this chapter is the way in which restrictions may unexpectedly channel human resources in directions of reduced benefit for conservation. This chapter has looked at groups for which wild Goshawks can become a hands-on exercise, as falconers, rehabilitators, hunters, poultry keepers, pest-controllers and researchers. In some countries, where bird-watching has become a major recreation, it has also become a business that employs many people to support that interest by working for birds. The hands-on people then tend to become a minority. Actions of the minority, which may be based on strong motives that can benefit or harm

Figure 127. A range of trapping equipment. Clockwise from left: a falling-lid trap; a 'Swedish' Goshawk trap; a Liljefors trap; a falling-end trap; a spring net; a bal-chatri noose trap; and a butterfly trap. All are set. See pages 116–118 for their use.

Goshawk populations, become subject to the lobbying power of the majority with the regulating authorities. If the actions of the active minority are to be beneficial, the majority needs understanding when applying its power.

Traditionally, falconers obtained raptors from the wild, trained them and later released them. A number of countries, notably including the USA, still practise falconry in this way. In Britain, few imports of Goshawks were permitted during the 1980s and 1990s. The lack of access to imported wild Goshawks stimulated the domestic breeding of short-winged hawks for hunting in wooded country. Effort concentrated initially on producing Northern Goshawks, until it was realised that Harris's Hawks were easier both to train and to breed.

The regulatory system in Britain has been liberal for falconry after adoption of domestic breeding. The strength of this approach is that it has permitted growth of an interest whose devotees are above average in environmental awareness and responsibility (Peyton *et al.* 1995), which provides substantial funding and volunteer effort for conservation and generates selective, low-impact use of game resources (Kenward 1987b, Fox 1995, Kenward & Gage in press). Development of expertise in domestic breeding is also an opportunity for conservation, by providing an insurance against problems that are sometimes detected so late in wild raptor populations that few individuals are left, as in the case of Mauritius Kestrels, Californian Condors and *Gyps* vultures in southern Asia. The kestrel and condor programmes, and six of seven major release projects for Peregrines, were all run by falconers (Saar 1988, 2000, Jones *et al.* 1994, Trommer *et al.* 2000, Wallace 2001, Cade & Burnham 2003).

The extensive flying of Harris Hawks instead of Goshawks does not risk introduction of an invasive alien species to Britain, because the species is frost-sensitive and, above all, so easily recognised that in the improbable event of opposite sexes surviving loss in the same area at the same time, they would be easily spotted and removed. However, obliging falconers to depend on domestic breeding does not use their funding and volunteer effort optimally to conserve wild stocks of species, such as the Goshawk, that are not rare in the wild but need monitoring to ensure the health of their populations.

CONCLUSIONS

1. Falconry started in eastern Asia two to four millennia ago and reached Europe with the Germanic tribes, remaining popular and protecting wild raptors there until the advent of modern firearms and game management turned a former social respect for raptors into persecution.
2. Falconry's history is more continuous in local agrarian societies from north Africa across Asia to northern China, where many early methods developed for training and veterinary care of raptors.
3. Falconry had a 20th century renaissance in Europe and transfer to North America. Falconers developed new techniques for domestic breeding and managing raptor behaviour to help restore populations that were decimated by pesticides, but also encountered concerns for raptor conservation.
4. Restrictions on obtaining raptors from the wild in some countries encouraged domestic breeding. Before 1970, Goshawk was the species most commonly used for hunting wooded country in Britain, but the species is difficult to breed domestically and is now trained by a minority.
5. Hacking, a soft-release method developed by falconers to encourage flying skills and fitness of young raptors by mimicking the post-fledging dependence period, is used widely to release young raptors and sometimes prior to training Goshawks.
6. Trained Goshawks can be useful for research and to manage species causing economic damage. Falconers are a rich source of volunteers for research and conservation. Regulations based on mentoring, with markers and DNA tests if needed, provide tools for managing this human resource.
7. Problems with predation by Goshawks can be managed by exclusion, diversion, pre-emption, deterrence and relocation. Goshawks at game feeding sites can be hard to discourage, but they can be trapped safely with spring nets and seldom return if released more than 30km away.
8. Relocation by falconers was an extremely cost-effective way to re-establish Goshawks in Britain. However, future benefits for conservation from 'hands-on' enthusiasts need better understanding by conservation majorities based on bird-watching.

CHAPTER 10

Conservation through protection and use

In a perfect world, this chapter would start with a heartening story about observing the first Goshawks breeding near my home in Dorset, hearing their calls and climbing the tree to mark the young. The idea of reaching agreement with enlightened and tolerant local landowners that Goshawks could be re-established here is seductive.

Sadly, it is only a dream, although the radio-tagging of Buzzards has proved that local land managers are enlightened and tolerant. Intolerance is applied to only a small minority of Buzzards that are most persistent at pheasant pens. A loss of 5% is far too small to reduce appreciably the 75% of birds which do not breed each spring, or to stop the trickle of eastward emigrants that helps to rebuild Buzzard numbers across England. However, the possibility of seeking agreement for having Goshawks locally is prejudiced by an unresolved issue with Hen Harriers and Red Grouse in Scotland.

Fifteen years ago, I walked across hills in Argyll with research friends from a large conservation organisation. They had asked me to help fit back-pack radios to

Golden Eagles. We talked with regret about how Goshawks were being killed at pheasant pens when they could be trapped and moved to the parts of Britain without Goshawks. There had recently been a courageous step by conservationists of agreeing to wider use of live-traps for corvids. Perhaps relocation of Goshawks could also be permitted by the authorities. We agreed that this would be worth examining within the following five years.

There was also in Scotland at that time a project to investigate the predatory impact of harriers on grouse. Grouse-shooting motivates preservation of moors and heaths in northern Britain. The study found that when harriers were carefully protected, their numbers increased and they took too many grouse for shooting to remain economic (references in Appendix 2). A proposal to remove some harriers to re-establish them in parts of Britain without grouse was not accepted. Fear of encouraging new persecution produced strong positions against permitting translocation of raptors. Another northern landowner reversed a decision to have Goshawks studied on his land, which prevented a project to see if Goshawks could be deterred by aversive chemicals on young pheasants.

Although grouse-shooting was no longer an economic use of the area where harriers had been studied, the landowner there was interested in what would happen if it was left without game-keepers (which could no longer be afforded) instead of immediately being turned over to forestry or for raising sheep. Without game-keepers, foxes were unconstrained and the ground-nesting harriers declined again in numbers, but the grouse did not recover.

For us to enjoy Goshawks in Dorset again, landowners and other conservation interests need trust-building agreements. There is already much cooperation at local level, notably to restore heathland. However, any management of raptors must be agreed at national level. It remains to be seen when that might be possible, but perhaps this book can help.

COOPERATION OR CONFLICT?

Goshawks are a key species for wildlife conservation. This is not only because they are vulnerable to many of the pressures that humans can exert on wildlife, but also because they are robust to those pressures. They can usually withstand pollution, persecution, predation pressures and prey deficit. People either cause these pressures directly, or we exacerbate them through our many different activities, especially our ever-changing use of land. Yet Goshawks remain widespread. As a well-studied and abundant species that interests birdwatchers, falconers, game-managers, scientists and landowners, the Northern Goshawk is a good test of our capability for pragmatic conservation.

This chapter examines the factors, including human activities, that can cause Goshawk populations to decline. It then considers how recent international conventions, which are implemented as national laws and management practices, can be used to improve the conservation of Goshawks and related wildlife.

PESTICIDES

Through the 1950s and 1960s, observers noted a decline in Goshawk numbers in parts of central Europe, followed in most cases by a recovery starting in the late 1970s. The best documented example is that of the Dutch Goshawk population, as reviewed over the years by Thissen *et al.* (1981), Bijlsma (1993) and Rutz *et al.* (2006a).

Thissen *et al.* (1981) considered that Goshawk numbers in The Netherlands had increased gradually before the 1950s, due partly to new woodlands, partly to reduced killing of hawks during the 1939–1945 war and partly to growth in Feral Pigeon populations. Then came a crash. Maarten Bijleveldt (1966) reported a decline from about 125 pairs in 1958 to as few as 20 pairs by 1963, with four of 20 clutches being broken. Dutch conservationists and a German falconry club (Deutscher Falkenorden) tried unsuccessfully to boost the population by releasing hawks from Germany (Bijleveld 1966). It was unlikely that persecution or food shortage had caused such a sudden, drastic population decline. However, there were simultaneous declines in Kestrel, Buzzard and Sparrowhawk populations, which were linked to high levels of DDT-metabolites and Dieldrin in eggs and dead raptors (Koeman *et al.* 1968, 1969, 1972).

DDT (Diethyl-Dichloro-Toluene), which was widely used as an insecticide in the 1950s and 1960s, causes shell-thinning, egg-breakage and death of embryos. A reduction in thickness of more than 17% was associated with population decline in several raptor species. This level of thinning occurs when eggs contain 15–20ppm (parts per million) of DDT-metabolites in fresh weight (Cook *et al.* 1982, Newton 1979, 1998). Organochlorines like DDT accumulate in fat and are lost from the body relatively slowly, except as egg lipids (Newton *et al* 1981). The contaminant levels in prey, which are passed to raptors, depend on what the prey has eaten. There is therefore concentration of organochlorines along food chains, with raptors getting the highest doses, especially those at the end of lengthy aquatic food chains.

Goshawks in other parts of north-central Europe showed signs of DDT contamination. During 1967–1979 in Schleswig-Holstein, 163 of 819 eggs (20%) failed to hatch and 49 of the eggs that failed were broken (Looft & Biesterfeld 1981). These egg losses were double the 10% that seems typical for Goshawks (Chapter 4). Moreover, some eggs in Schleswig-Holstein had noticeably thin shells and at least one was eaten by a female (Looft & Biesterfeld 1981).

Although material to analyse for DDT and its metabolites (DDE & DDD) was not collected until DDT was banned from agricultural use in West Germany in 1972–4, data from Looft's study area in Schleswig-Holstein showed that some eggs still contained more than 20ppm in the 1970s (Figure 128). Moreover, while average DDT levels halved in the six years following the ban (Baum & Conrad 1978), the average levels at the start of the decline suggest that earlier DDT levels in Goshawks had probably been at least around this threshold for population decline (Figure 128).

280   The Goshawk

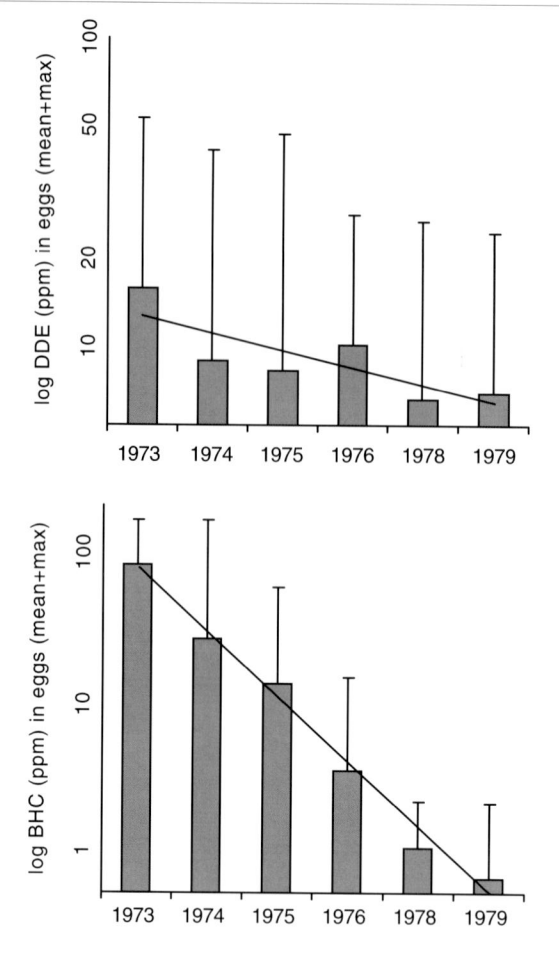

Figure 128. Levels of DDE (the main metabolite of DDT) and benzene hexachloride (BHC) in Goshawk eggs from Schleswig-Holstein declined during the 1970s from thresholds for population impacts, 15–20ppm for DDT and 100 ppm for BHC. Data from Baum & Conrad 1978.

Another organochlorine measured in the Goshawk eggs from Schleswig-Holstein was benzene hexachloride (BHC), also sometimes called hexachlorobenzene (HCB). This was widely used as a dressing for cereal seeds, and would have reached Goshawks through feeding on pigeons. BHC is not acutely toxic below 100 ppm fresh weight, and is thought not to have affected raptor populations adversely, but was clearly prevalent in Goshawks before it too was banned in the early 1970s (Figure 128).

During the 1970s, average shell-thinning among Goshawk eggs from Schleswig-Holstein was only 7% (Conrad 1981), not enough to cause population decline, and the same was true in Scandinavian countries (Anderson & Hickey 1974, Nygård

1991). Goshawk eggs generally contained lower residues of DDE than smaller accipiters that eat insectivorous birds (Snyder *et al.* 1973, Marquiss & Newton 1982b, Frøslie *et al.* 1986, Elliott & Martin 1994). Sparrowhawk eggs in Germany contained 10–20 times the levels of DDT found in Goshawk eggs from similar areas (Bednarek *et al.* 1975).

DDT had most impact in areas with much agriculture. In Britain, Sparrowhawks and Peregrines were more contaminated and less successful at nests close to farmland than in more remote areas (Newton 1979), and the same was true for German Sparrowhawks (Bednarek *et al.* 1975). DDT was also used most intensively as an insecticide in the orchards and vineyards south of Schleswig-Holstein (Ellenberg & Dietrich 1981). A study in Hessen experienced total brood failure of Goshawks in several years, in some cases because of broken eggs (Kollinger 1975). The situation was similar in Poland, where tree-nesting Peregrines, which would have shared many prey species with Goshawks, rapidly became extinct. Pielowski (1968) described deterioration in breeding success in the Kampinoski National Park, near Warsaw, from two young Goshawks per pair at ten nests in 1956–7 to only 0.6 young at the same nest sites in 1964–5. Goshawks almost disappeared by 1967 in the agricultural parts of central Switzerland, but had recovered somewhat by 1981 (Oggier 1981), whereas in the Alpine valleys the populations remained fairly stable throughout.

Nevertheless, the evidence for severe impacts of DDT on Goshawks is inconclusive. Eggs were not sampled widely until populations had started to recover. Some collected in Belgium had their shell-thickness reduced by only 13% in the 1950s and 10% in the 1960s (Joiris and Delbeke 1985). No study found Goshawk eggs thinned by as much as 17% and thus showed that DDT was entirely responsible for a population decline. However, residual high levels of DDT metabolites during the 1990s in hawks from the former East Germany (Kenntner *et al.* 2003) suggest appreciable impacts there during the height of DDT usage.

The other important organochlorine contaminants in Goshawks were cyclodienes. Dieldrin and the related cyclodiene Aldrin were introduced in the mid-1950s as seed dressings, and were therefore likely to have reached Goshawks in quantity via pigeons. Cyclodienes are more toxic than DDT: crashes in British Sparrowhawk and Peregrine populations were associated with their introduction rather than with DDT alone. Liver levels of cyclodienes above 10ppm in fresh weight are considered lethal and lower levels can reduce egg hatchability (Cook *et al.* 1982). The most telling evidence from The Netherlands was the finding of 17 to 44ppm of Dieldrin in the livers of three Goshawks (Koeman & van Genderen 1975), well above the 10ppm lethal level. The average Dieldrin level in livers of 24 Dutch Buzzards in 1968/69 was 7.6ppm (Fuchs & Thissen 1981), which shows that there was serious contamination of the Dutch environment with this seed dressing. It would be interesting to see if the severity of declines of Goshawk populations in different European countries can be linked to the volume of cyclodiene seed-dressings used in each.

In Scandinavia, the main pesticide problem for Goshawks from agricultural chemicals appeared to be poisoning with methyl mercury seed dressings, which were used against fungal attack. The initial suspicion that these caused neurological symptoms and death in Swedish raptors (Borg *et al.* 1969) was proven experimentally using captive Goshawks (Borg *et al.* 1970, Johnels *et al.* 1979). There was a story that a Swedish taxidermist had been eating the bodies of hawks he had prepared and suffered tunnel vision before the secondary poisoning was recognised, but this tale may have been apocryphal. Levels of water-soluble methyl mercury declined rapidly in agricultural areas after bans on use in seed dressings, but run-off from agriculture led to accumulation in aquatic food chains (e.g. Nygård 1997). Thus, although very high levels were found in a few juvenile Norwegian Goshawks that may have been submitted for analysis from agricultural areas, larger samples of adults showed maximum levels later in coastal areas (Figure 129).

Thus, cyclodienes and mercury were an additional source of mortality for Goshawks in Europe mainly during the 1950s and 1960s, with DDT also causing brood failures in some areas. However, residue levels fell after restrictions were placed on widespread use of these substances. Raptor populations recovered rapidly in the worst affected areas (Thissen *et al.* 1981, Rutz *et al.* 2006a). The impacts of organochlorines on Goshawk populations were less extreme than for Sparrowhawks and Peregrines, which were extirpated by these pesticides in large parts of Europe (Ratcliffe 1980, Newton 1986).

The dangers from organochlorines and some other pesticides are now well recognised, but there remains a danger of environmental toxins from new sources. From the 1980s, it took the *Gyps* vultures on the Indian subcontinent little more than a decade to decline to perhaps one thousandth of their former abundance. It took several more years to identify the cause as a non-steroidal anti-inflammatory drug (NSAID) called Diclofenac, which was being used widely to treat domestic animals but caused lethal kidney damage in vultures (Oaks *et al.* 2004, Riseborough 2004). Restoring the vulture populations will take much longer.

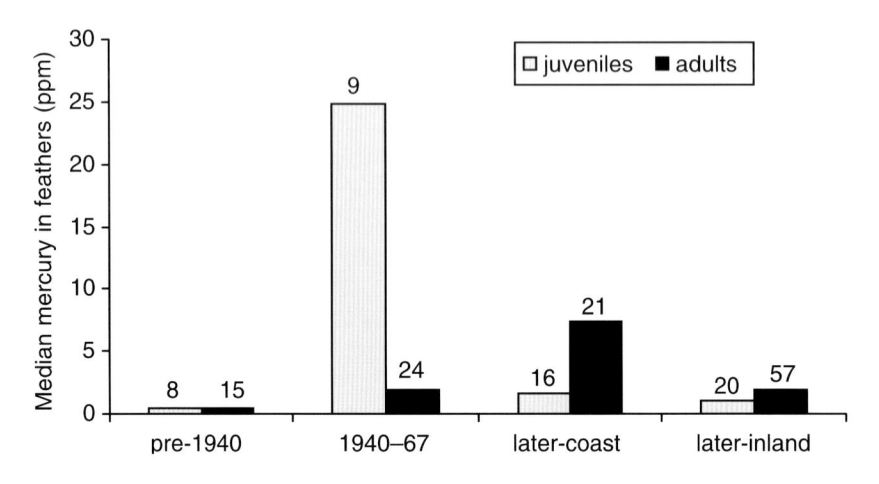

Figure 129. Mercury in feathers of juvenile Norwegian Goshawks peaked in 1940–67 but later in adults from coastal areas. Values for *n* are included above each bar. Data from Nygård 1997.

Two lessons from Diclofenac are that substances can vary greatly in toxicity between species and that unmonitored populations may decline greatly before problems are detected. The same lessons applied with DDT. A third lesson is that careless attribution of blame diverts attention and causes conflict instead of cooperation to solve problems. Accusations of deliberate killing of vultures are still being levelled against several groups (Satheesan 2005), just as the decline of Peregrine populations was initially attributed in error to falconry rather than to organochlorines. This mistake has been continued for years (Rockenbauch 1998), apparently to give the impression that protection efforts rather than restrictions on DDT enabled recovery of Peregrines in Germany and to deny credit to falconers for restoring northern and tree-nesting populations (Müller 2000).

To reduce risks of a 'Diclofenac for Goshawks' (or other raptors), it would seem wise to ensure both widespread monitoring and the maintenance of expertise in domestic breeding. The focus should probably be on monitoring, now that the basics of breeding Northern Goshawks are understood. Maintenance of hands-on expertise might best be applied for *ex-situ* populations of closely related species with small and localised populations, such as Henst's and Meyer's Goshawks.

Dark clouds often have silver linings, and this may be the case for past problems with pesticides. After the environmental threats from persistent organic pollutants and heavy metals had been recognised, interest developed in long-term monitoring of their presence. Hermann Ellenberg and Jörg Dietrich (1981) pointed out that the Goshawk was an ideal bio-monitor, because it was sedentary, widely distributed geographically, accumulated pollutants but was robust to them and therefore reasonably common, and foraged over large areas for many species, thereby sampling pollution widely. Samples could be obtained from eggs for fat-soluble organic pollutants, and from moulted feathers for heavy metals that bind to the sulphur-rich amino acids in plumage (Ellenberg *et al.* 1986).

The research group found that variation between age, sex and moult dates of feathers could be minimised by analysis of the first three primaries of breeding females, which are moulted in close succession (Chapter 4). Eddie Hahn (1991) then showed that metals in feathers were accumulated mainly from the air rather than through the diet, such that variation in metals between feathers was explained by differences in their exposure. For example, middle primaries were most contaminated because they sweep through most air. Concentrations in primaries 1–3 correlated very strongly with annual mean data from atmospheric sampling stations for lead (Figure 130), and also for cadmium and copper.

Recent work from several parts of Europe shows that Goshawks maintain low but readily measurable levels of organochlorine contamination. This includes studies in Germany (Kenntner *et al.* 2003, Scharenberg & Looft in press), Norway (Herzke *et al.* 2002) and Spain (Mañosa *et al.* 2003). Local variability, linked to likely soil residue levels (Kenntner *et al.* 2003), confirms that the species is suitable for biomonitoring of these pollutants as well as heavy metals.

Pollution may also have indirect effects on Goshawks. For example, trees in some parts of Europe have been killed by acid rain, a product of atmospheric emissions. In

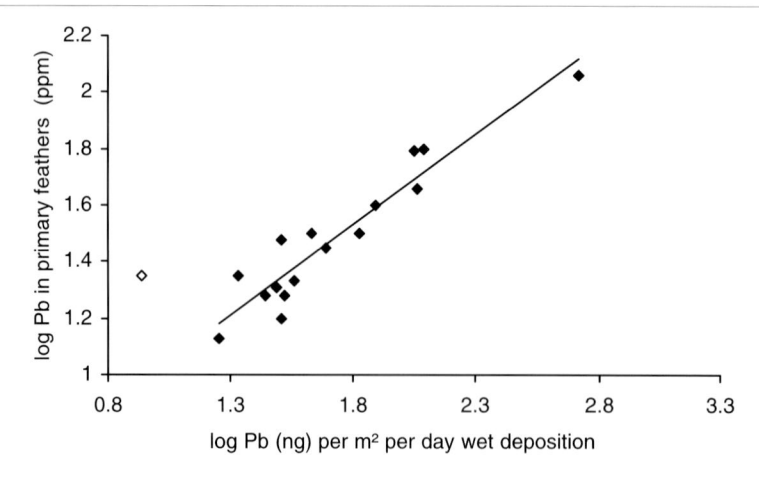

Figure 130. Lead (Pb) in primary feathers 1–3 of breeding female Goshawks was strongly related to records at nearby wet deposition monitoring sites. Data from Ellenberg *et al.* 1986; trend line omits a nest in mountains 30km from the nearest monitor site.

such areas, Anita Gamauf (1988) found that Goshawks tended to avoid use of damaged trees to a greater extent than Buzzards but not as much as Sparrowhawks. The impending effects of other atmospheric emissions on global climate may impact Goshawks mainly through changes in prey populations, unless land areas are reduced greatly by rising sea-level.

PAST PERSECUTION AND PRESENT ILLEGAL KILLING

We saw in Chapter 8 how the impact of Goshawk predation could vary from the killing of a few prey individuals to the taking of more than the annual production, thereby causing population decline and, in heavily modified environments, perhaps even local extinction. The same can apply to the killing of Goshawks by humans. Persecution, as a term applied to human populations, implies malicious intent and is appropriate for times when there was a desire to extirpate predators rather than merely defend against depredation.

Sadly, extirpation is a cost-effective management technique. Far less work is needed to kill occasional colonists than to reduce numbers in a healthy breeding population. During the era of intense persecution, there were official bounty schemes for killing Goshawks and other raptors in many countries, and gamekeepers proudly recorded their prowess. State bounties continued to be paid for killing some raptors until 1952 in Alaska, 1957 in The Netherlands, 1963 in Norway and 1968 in Austria and Australia (Newton 1979).

Despite generally heavy persecution, the Goshawk became extinct only in the British Isles. It probably became extinct as a regular breeder in England, Wales and Ireland not long after 1800, as 19 records in the *Zoologist* (now the *Journal of Zoology*) between 1841 and 1888 mark it as a great rarity. These records (Appendix 2) include only three adults, or five of 20 birds if credence is given to a second-hand report of blue eggs from raptors nesting in an ivy-covered oak on the edge of a plantation (Ranson 1863). All but two records were from east-coast counties, with half from Norfolk (including two from boats off that coast) and suggest vagrants from the European mainland. Breeding Goshawks may have lingered longer in Scotland, because records from the estate of Glengarry claim the killing of 63 Goshawks during 1837–40, along with 98 Peregrines, 275 Red Kites, 27 White-tailed Eagles and 18 Ospreys (Richmond 1959).

With Goshawks present throughout Great Britain (England, Scotland and Wales) at least in Mediaeval times, it is unlikely that they would not have crossed to Ireland, and the falconry literature seems to confirm this (Blome 1686, Campbell 1780, Harting 1867). However, they were probably gone by 1850. Records from Wicklow and Tipperary in 1870 (Brooke 1870, 1871) and Tyrone in 1919 (Williams *et al.* 1919) were reported as representing vagrants from North America, although the distance is large compared with that to mainland Europe, and three English records for 1869–71 suggest an influx at that time.

There is no evidence that the Goshawk was more than an occasional breeder in the British Isles (including Ireland) between 1900 and 1965. There is a surprising (and therefore perhaps mistaken) report of breeding in Gloucestershire in 1904, with continuing records of vagrants in Norfolk (Vincent & Wormald 1943, Nisbet & Smart 1959). A nest was also reported for Cheshire, and a displaying pair in Nottinghamshire, although no nest was found (Raines 1946, 1956). Recolonisation by two breeding pairs in Sussex in the late 1940s failed, with the disappearance of the hawks shortly after it was reported that they were eating mainly pigeons and pheasants (Meinertzhagen 1950, 1959).

The White-tailed Eagle, Osprey, Honey Buzzard and Marsh Harrier joined the Northern Goshawk in extirpation from the British Isles during the 19th century and many other raptor populations were reduced to remnants (Newton 1979). Although persecution and collecting were critical for them, loss of habitat was probably a major predisposing factor, for Goshawks at least. By 1900, only 2–4% of Britain remained wooded.

In more wooded parts of Europe, persecution may have reduced Goshawk breeding populations but did not cause extinction. The Norwegian Goshawk population was large enough to sustain the shooting of some 4,000 hawks a year between 1881 and 1931 without obvious decline (Sollien 1979). The recorded destruction of Finnish Goshawks, some 5,800 annually between 1964 and 1975, was even greater than in Norway but again was not associated with any obvious population decline (Moilanen 1976). Saurola (1976) estimated that 30.8% of Finnish Goshawks were killed in their first year, with first-year mortality of 63–64% (Haukioja & Haukioja 1970, Saurola 1976). The population model for Gotland

(Chapter 8) was able to accommodate greater first-year mortality (67%) with a breeding rate in older hawks of only 85% (Kenward *et al.* in press).

Goshawks were probably less able to tolerate historical levels of persecution when also faced with pesticide contamination. In Bavaria, Rust (1977) and Link (1977, 1986) noted declines in the number of breeding pairs, especially in the least wooded areas (i.e. with most agriculture), but there was recovery in the late 1970s (see Figure 110 in Chapter 8).

Since the complete protection of German Goshawks in 1971 was followed within three years by bans on the use of organochlorines in agriculture, and then by increases in Goshawk populations, there has been controversy about whether persecution or pesticides caused the original decline (Kalchreuter 1981b). Deliberate destruction of nests was apparently as high as 43% during the 1970s in at least one study area (Link 1986), and killing of adults continues (Bezzel *et al.* 1997a). However, Goshawk population declines during the 1960s in the less wooded areas of Bavaria and elsewhere in Germany were probably not due solely to deliberate killing.

The shocking population declines of some raptor populations during the pesticide era initiated a marked change in public attitudes to raptors. Protective legislation and education through books, films and displays of live raptors have made extirpation undesirable in modern societies. There are no longer bounty schemes. Hunters in most of the developed countries resent continued accusations of persecution and seek to dissuade their less informed colleagues who use raptors for target-practice. However, there is still appreciable illegal killing, aimed at preserving game and poultry. This can involve use of poisoned bait that may be targeted at other predators, such as foxes, but may also involve deliberate removal of individual raptors at sites where game-birds are fed.

Unfortunately, both these processes still have an impact on British Goshawks. Following the reintroductions in the 1960s and 1970s, the British Goshawk population did not expand as fast as during the post-pesticide recovery in The Netherlands (Figure 131). By 1977 there were 18 separate areas occupied by Goshawks in England, Scotland and Wales, all at least 80km apart, but hawks did not persist in seven areas. Areas settled by only one pair were not secure, because only three of nine 'single-pair sites' were still occupied in 1980. At least 49 fledged hawks had been killed since 1971, by shooting (25), trapping (8), poisoning (4), and unspecified methods (12). Thirteen of these birds were breeding adults and proven or probable interference had destroyed 29% of 171 broods (Marquiss 1981, Marquiss & Newton 1982b).

Two decades later, most British Goshawks are still in northern and western areas, with Wales and bordering parts of England especially well populated. Populations elsewhere have not expanded greatly. A population in north-east Scotland has been particularly constrained, expanding by only one pair for every 9.5 young that were fledged since the early 1970s, to reach 17 pairs by 1996. In contrast, the population impinging on Kestrels on the English border with Scotland (Petty *et al.* 2003a) grew to 87 pairs during the same period, by one pair per six young fledged (Marquiss *et al.* 2003), which is close to the proportion of males that survived two years on Gotland (Chapter 8).

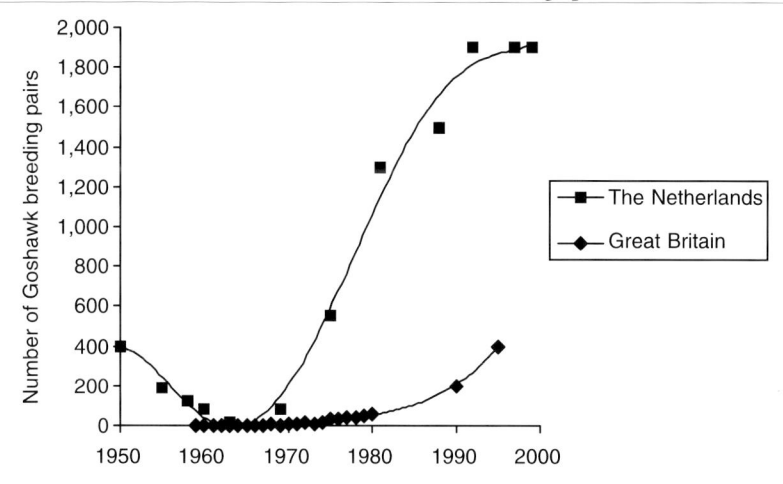

Figure 131. Goshawk numbers increased much less rapidly after reintroduction in the UK than during recovery from organochlorine impacts in The Netherlands. Data from Marquiss 1981, Marquiss *et al.* 2003, Thissen *et al.* 1981, Rutz *et al.* 2006a.

There may be reasons other than illegal killing behind the difference in growth rates for Britain and The Netherlands. Immigration from neighbouring countries and survival of single, less-contaminated hawks, may have aided recovery of the Dutch Goshawk population. Moreover, a different approach to conservation might have helped Goshawks to recolonise Britain faster. Of 23 birds killed illegally, 17 were juveniles and 12 were at pheasant pens. The problem is that Goshawks disperse little from natal areas in which food is abundant (Chapter 5). They may therefore tend to accumulate in a suitable area before 'overflowing' into new areas. Unfortunately, a high density of hawks can lead to appreciable impacts on game, especially at pheasant release sites (Chapter 7). If human livelihoods are felt to be threatened by a high level of predation on game, the law is less likely to be kept. Illegal killing may then make the local Goshawk population a 'sink' rather than a 'source' of hawks. However, what if gamekeepers had been encouraged to live-trap Goshawks at pheasant pens for release in areas with less game rearing? What if the imports that helped spread Goshawk colonisation of Britain had been allowed to continue?

PREDATION

Deliberate killing by humans is not the only predation problem that faces Goshawks in future. Another species, the Eagle Owl, became rare in Europe during the last two centuries. However, growing interest in raptors and in the ability to breed them in captivity resulted in reintroduction projects in Germany (Herrlinger 1973), followed by Sweden after evidence of a sharp decline during the pesticide era (Broo 1978). In Schleswig-Holstein, Eagle Owls have taken to nesting in trees, like their smaller American cousin the Great Horned Owl. They need large stick nests, and in southwest Schleswig-Holstein 59% of the nests adopted by these owls were originally built by Goshawks (and 30% by Buzzards). This effectively takes out Goshawk breeding sites, because no Goshawk breeds within 500m of an active Eagle Owl nest (Busche *et al.* 2004). Moreover, among 24 Goshawk nest failures with known causes, nine (18%) were definitely due to these owls and four (8%) were probably due to them, compared with five losses attributed to humans.

Goshawk numbers are now a third of what they were in Schleswig-Holstein in the year after Eagle Owls first nested (Figure 132). That also was the publication year for the concept of 'intra-guild predation', for 'potential competitors that also eat each other' (Polis *et al.* 1989). In Italy, Black Kite productivity was severely reduced within two kilometres of Eagle Owl nests and were therefore confined to interstitial sites between owl nests, which were 3–4 km apart (Sergio *et al.* 2003). In areas with good prey for owls (for which Brown Rats and European Hedgehogs were staple biomass in Italy) and limited nest sites for large raptors, safe areas might become rare. Further evidence of Eagle Owls impacting on raptors can be anticipated as the owls again become widespread (Rutz *et al.* 2006a).

Although Goshawks are now abundant in European areas of fragmented woodland and farmland, it has been suggested that the abundance of tree-nesting Great Horned Owls is what prevents a similar expansion of their distribution outside extensive forest in North America (Kenward 1996). We now have the question of what will happen to densities of Goshawks and other medium-sized raptors if

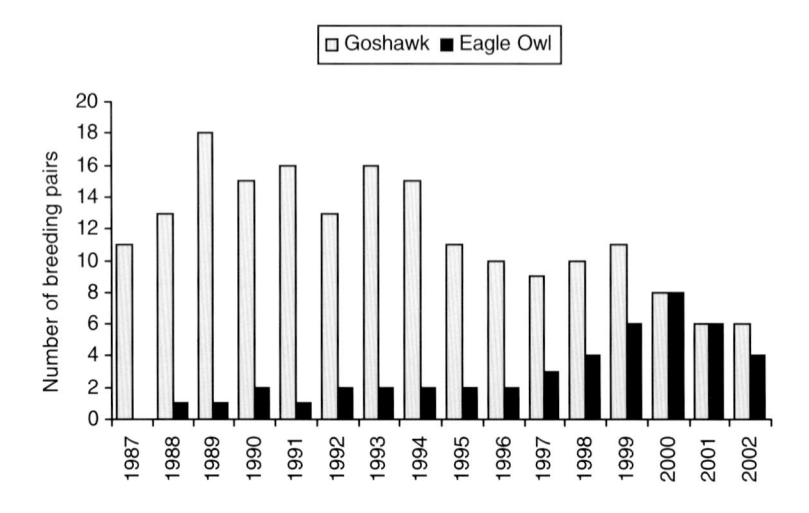

Figure 132. Goshawk numbers decreased as an Eagle Owl population re-established itself in southwest Schleswig-Holstein. Data from Busche *et al.* 2004.

tree-nesting Eagle Owls become widespread in Europe. Will intra-guild predation benefit species such as gamebirds (Tapper 1999), which may be more affected by diurnal raptors than by Eagle Owls? Do Goshawks breed well in some fragmented woods of the western USA because an even bigger predator, the Golden Eagle, likes the habitat and suppresses numbers of Great Horned Owls?

PREY DEFICITS

Evidence is growing that Goshawks have problems with finding adequate food supplies in several parts of Europe. In The Netherlands, Goshawks increased in numbers after the pesticide era to levels not seen earlier in the 20th century (Figure 131) and spread into areas that previously lacked tree cover (Bijlsma 1993). Hawk numbers, diet and prey numbers were recorded in a small study area (20km^2) called Planken Wambuis, in Veluwe province. Christian Rutz and Rob Bijlsma (2006) recorded a decline from 6–7 Goshawk pairs during 1976–86 to only 3–5 pairs during 1990–2000 (Figure 133). This decline was associated with an 80% reduction in biomass of avian prey during 1975–2000, which was thought to be due to changes in land management and acidification. There was an especially marked fall in numbers of Woodpigeons. Moreover, there was a 95% reduction in rabbit numbers due to viral haemorrhagic enteritis and from vegetation change resulting from nitrogen deposition.

Searching of this area and surrounding woodland was thorough enough to find many feathers moulted by hawks that were not recognisable as breeders, and thus to show that non-breeder numbers declined in parallel with the number of occupied

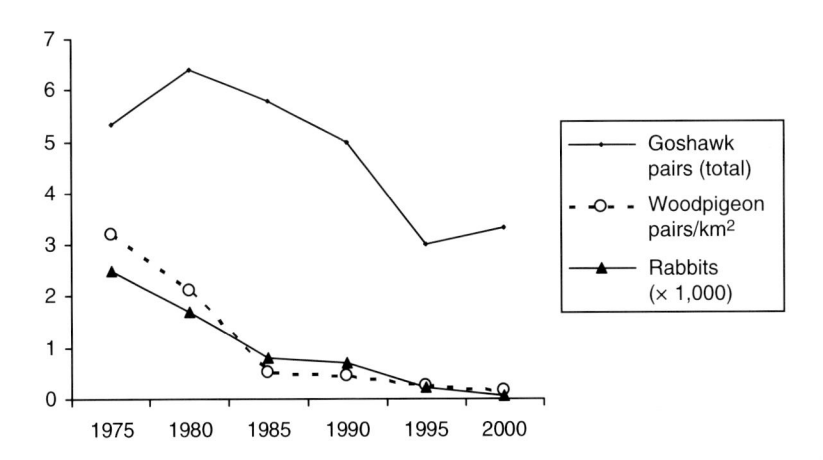

Figure 133. Goshawk numbers in Planken Wambuis declined following declines in numbers of important prey species. Data from Rutz & Bijlsma 2006.

sites and of pairs that laid eggs. There was no increase in adult turnover, breeding by first-year hawks or physical evidence of illegal killing to indicate increased mortality. Nor were there fewer young per clutch. There was also no loss of nesting habitat, but fewer pairs of hawks laid eggs and their diet diversity increased, notably including more raptors (Rutz & Bijlsma 2006). Moreover, the drop in the number of Woodpigeons in the area, from more than three pairs per km^2 in 1975 to 0.5 per km^2 or fewer from 1985, reduced density of this favoured prey to a level at which it might be suppressed by Goshawks.

In other areas, declines in prey and consequent reduction in numbers of Goshawks appear to reflect impacts on prey species of intensified land management. Thus in Finland, declining numbers of Goshawks follow declines in numbers of woodland grouse and their proportion in nesting diet (Wikman & Lindén 1981, Tornberg *et al.* 2005, 2006). Data on predation rates (Tornberg 2001) suggest that Goshawks may be able to suppress numbers of their woodland grouse prey when predator numbers peak, which occurs at intervals of two years after peaks in grouse numbers, and that this delay could contribute to driving cycles in the grouse populations (Tornberg *et al.* 2005, 2006). Goshawks may also be able to suppress numbers of another important winter prey, the Red Squirrel. In Swedish taiga, squirrel densities were equivalent to 30–90 per hawk in most years (Andrén 1996), so that predation by Goshawks could account for most of the population (Kenward 1996). In such habitats, prey population spikes might represent escape from a predation-trap (Chapter 7).

Goshawk numbers have declined recently in two other European areas with long-term studies. At Vendsyssel in Denmark, a decline since 1994 has been associated with an increase in release of pheasants, from 6,000 to 35,000 annually, with an associated increase of pheasants in the Goshawk winter diet (Figure 134) and with increased illegal killing (Nielsen 2003a, b). Thus, the Goshawk decline might be due to the latter. However, as there has also been a substantial decline since the 1980s in numbers of local Feral Pigeons (Nielsen 1998), which were a staple prey when breeding and are now much less prevalent in the winter diet, an alternative explanation would be a fall in food supply.

Similarly, there have been declines since 1990 in Goshawk populations in Bavaria (Figure 110 in Chapter 8), in a series of steps preceded by high numbers of breeding juveniles (Bezzel *et al.* 1997a). This suggests that there have been bursts of high adult mortality, which could be explained by intermittent bouts of shooting adults at nests. However, fluctuating food supplies with a downward trend could produce the same pattern, if several years of high mortality were followed by a couple of years of abundant food, which would permit good breeding followed by good recruitment.

When several factors may be involved in a population decline, it can be very difficult to separate them. Change in productivity, caused by pesticides and other failures of nests, is relatively easy to detect. Indeed, there was increased deliberate destruction of nests in Denmark, which could have caused the reduced survival of breeders that was modelled to fit the population decline (Drachmann & Nielsen 2002). However, although the age of first breeding had increased after a preceding

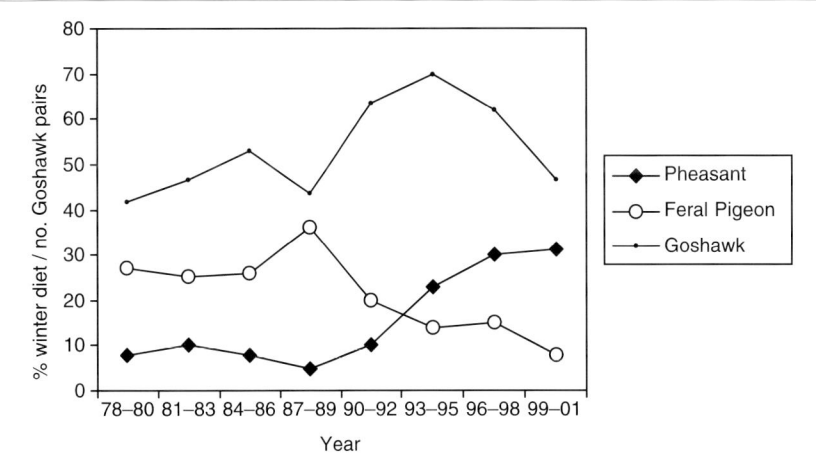

Figure 134. Danish Goshawk numbers decreased at Vendsyssel as pheasants replaced pigeons in the winter diet. Data from Nielsen 2003a, b.

period of population growth (see Chapter 8), the proportion of young breeders did not increase again to compensate for a reduction in survival of adults, which suggests that food shortage may also have been involved.

It is especially important to understand the likely symptoms of a Goshawk population decline caused by food shortage. If the decline is blamed erroneously on illegal killing, it will lead to disputes between hunters and protection organisations at precisely the time when they should be cooperating to identify and help remedy the causative changes in land-use. A decline caused by overall reduction in food supply may show no marked accompanying reduction in breeding age (Figure 135a), merely a cessation of breeding by some pairs, as reported by Rutz & Bijlsma (2006). There might even be less recruitment of young birds initially, due to established breeders being best able to hold sites that remain adequately provisioned (Figure 135b). In contrast, if the increased mortality of adults is caused by illegal killing at nests and is severe enough to reduce numbers of breeding pairs, it will increase the numbers of young breeders. However, if a population has many non-breeders, there will be some delay before numbers of breeders decrease, even with annual destruction of 20% of nests and adult females (Figure 135c). This model allows up to 25% of first-year birds to occupy breeding areas vacated by adults.

The resilience of populations to removal of juveniles, combined with the practical difficulty of recording deaths away from nests, make it difficult to distinguish the impacts of food shortage and illegal killing in winter. However, killing focussed on juveniles at pheasant pens will need to be severe in order to reduce breeding numbers, due to each death having relatively less effect than that of a breeding adult (Drachmann & Nielsen 2002). This is partly because some juveniles that are killed would otherwise have died of other causes before adulthood ('compensatory mortality'), but also because an increase in turnover of adults increases the

Figure 135. The structure of a Goshawk population in spring, modelled from the Gotland population in Figure 117 but permitting up to 25% of first-year birds to occupy breeding areas vacated by adults, would show no change in the proportion of young breeders recruited if food supplies halved during ten years (a), or perhaps a temporary decrease if young breeders are most affected (b). Killing 20% of breeding adults for ten years (c) produces an increased proportion of first-year breeders, (a decline in the number of nests is delayed by non-breeder abundance), whereas killing 20% of juveniles does not reduce breeding (d).

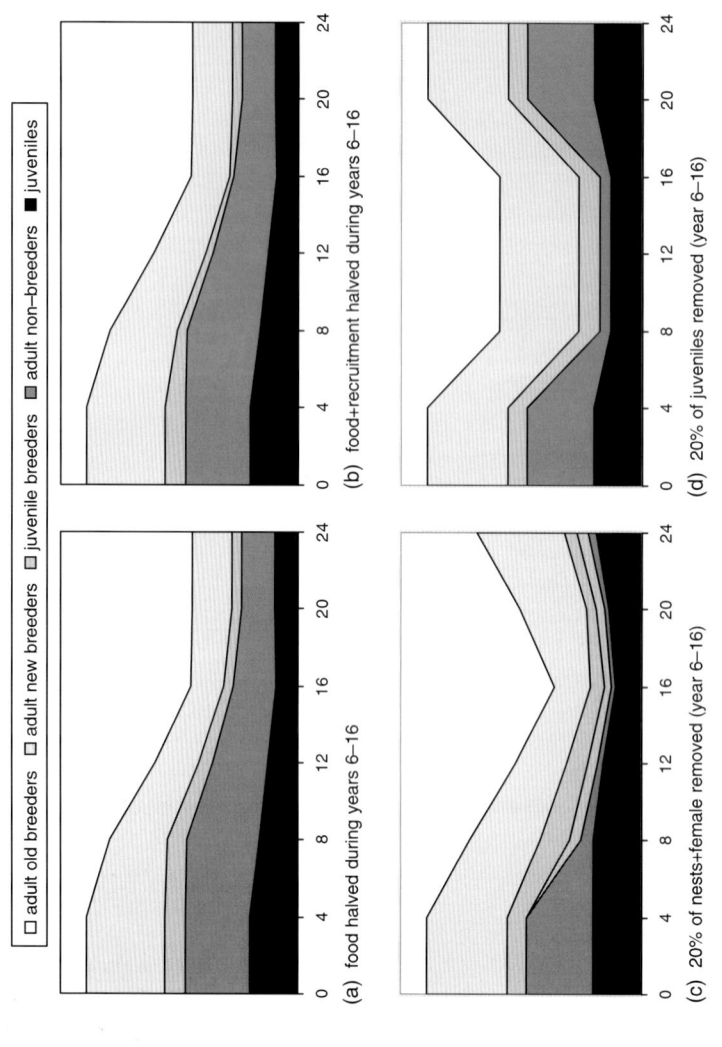

proportion of new breeders, which are least successful. Removing 20% of juveniles can be sustained by an increase from 60% to 90% of adults breeding without reduction in the number of breeding pairs (Figure 135d).

Radio-tracking can be used to link dispersal and mortality to food supplies, which may help to reveal and hence remedy land-use problems, and also to confirm or refute suspicions that killing is severe enough to have an impact on populations (Kenward 2002b). However, this is too costly a solution to be applied to investigate every population decline. A better solution for discriminating between deliberate killing and indirect human activities as causes of decline would be to ensure that any removal of Goshawks is regulated, and therefore measurable, not an 'underground' illegal activity.

GOSHAWK HABITATS

Habitats are important to Goshawks for supporting prey populations, for structure when foraging (e.g. for perches) and nesting, and for cover from predators. The ability of North American Goshawks to breed very successfully in strips of woodland (Younk & Bechard 1994, Bechard *et al.* 2006) indicates that extensive mature woodland is not more important for prey and structure there than in Europe, where Goshawks nest in woodlots and town parks. Although rural Goshawk nests are found in some studies to be more remote from humans than expected by chance (Bosakowski 1999) and sometimes to be more successful when remote from human disturbance (Krüger 2002b), this does not mean that remoteness is any more essential to them than extensive woodland. In North America too they can adapt to human proximity. One pair tolerated a house being built 90m from the nest, and another tolerated skiers and snow-scooters passing underneath the nest about twice a day (Lee 1981b). As stressed repeatedly, a preference of wildlife for a particular habitat does not imply that it is critical (Kenward & Widén 1989, White & Garrott 1990). Perhaps, like Cooper's Hawks, Northern Goshawks could also adapt to nest in American cities, although their aggressiveness to humans in America might make this undesirable.

In North America, the result of attempts to use Goshawks to preserve mature woodland has been an emphasis on managing woodland to maintain prey populations (Reynolds *et al.* 1992, 2006 a, b). Although this may not yet have changed forestry practices that would have enhanced Goshawk prey availability range-wide, as was the original aim, the integrated management of habitats to maintain biodiversity is an important advance (Graham *et al.* 1994, Boyce *et al.* 2006) that may benefit species other than Goshawks (Caroll *et al.* 2006). Moreover, the controversy has helped educate scientists about socio-economic issues (White & Kiff 1998) and lawyers about science. Above all, it has taken forward the study of Goshawk biology and provided training for many biologists. Training to study predators is valuable because the special vulnerabilities of such species require biologists to gain extensive field skills and cover many aspects of science.

Moreover, Goshawks have stimulated important re-examination of methods for assessing status and trends in a sparsely distributed species. The importance of systematic and accurate density estimation has been stressed (Kennedy 1997, Smallwood 1998, Hargis & Woodbridge 2006), as has the need for caution about indices of abundance that are based primarily on visual observations (Kennedy 1998, Doyle 2006).

Finally, American studies have indicated where Goshawks genuinely appear to be vulnerable and at risk of losing genetic resources. For offshore islands in the Pacific Northwest, where there are no lagomorphs or other mammals that provide a substantial meal (because the Red Squirrel in North America is much smaller than its Eurasian counterpart), Goshawks have become dependent on avian prey. These island hawks, and their conservation, may therefore differ in subtle ways from hawks elsewhere in North America. Frank Doyle (2006) notes that felling might benefit mainland Goshawks if it creates open spaces that favour hares and ptarmigan as prey, and that felling acts most negatively when it removes habitat on islands that lack these prey. On two islands off the Canadian coast, Goshawk nests were 6–8km and 9–15km apart, compared with 4–5km at two mainland sites. The conservation of Goshawks on islands of the northwest Pacific needs careful attention, again involving management of their prey-base (Reynolds *et al.* 1992).

In Europe there are serious issues about intensification of northern forestry (Widén 1997, Selås 1998b), for which the impacts on wildlife in general have been less addressed than has been the case with assessing the damage done by intensive farming further south. Both types of intensification already appear capable of having a negative impact on Goshawks by reducing prey populations. Impacts would be especially severe if the food or cover available for prey species are reduced to levels at which Goshawks can put important species in a predation trap or even extirpate them, because this substantially reduces food for the hawks. Perhaps the original decline of Goshawk populations in the eastern USA was triggered by the extinction of the Passenger Pigeon which coincided with massive deforestation (P. Kennedy, pers. comm.). Recent growth of new woodland has enabled the predator to recolonise (Speiser & Bosakowski 1984, Veit & Peterson 1993, DeStefano 2005).

Although the interest in Goshawk habitats was kindled by concerns about logging (Crocker-Bedford 1990), removal of up to 30% of mature trees has had little effect on numbers of breeding Goshawks in several mainland studies (Penteriani & Faivre 1997, 2001, Penteriani *et al.* 2002, Mahon & Doyle 2005). Goshawks may move from some traditional nest areas when timber is harvested (Patla 2005), but are also colonising new areas elsewhere. Overall, there is no good evidence of Goshawk population declines in North America (Braun *et al.* 1996, Kirk & Hyslop 1998, Andersen *et al.* 2005, Squires and Kennedy 2006). The move towards integrated management of landscapes may help prevent declines in future, but now needs application in Europe too, and eventually across Eurasia as a whole. Many interests will have a role to play, with monitoring of diet at nests a possible inexpensive way to obtain early warning of problems (Chapter 8).

GOSHAWKS REMAIN WIDESPREAD

Goshawk population density at national level tends to be higher in central parts of Europe than in the most northern and southern countries (Figure 136), as was the case for average densities across studies within 5° quadrats of latitude and longitude (Rutz *et al.* 2005). The most noticeable exception to this trend are in Britain, where

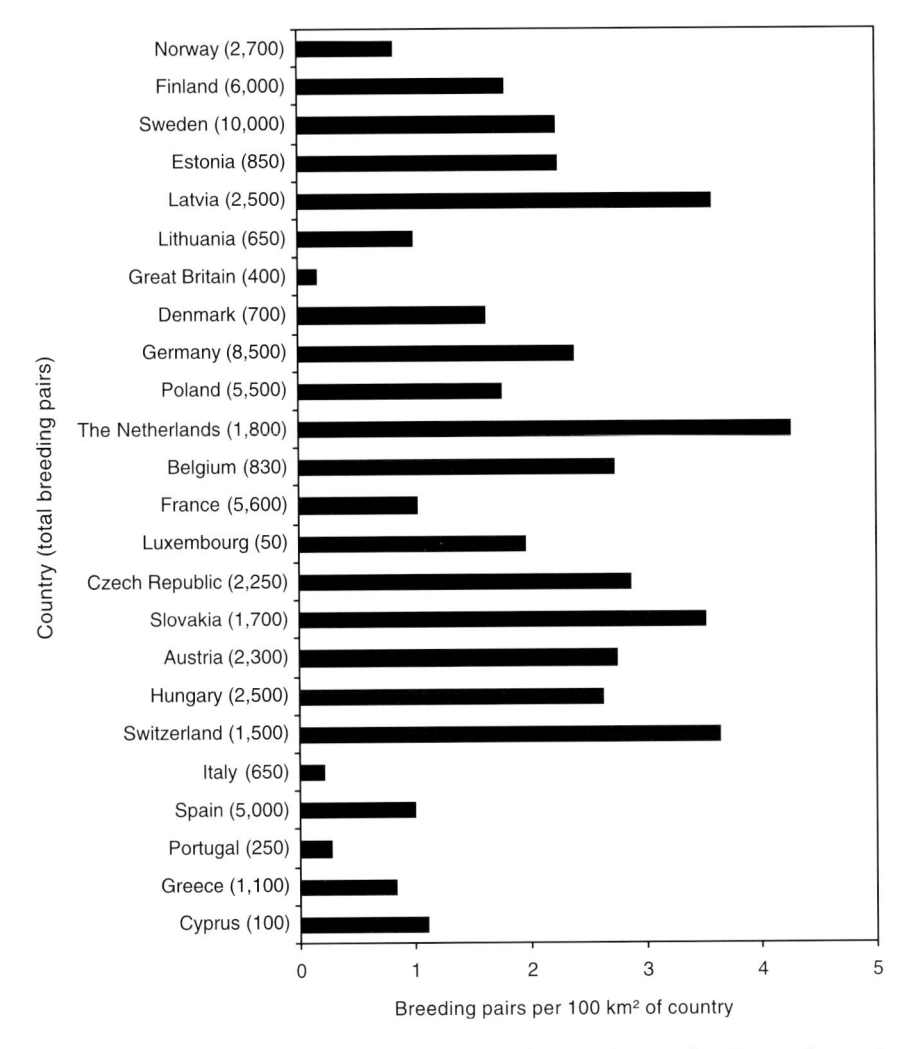

Figure 136. Goshawk breeding density at national level from north to south in Europe. Sources in Appendix 2.

low numbers reflect incomplete recolonisation (Ireland is not shown but may have a small number of pairs). The population estimate for France also seems relatively low and perhaps needs revision.

The total for these country estimates exceeds 60,000 breeding pairs. The most recent assessment by BirdLife International (Burfield & van Bommel 2004) estimates a total Goshawk population for Europe, including Belarus, Ukraine and European Russia, of at least 160,000 breeding pairs. This makes the Northern Goshawk the fourth most abundant diurnal raptor in Europe, after the Common Buzzard (>710,000 pairs), the Eurasian Sparrowhawk (>340,000 pairs) and the Common Kestrel (>330,000 pairs), with the Marsh Harrier (>140,000 pairs) the next most abundant. The Northern Goshawk is considered secure in Europe and is listed in Annex II of the European Community Birds Directive, although the race *A. g. arrigonii* is in Annex I and is the subject of a conservation action plan for Corsica.

There are few data from which to estimate population size in the vast wooded northern areas of Asia and North America. Mean densities of 1.7 and 2.4 pairs/100 km^2 have been recorded in the extensive northern forests of Alaska and the Yukon (McGowan 1975, Doyle 2000), about half the values from Fennoscandian Taiga (Widén 1985, Tornberg 2000). If we take 1–2 pairs/100 km^2 to estimate values that are likely to be conservative, bearing in mind that median values for both the American and European study areas are 3.4 (Figure 118), there would be 150,000–300,000 breeding pairs in the 15,000,000 km^2 of northern American forests, and at least as many again in the larger forests of Siberia & China. With a European population of at least 160,000 pairs, and other Goshawk populations south of the taiga in America and Asia, the global Goshawk population is unlikely to be less than 500,000 pairs.

With such a large global population, the Northern Goshawk is about as far from being an endangered species as is possible for a raptor. On the one hand, it thrives in landscapes developed by man, provided that the cultivation does not become so intensive that prey populations decline or become contaminated with toxic chemicals. On the other hand, much of the species' large global population is in areas remote from direct or indirect human impact. Why, then, should its conservation be controversial? The answer lies in the way in which conservation has developed, as illustrated by international conventions.

THE APPLICATION OF SOME CONVENTIONS

Wildlife conservation, like other human activities, is based increasingly on international conventions. Pre-existing national laws and practices, which at their best were important guidance for those conventions, are gradually becoming aligned with the conventions.

An early wildlife convention in Europe was agreed in 1950 for the protection of birds. This was followed in 1979 by the Bern Convention on Conservation of European

Wildlife and Natural Habitats, which is reflected in the Birds Directive of the European Union (1979). All raptors are protected, but countries may derogate to permit taking of raptors that cause problems and for science or falconry. The Bern Convention embraces Belarus, Russia and Ukraine and therefore applies protection for much of the Goshawk distribution across Eurasia, but falconry regulations still differ in every state.

The Goshawk is given an equivalent level of protection across North America. In contrast to the European Union, there are also federal regulations for falconry under the Migratory Birds Act in the United States. Another difference from Europe is the Endangered Species Act, which obliges restoration activities for listed species, including habitat conservation. Unsuccessful attempts were made for 12 years to list the Goshawk under this act and to use the species in other ways to preserve old-growth timber. The litigation and resulting controversy stimulated much research and debate on the habitats and prey that Goshawks need, as concisely reviewed and tabulated by Squires and Kennedy (2006). The listing of the species as 'of concern' by the US Fish and Wildlife Service and 'sensitive' by the Forest Service now obliges its requirements to be considered in forest management.

A more recent international agreement, the Convention on Biological Diversity (1992), combines protection of species and habitats '*in situ*' (Article 8) with requirements for '*ex-situ*' conservation (Article 9), as in breeding for release, and to sustain components of biodiversity through use. In particular, there are requirements to 'Protect and encourage customary use of biological resources in accordance with traditional cultural practices that are compatible with conservation or sustainable use requirements' (Article 10) and to 'adopt economically and socially sound measures that act as incentives for the conservation and sustainable use of components of biological diversity' (Article 11). Sustainable use is mentioned in 13 of the 19 principle Articles; protection in two of them.

Two other international agreements with some relevance to conservation of Goshawks are the Convention on Trade in Endangered Species (CITES) and the Convention on Migratory Species (CMS). Both are applied mainly to rarer species than the Northern Goshawk, which is neither endangered nor migratory over long distances. However, all raptors are included as if endangered for implementing CITES in the European Union.

Although the Convention on Biological Diversity (CBD) is the most widely applicable global convention on wildlife, it has not yet been widely applied in national laws and regulations, except for the pre-existence of protection measures as required in Article 8. The earlier conventions had placed an emphasis on protecting species and creating reserves of rare habitat, which has helped change popular attitudes to wildlife. Protection of particular species shows that society values them. Protection of areas has been crucial for holding back intensive cultivation and other development. Currently, some 11.5% (17,100,000 km^2) of land on Earth is in protected areas (Chape *et al.* 2003), of which about 36% exclude some or all extractive use of wild resources.

However, a 'protect and reserve' approach can lead to polarised attitudes and landscapes. For example, people resent protection that stops valuable and previously legal activities or the control of species that eat crops or kill game. Generalised protection has failed to prevent wildlife populations declining in northern countries and can be hard to implement in developing countries (Getz *et al.* 1999, Adams *et al.* 2004). The CBD offers a second approach, of using wild resources to give value to land, which can motivate conservation through less intensive agriculture, forestry or other development. This seems especially important, because 27 of the 29 most vulnerable bird species in Europe are adversely affected by change, loss or fragmentation of their habitats (Stroud 2003). An incentive-driven or 'conserve-by-use' approach goes with the flow of socio-economics, instead of blocking it by constraints or bans on use (Hutton & Leader-Williams 2003).

Where land is relatively unproductive, sustainable use of wild resources can completely replace value from intensive use. For instance, wildlife tourism and especially hunting can be more cost-effective than cattle on African rangelands (Child 1995, Prins *et al.* 2000). Although intensive cultivation creates far more value than sustainable use of wild resources on fertile land, de-intensification measures provide remedies. For instance, Newton (2004) identified six main factors that were associated with declines of 30 bird species in Britain, namely (i) weed control, (ii) early ploughing, (iii) grassland management, (iv) intensive stocking, (v) hedgerow loss and (vi) predation. All five habitat factors can be addressed in ways that produce fractional reductions in yield, which can be offset by private 'pay-for-use' and (where available) state subsidy (Kenward & Garcia Cidad 2002). The vision is a 'more biodiversity-friendly mosaic of land driven by the livelihoods that are derived from the sustainable use of wild living resources, instead of landscapes with small islands of biodiversity in a sea of agriculture' (Hutton & Leader-Williams 2003).

The vision from CBD is of protection ('sticks') complemented by incentive ('carrots') to conserve more land. For example, protected areas for fish may support core populations that safeguard a conserve-by-use approach in surrounding zones (Roberts *et al.* 2002). Similar zoning approaches are being designed for predatory mammals. However, the challenge of integrating the previous protection-based approach with the new incentive-driven approach is especially demanding in the case of charismatic species, like the Northern Goshawk, that have proved vulnerable to pesticides and persecution.

On the one hand is an understanding that it is in everyone's interest to work for land-use that supports wildlife as well as farm and forest products, if only to maximise income from recreation as well as industrial output. Landscapes rich in wildlife benefit hunters and predators. There is also appreciation that achieving such landscapes is a complex task and that, as the number of people in land management declines, cooperation becomes increasingly important (Galbraith *et al.* 2003). Understanding of raptor-human conflicts has resulted in a resolution by the two main global organisations devoted to raptor research and conservation (the Raptor Research Foundation and the World Working Group on Birds of Prey and Owls). Both resolutions accept management of raptors to resolve conflicts, including (as a last resort) the selective removal of problematic individuals.

On the other hand, there is also concern that any deviation from strict protection will mean a return to persecution that extirpates raptors like the Goshawk. Many organisations devoted to a 'protect and reserve' approach may also fear that a more pragmatic approach would cost them support, especially in view of public attraction to 'animal rights' campaigns that seek to safeguard individual animals.

The result of these conflicting viewpoints is something of an impasse. Where protection is unyielding, Goshawks are killed illegally. This slows their repopulation of Britain, makes it hard to assess the importance of other threats to the species and may encourage Goshawks to avoid humans and hence indirectly reduces the habitat available for them (Rutz et al. 2006a). In Sweden, a period of unyielding protection of Goshawks discouraged work to maintain pheasant populations and will have harmed hawk populations in areas where wild pheasants disappeared, since these formerly provided up to 43% of the biomass for male hawks (Marcström & Widén 1977, Kenward et al. 1981a). If falconers are not permitted access to wild Goshawks, they spend large sums on domestic breeding and may lose interest in local raptors (Chapter 9).

Protection can even prevent conservation research, for example by creating excessive administrative effort to import blood and feather samples (Cooper 2000). Research that shows raptors having an adverse impact on human interests may be unwelcome, as when a raptor enthusiast disapproved of further work on Goshawk predation in case they showed an impact on game (Chapter 8). When protection organisations campaign against other groups instead of cooperating for practical conservation, human resources go to waste and attention is diverted from more insidious threats like pollution and habitat loss.

COOPERATIVE CONSERVATION

Ian Newton (1998) reminds us that although recent animal extinctions have mostly followed human colonisation and 'been attributed mainly to overhunting', nevertheless 'habitat destruction and fragmentation have also caused extinctions and are likely to cause greater extinctions in future'. So how can the problems of habitat change best be addressed? One solution is to create protected areas. However, this is likely to have quite limited benefit for species as widespread and wide-ranging as the Northern Goshawk. Likewise, protection of individuals does not safeguard hawks when changing land-use reduces their food supply or when they are affected by larger raptors.

The previous consensus, that predation rarely if ever has an impact on prey populations and that raptors are so vulnerable that they need rigid protection, was not wrong. It was a good way to reach an accommodation about raptors when their populations were reduced. However there is now scope and need for changes, in order to encourage benefits from conservation from as many interests as possible. If the hunting of game is a way to preserve habitats and feed hawks, might it not be best to help hunters and hawks to live together, through measures that reduce predation without also reducing the numbers of hawks breeding? If falconers already contribute management skills and education for conservation, is it not appropriate to seek further benefits from them in exchange for access to wild hawks? If birdwatchers wish to maintain their non-consumptive use of wildlife resources, can they contribute more to offset a decline in agri-environment subsidies? How can scientists and governments help to bring all these interests together?

A first step is to reduce the conflicts between different interest groups. Everyone can participate in simple measures to help rebuild trust, such as replacement of terms like 'persecution', which reflects past activities but has the insulting implication of a continuing drive for extirpation. Nowadays, terms like 'illegal killing' and 'selective removal' are more objective and accurate (Rutz *et al.* 2006a). Further steps could be the development of formal agreements between governments, protection societies and 'hands-on' groups, concerning rights, responsibilities and practical cooperation.

Among the tasks that provide scope for cooperation is the need to monitor Goshawk populations widely and at minimal expense. Work in Scandinavia indicates that mark-recapture methods estimate similar numbers of breeding pairs to extrapolations from known nest densities. On Gotland, mark-recapture data estimated a production of 227 hawks in their first year (excluding an estimated 25 from Finland) or 132 clutches at an observed productivity of 1.7 young per clutch (Kenward *et al.* 1999). In comparison, records of 31 clutches in the 846 km^2 study area in each of the last two study years estimated 115 clutches for the 3,100 km^2 island. For Sweden as a whole, Marcström and Kenward (1981) estimated 3,500–13,600 pairs (with regional variation in recovery rates), compared with an estimate of 10,000 pairs from quadrat surveys (Svensson 2002). In Finland, if an annual sample of 5,800 pairs (Moilanen 1976) contained 73% in their first year and the recovery rate was 30.8% (Saurola 1976), annual production would have been 13,750 young Goshawks (5,800 x 0.73/0.308). With 1.8 young per occupied nest site in Finland (Chapter 8), the breeding population would have been 7,600 pairs. This compares with a density-based estimate, after recent declines, of 6,000 pairs (Väisänen *et al.* 1998).

A task of monitoring by mark-recapture could involve most interest groups. Volunteers, or falconers in exchange for hawks, could mark nestlings, record productivity and prey remains, and collect feathers for analysis of first breeding age and turnover of breeders (Figure 137). Live trapping of hawks could be licensed at sites with a predation problem, on condition that (i) selective spring nets are used, (ii) numbers of marked and unmarked hawks are reported and (iii) they are taken to falconers or released elsewhere (Chapter 9). Scientists would be needed for planning and for quality control of data collection to ensure robust estimations. Use of visual markers (e.g. colour rings) could enable complementation and cross-checking by hawk-watchers (Figure 137). Surely such a cooperative system is better than hawks being shot and buried unseen or, worse still, poisons being used. Social and economic rewards from falconers could even make Goshawk nests more valued by landowners.

Another task for ornithological volunteers and scientists could be to establish urban Goshawks. Hawks tamed by falconers, or hacked with falconry techniques, might be especially tolerant of human activities (Waardenburg 1977b) and would be welcome where there are problems with aggregations of Feral Pigeons. The same voluntary and professional interests might cooperate to gain data for population modelling by tracking radio-tagged birds. With careful training and planning (and noting that falconers routinely use radio-tracking), it is practical to measure survival as well as productivity with just three location checks a year, and hence to build geographically specific predictive models in the growing number of areas where there are maps (e.g. from remote sensing) of relevant habitats (Kenward 2002b). Pan-European analyses, as in Rutz *et al.* (2005) could become global if researchers and volunteers from Europe and America can agree to standardise collection of data on Goshawk nests, densities, foraging areas, diet and methods for recording recruitment, adult turnover and composition of landscapes.

The Goshawk has proved a superb subject for scientific study, with nests that are found easily enough for studies of breeding work and the hawks large enough to tag for tracking outside the breeding season. Science that was descriptive and correlative has progressed to experimental studies, modelling and, thanks to a huge distribution of comparable field studies, to analyses that combine data from large numbers of separate studies (meta-analyses). There is now also opportunity for bio-socio-economic projects that encourage cooperation of different interests for incentive-based conservation.

Nevertheless, there is concern that permitting the management of Goshawks could put raptors back on a slippery slope towards extirpation through persecution. This concern depends on how people view human societies. My view is that human societies change with time. For example, they learn that wars between and within nations do not pay, and hence to curb tribalism (which also lies at the heart of

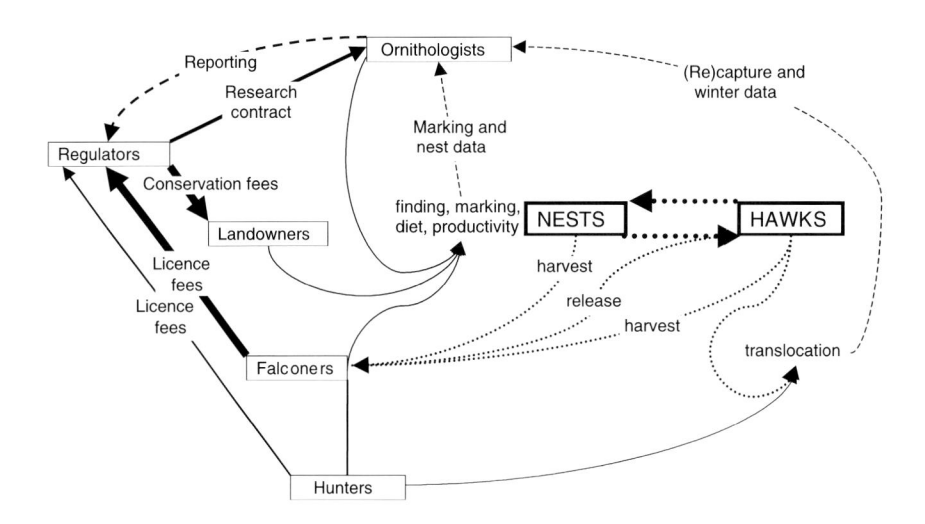

Figure 137. Conservation of Goshawks through sustainable use, with stakeholder contributions through actions (thin lines) on resources (dotted lines) that produce monitoring data (dashed lines) and economic motivations (thick lines).

campaigns by special-interest groups). Human societies have historically found sustainable paths between unwise exploitation and uninformed adulation of their environment, and should be able to do so again. Quite recently, deer were hunted almost to extinction in many parts of Europe, then they were protected and now they are again common enough to be widely harvested and sometimes treated as pests. When protection was relaxed on Goshawks in Schleswig-Holstein in the 1980s, it reduced Goshawk breeding numbers, but only by 20–30% (see Figure 110). The protection was then increased again – at the request of hunting organisations (Looft 2000).

Are people ready for innovative cooperation in conservation? Can protection groups widen their view of conservation? Can animal welfare interests accept that death is part of life's cycle and re-focus on welfare instead of preventing predation by humans? Can governments resist the temptation to play politics with such issues? Can falconer and hunter groups persuade their members to contribute more to conservation in exchange for restoration of past privileges? Can everyone accept that the more ways in which an animal is loved, as a free-spirit or a companion or even as a delicacy, the more human resources can be available for its conservation? I don't know. However, cooperation requires tolerance, which starts with knowledge. I hope this book will bring some knowledge to help the Northern Goshawk.

CONCLUSIONS

1. Goshawks were killed by organochlorines and methyl mercury compounds during the 'pesticide era' and suffered a large population reduction in The Netherlands, possibly through eating pigeons contaminated by seed dressings.
2. Goshawks still accumulate environmental pollutants, including heavy metals deposited on feathers, but residues are generally low. The integrating effect of this widespread top predator makes it a good species for biomonitoring.
3. Humans eliminated Goshawks from Britain while there was little woodland. Goshawks have tolerated high levels of killing elsewhere in Europe, but this can still have an impact on recovering populations and hinder detection of other environmental stresses.
4. There are increasing signs of Goshawk populations suffering from anthropogenic change that influences their prey-base in forests and woodland-farmland areas. They may also be vulnerable to decrease in breeding densities as Eagle Owls repopulate Europe.
5. The Goshawk population is estimated to exceed 60,000 breeding pairs in Europe. Based on 1–2 pairs/100km^2 in northern forests, there may be at least 500,000 pairs worldwide.
6. Recommendations, to reduce risk of extreme population decline from future pollutants are (i) widespread monitoring and (ii) maintenance of expertise in domestic breeding.

7. To reduce illegal killing, non-lethal measures such as live-trapping and translocation could be encouraged and used to increase cooperation between different interest groups, including monitoring through mark-recapture schemes.
8. Falconers could be encouraged to contribute to preserving nests, marking young and collecting data on turnover, first-year breeding, productivity and diet, in order to detect impacts of changing land-use.
9. By promoting cooperation for incentive-based conservation, governments could make best use of human voluntary resources, better implement the Convention on Biological Diversity and encourage landscapes rich in resources for Goshawks and other wildlife.

APPENDIX 1

Scientific names of vertebrates mentioned in the text

BIRDS
American Robin	*Turdus migratorius*
Bald Eagle	*Haliaeetus leucocephalus*
Barnacle Goose	*Branta bernicla*
Blackbird	*Turdus merula*
Black Goshawk	*Accipiter melanoleucus*
Black Grouse	*Tetrao tetrix*
Black Kite	*Milvus migrans*
Blue Tit	*Parus caeruleus*
Booted Eagle	*Hieraaetus pennatus*
Bonelli's Eagle	*Hieraaetus fasciatus*
Brown Goshawk	*Accipiter fasciatus*
Bullfinch	*Pyrrhula pyrrhula*
California Condor	*Gymnogyps californianus*
Capercaillie	*Tetrao urogallus*
Carrion Crow	*Corvus corone*
Coal Tit	*Parus ater*
Common Buzzard	*Buteo buteo*
Common Kestrel	*Falco tinnunculus*
Common Redstart	*Phoenicurus phoenicurus*
Common Starling	*Sturnus vulgaris*
Cooper's Hawk	*Accipiter cooperi*
Eagle Owl	*Bubo bubo*
(Eastern) Imperial Eagle	*Aquila heliaca*
Eurasian Sparrowhawk	*Accipiter nisus*
Eurasian Treecreeper	*Certhia familiaris*
Feral Pigeon (Roch Dove)	*Columba livia*
Fieldfare	*Turdus pilaris*
Harris's Hawk	*Parabuteo unicinctus*
Hawk Owl	*Surnia ulula*

Hazel Grouse	*Bonasa bonasia*
Heath Hen	*Tympanachus cupido cupido*
Herring Gull	*Larus argentatus*
Hobby	*Falco subbuteo*
Hooded Crow	*Corvus cornix*
Goldcrest	*Regulus regulus*
Golden Eagle	*Aquila chrysaetos*
Grey Goshawk	*Accipiter novaehollandiae*
Grey-bellied Goshawk	*Accipiter poliogaster*
Great Horned Owl	*Bubo virginianus*
Great Tit	*Parus major*
Grey Partridge	*Perdix perdix*
Gyr Falcon	*Falco rusticolus*
Henst's Goshawk	*Accipiter henstii*
Hen Harrier	*Circus cyaneus*
Honey Buzzard	*Pernis apivorus*
Jackdaw	*Corvus monedula*
Jay	*Garrulus glandarius*
Long-eared Owl	*Asio otus*
Magpie	*Pica pica*
Mallard	*Anas platyrhynchos*
Marsh Harrier	*Circus aeruginosus*
Mauritius Kestrel	*Falco punctatus*
Merlin	*Falco columbarius*
Meyer's Goshawk	*Accipiter meyerianus*
Mistle Thrush	*Turdus viscivorus*
Moorhen	*Gallinula chloropus*
(Northern) Goshawk	*Accipiter gentilis*
Northern Harrier	*Circus (cyaneus) hudsonicus*
Osprey	*Pandion haliaetus*
Passenger Pigeon	*Ectopistes migratorius*
Peregrine Falcon	*Falco peregrinus*
Raven	*Corvus corax*
Red-breasted Goose	*Branta ruficollis*
Red Grouse	*Lagopus lagopus scoticus*
Red Kite	*Milvus milvus*
Ring-necked Pheasant	*Phasianus colchicus*
Rook	*Corvus frugilegus*
Ruffed Grouse	*Bonasa umbellus*
Sharp-shinned Hawk	*Accipiter striatus*
Short-toed Eagle	*Circaetus gallicus*
Steller's Jay	*Cyanocitta stelleri*
Stock Dove	*Columba oenas*
Tawny Owl	*Strix aluco*

Tufted Duck — *Aythya fuligula*
White-tailed Eagle — *Haliaetus albicilla*
Woodcock — *Scolopax rusticola*
Woodpigeon — *Columba palumbus*
Willow Grouse — *Lagopus lagopus lagopus*

Mammals
Abert's Squirrel — *Sciurus aberti*
Cottontail Rabbit — *Sylvilagus floridianus*
Brown Rat — *Rattus norvegicus*
(European) Hare — *Lepus europaeus*
(European) Rabbit — *Oryctolagus cuniculus*
Field Vole — *Microtus agrestis*
Fisher — *Martes pennanti*
Flying Squirrel — *Glaucomys* spp.
Grey Squirrel — *Sciurus carolinensis*
(European) Hedgehog — *Erinaceus europaeus*
Lynx — *Felis lynx*
Muskrat — *Ondatra zibethicus*
Porcupine (North American) — *Erethizon dorsatum*
Raccoon — *Procyon lotor*
Red Squirrel (American) — *Tamiasciurus hudsonicus*
Red Squirrel (Eurasian) — *Sciurus vulgaris*
Snowshoe Hare — *Lepus canadensis*
Varying Hare — *Lepus timidus*
Wolverine — *Gulo gulo*

Fish
Pike — *Esox lucius*

APPENDIX 2

Sources for figures that used data from more than 10 publications

CHAPTER 3

Figure 23. Nest and tree heights in European and North American studies:

Allen (1978), Anonymous (1989), Bartelt (1977), Becker *et al.* (in press), Bijlsma (1993), Bosakowski & Rithaler (1997), Bosakowski *et al.* (1992), Bosakowski *et al.* (1999), Bull & Hohmann (1994), Daw *et al.* (1998), Fischer (1986), Fleming (1987), Hall (1984), Hargis *et al.* (1994), Hayward & Escaño (1989), Kennedy (1988), Kimmel (1995), Kostrzewa (1987), La Sorte *et al.* (2004), Looft & Biesterfeldt (1981), Mañosa (1993), McGrath *et al.* (2003), Patla (1997), Penteriani (2002), Penteriani & Faivre (1997), Penteriani *et al.* (2001), Reynolds *et al.* (1982), Rosenfield *et al.* (1998), Saunders (1982), Schaffer (1998), Shuster (1980), Speiser & Bosakowski (1984, 1989), Squires & Ruggiero (1996).

Figure 34. Laying date in European studies:

Altenkamp (2002), Anonymous (1990), Bezzel *et al.* (1997a), Bijlsma (1989), Bijlsma (1993), Bijlsma (1998–2003), Byholm *et al.* (2002a), Dekker *et al.* (2004), Diviš (2003), Gedeon (1984), Huhtala & Sulkava (1976, 1981), Ivanovsky (1998), Kalaber (1984), Kenward *et al.* (unpublished), Looft (1984), Mañosa (1991), Mañosa *et al.* (1990), Marquiss & Newton (1982b), Möckel & Günther (1987), Pugacewicz (1996), Schönbrodt & Tauchnitz (1991), Tornberg (2000), Tornberg *et al.* (in press).

Figure 37. Percentage woodland in European and American studies:

Altenkamp & Herold (2001), Anonymous (1990), Bakker (1996), Bijlsma (1989, 1993, 1998–2003), Boal *et al.* (in press), Bosakowski *et al.* (1999), Bühler *et al.* (1987), Byholm (2003), Czuchnowski (1993), Dekker *et al.* (2004), DeStefano *et al.* (1994a), Dietrich (1982), Dietzen (1978), Dobler (1991), Doyle (2000), Draulans (1984, 1988), Drazny & Adamski (1996), Ehring (2004), A. Gamauf (unpublished),

Goszczyński (1997, 2001), Hausch (1997), Heise (1986), Huhtala & Sulkava (1976, 1981), Ivanovsky (1998), Jędrzejewska & Jędrzejewski (1998), Joubert & Margerit (1986), Joubert (1994), Kenward *et al.* (1981a), Kenward *et al.* (unpublished), Knüwer (1981), Koning (2000), Kos (1980), Kostrzewa *et al.* (2000), Krol (1985), Krüger & Stefener (2000), Lelov (1991), Link (1986), Lõhmus (2004), Looft (1984), Mahon *et al.* (2003), Mañosa (1994), Mañosa *et al.* (1990), Möckel & Günther (1987), Nielsen (2003b), Nore (1979), Oelke (1981), Olech (1997), Opdam (1975), Penteriani (1997), Penteriani *et al.* (2002), Petronilho & Vingada (2002), Petty *et al.* (2003a), Raddatz (1997), Rassmussen & Storgård (1989), Reynolds & Joy (1998), Reynolds & Meslow (1984), Reynolds & Wight (1978), Schönbrodt & Tauchnitz (1987), Selås (1997b), Šotnár (2000), Speiser & Bosakowski (1991), Staude (1987), Steen (2004), Steiner (1999), Sulkava (1964), Thiollay (1967), Tornberg (1997), Toyne & Ashford (1997), van Lent (2004), Warnecke (1961), Weber (2001), Widén (1987), Wikman & Tarsa (1980), Zawadzka & Zawadzki (1998), Ziesemer (1983).

CHAPTER 6

Figure 76. Diet data from central European studies:

Altenkamp & Herold (2001), Bezzel *et al.* (1997b), Bijlsma (1998–2003), Drazny & Adamski (1996), Goszczyński & Pilatowski (1986), Ivanovsky (1998), Jacob & Witt (1986), Kayser (1993), Lõhmus (1993), Looft & Biesterfeld (1981), Mañosa (1994), Marquiss & Newton (1982b), Nielsen & Drachmann (1999b), Olech (1997), Padial *et al.* (1998), Penteriani (1997), Petronilho & Vingada (2002), Petty *et al.* (2003a), Rutz (2004), Šotnár (2000), Steiner (1999), Storgård & Birkholm-Clausen (1983), Thiollay (1967), Toyne (1998), Würfels (1994, 1999), Zawadzka & Zawadzki (1998).

Figure 84. Mammals in diet for European and North American studies:

Allen (1978), Altenkamp (2002), Arntz (1998), Bakker (1996), Becker *et al.* (in press), Bezzel *et al.* (1997a), Bijlsma (1993, 1998–2003), Bloom *et al.* (1986), Bloxton (2002), Boal & Mannan (1994), Bosakowski & Smith (1992, in press), Bosakowski *et al.* (1992), Bosakowski *et al.* (1999), Bull & Hohman (1994), Craighead & Craighead (1956), Deppe (1976), DeStefano & Cutler (1998), DeStefano *et al.* (1994a), Dietrich (1982), Diviš (2003), Doerr (1968), Doucet (1989), Doyle (2000), Doyle & Smith (1994), Drazny & Adamski (1996), Erkens & Hendrix (1984), Fleming (1987), Good *et al* (2001), Goszczyński (1997, 2001), Grønnesby & Nygård (2000), Grzybowski & Eaton (1976), Ivanovsky (1998), Jacob & Witt (1986), Joubert & Margerit (1986), Kalabér (1984), Kayser (1993), Keane (1999), Keane *et al.* (in press), Kennedy (1991), Kenward *et al.* (unpublished), Klaas (1967), Koning (2000), Lee (1981a), Lewis (2001), Lewis *et al*

(2004), Lõhmus (2004), Looft & Biesterfeld (1981), Looft (1984), Mañosa (1991), Mañosa *et al.* (1990), Marquiss & Newton (1982b), Maurer (2000), McCoy (1999), McGowan (1975), Meng (1959), Morillo & Lalanda (1972), Myrberget (1989), Nielsen & Drachmann (1999b), Nore (1979), Olech (1998), Padial *et al.* (1998), Patla (1997), Penteriani (1997), Petronilho & Vingada (2002), Petty *et al.* (2003a), Reynolds & Meslow (1984), Reynolds *et al.* (1994), Rogers *et al.* (in press), Root & DeSimone (1980), Rosendaal (1990), C. Rutz (unpublished), Schaffer (1998), Schnell (1958), Schnurre (1934, 1956, 1973), Smithers (2003), Snyder & Wiley (1976), Šotnár (2000), Squires (2000), Steiner (1999), Stephens (2001), Storer (1966), Storgård & Birkholm-Clausen (1983), Sulkava *et al.* (1994), Sutton (1931), Thissen *et al.* (1981), Thrailkill *et al.* (2000), Tornberg (1997, 2000), Toyne & Ashford (1997), van Lent (2004), Veiga (1982), Verdejo (1994), Warncke (1961), Watson *et al.* (1998), Widén (1985), Wikman & Lindén (1981), Woets (1998), Wood (1938), Woodbridge *et al.* (1988), Würfels (1994, 1999), Younk & Bechard (1994), Zachel (1985), Zawadzka & Zawadzki (1998).

Mammals in Goshawk diet in Russian studies since 1985:

Arkhipov (2003), Belik (2003), Chumankin (2003), Domashevsky (2003), Ivanovsky (1991), Ivanovsky *et al.* (2003), Krechmar & Probst (2003), Melnikov & Buslaev (2003), Mitrofanov (2003), Novak (1998), Petrov & Gusev (1995), Shepel (1992), Trofimenko (2002), Vitovich (1985), Voronin (1995)

CHAPTER 8

Figure 106. Breeding success in European and North American studies:

Altenkamp & Herold (2001), Altenkamp (2002), Anonymous (1990), Arbeitsgruppe Greifvögel NRW (2002), Austin (1993), Bartelt (1977), Bechard *et al.* (in press), Bijlsma (1989, 1993), Bloom *et al.* (1986), Boal & Mannan (1994), Boal *et al.* (in press), Bühler *et al.* (1987), Byholm (2003), Byholm *et al.* (2002a), De Fraine & Verboven (1997), Dekker *et al.* (2004), Diviš (2003), Doucet (1989), Doyle (2000), Draulans (1984, 1988), Drazny & Adamski (1996), Ehring (2004), Erdman *et al.* (1998), A. Gamauf (unpublished), Gedeon (1984), Goszczyński (1997, 2001), Hanauska-Brown *et al.* (2003), Hausch (1997), Heise (1986), Hennessy (1978), Hillerich (1978), Ivanovsky (1998), Joubert (1994), Keane (1999), Keane *et al.* (in press), Kehl & Zerning (1993), Kenward *et al.* (1999), Knüwer (1981), Kostrzewa *et al.* (2000), Lapinski *et al.* (2000), Link (1986), Lõhmus (2004), Looft (1984), Mahon *et al.* (2003), Mañosa (1991), Mañosa *et al.* (1990), Marquiss *et al.* (2003), McGowan (1975), McGrath *et al.* (2003), Möckel & Günther (1987), Nielsen & Drachmann (1999c), Nore (1979), Oakleaf (1975), Olech (1998), Patla (1997), Pugacewicz (1996), Rassmussen & Storgård (1989), Reynolds & Joy (in press), Root & Root (1978), C. Rutz (unpublished), Schönbrodt & Tauchnitz (1991),

Šotnár (2000), Speiser (1992), Steen (2004), Storgård & Birkholm-Clausen (1983), Stubbe *et al.* (1991), Sulkava *et al.* (1994), Thissen *et al.* (1981), Titus *et al.* (1997), Tornberg (1997, 2000), Verdejo (1994), Widén (1985), Wikman & Lindén (1981), Woodbridge & Detrich (1994), Zawadzka & Zawadzki (1998), Zinn & Tibbitts (1990).

Figure 118. Goshawk density for study areas in Europe and North America:

Albig & Schreiber (1996), Altenkamp & Herold (2001), Austin (1993), Bartelt (1977), Bijlsma (1989), Bosakowski *et al.* (1999), Bühler *et al.* (1987), Czuchnowski (1993), DeStefano *et al* (1994a), Dietrich (1982), Dobler (1991), Doyle (2000), Drazny & Adamski (1996), Forsman & Solonen (1984), A. Gamauf (unpublished), Goszczyński (1997, 2001), Heise (1986), Ivanovsky (1998), Keane (1999), Keane *et al.* (in press), Kehl & Zerning (1993), Kenward *et al.* (1999), Kimmel & Yahner (1994), Knüwer (1981), Kos (1980), Krüger & Stefener (2000), Lelov (1991), Lõhmus (1993), Looft (1984), Looft & Biesterfeld (1981), Mahon *et al.* (2003), Mañosa (1991), Mañosa *et al* (1990), McGowan (1975), McGrath *et al.* (2003), Nore (1979), Oelke (1981), Olech (1998), Penteriani *et al.* (2002), Petty *et al.* (2003a), Pugacewicz (1996), Raddatz (1997), Rasmussen & Storgård (1989), Reynolds & Joy (in press), Reynolds & Meslow (1984), Reynolds & Wight (1978), Sarkanen (1971), Schönbrodt & Tauchnitz (1991), Shuster (1976), Šotnár (2000), Tornberg (1997, 2000), Verdejo (1994), Weber (2001), Widén (1985), Wikman & Lindén (1981), Woodbridge & Detrich (1994).

CHAPTER 10

Key references for raptor predation on Red Grouse in Scotland:

Redpath & Thirgood (1997, 1999), Redpath *et al.* (2001, 2003), Thirgood & Redpath (1997), Thirgood *et al.* (2000a, b, c), Valkama *et al.* (2005).

Records of Goshawks in Britain during 1841–1960:

Willoughby (1843), Fisher (1844), Horn (1846), Bree (1850), Bold (1850), Gurney (1851), Stevenson (1859a, b), Ranson (1863), Harting (1867), Gunn (1870), Brooke (1870, 1871), Cordeaux (1876), Tuck (1877), MacPherson (1883), Gunn (1888), Williams *et al.* (1919), Vincent & Wormald (1943), Raines (1946, 1956), Nisbet & Smart (1959).

Figure 136. Goshawk populations in European countries:

Cyprus: Nicalaos Kassinis (pers. comm.), Greece: Burfield and van Bommel (2004), Portugal: Palma *et al.* (1999), Spain: Martí and Del Moral (2003), Italy: Brichetti and Fracasso (2003), Switzerland: Oggier and Bühler (1998), Hungary: Bagyura and Haraszthy (2004), Austria: Dvorak *et al.* (1993), Czech Republic: Danko *et al.* (2002), Slovakia: Danko *et al.* (1994), Luxembourg: Melchior *et al.* (1987), France: Thiollay *et al.* (2003), Belgium: Rutz *et al.* (in press), The Netherlands: Bijlsma *et al.* (2001), Poland: Tomialojč and Stawarczyk (2003), Germany: Kostrzewa and Speer (2001), Denmark: Grell (1998), Great Britain: Petty (1996a), Lithuania: Burfield and van Bommel (2004), Latvia: Strazds *et al.* (1994), Estonia: Elts *et al.* (2003), Sweden: Svensson (2002), Finland: Väisänen *et al.* (1998), Norway: Bergo (1996).

References

ABULADZE, A.V. 1999. *[On the birds of prey passage in Georgia in autumn 1997]*. [Proceedings of the Conference on Birds of Prey of Eastern Europe and Northern Asia] 3(2)13–16. [Russian]

ABULADZE, A.V. & EDISHERASHVILI, G.V. 2003. *[Raptor migration in Georgia in winter and autumn 1998]*. [Proceedings of the North-Eurasian Raptor Conferences] 4 : 113–117. [Russian]

ADAMS, W.M., AVELING, R., BROCKINGTON, D., DICKSON, B., ELLIOTT, J., HUTTON, J., ROE, D., VIRA, B. & WOLMER, W. 2004. Biodiversity conservation and the eradication of poverty. *Science* 306 : 1146–1148.

ALBIG, A. & SCHREIBER, A. 1996. Bestandsentwicklung von Habicht, Sperber und Mäusebussard auf einer Fläche in der Stader Geest (Nord-West-Niedersachsen). *Seevögel* 17 : 15–19.

ALLEN, B.A. 1978. Nesting ecology of the goshawk in the Adirondacks. MSc thesis, State University of New York, Syracuse, New York, USA.

ALLEN, M. 1980. *Falconry in Arabia*. Orbis, London, UK.

ALTENKAMP, R. 2002. Bestandsentwicklung, Reproduktion und Brutbiologie einer urbanen Population des Habichts *Accipiter gentilis* (Linné 1758). Diploma thesis. University of Berlin, Berlin, Germany.

ALTENKAMP, R. & HEROLD, S. 2001. Habicht (*Accipiter gentilis*). Pp. 175–179 in ABBO. Die Vogelwelt von Brandenburg und Berlin. Verlag Natur and Text, Rangsdorf, Germany.

AMADON, D. 1959. The significance of sexual differences in size among birds. *Proceedings of the American Philosophical Society* 103 : 531–536.

AMADON, D. 1964. Taxonomic notes on birds of prey. *American Museum Novitatae* 2166:1–24.

ANDERSEN, D.E., DESTEFANO, S., GOLDSTEIN, M.I., TITUS, K., CROCKER-BEDFORD, C., KEANE, J.J., ANTHONY, R.G. & ROSENFIELD, R.N. 2005. Technical review of the status of Northern Goshawks in the western United States. *Journal of Raptor Research* 39: 192–209.

ANDERSON, D.W. & HICKEY, J.J. 1974. Eggshell changes in raptors from the Baltic region. *Oikos* 25 : 395–401.

ANDERSSON, M. 1981. On optimal predator search. *Theoretical Population Biology* 19 : 58–86.

ANDERSSON, M. & NORBERG, R.A. 1981. Evolution of reversed size dimorphism and role partitioning among predatory birds with a size scaling of flight performance. *Biological Journal of the Linnean Society* 15 : 105–130.

ANDRÉN, H. 1996. Populationsfluktuationer och biotopval hos ekorre – ett viktigt bytesdjur för duvhök. Pp. 31–33 in NYGÅRD, T. and WISETH, B. (eds) Hønsehauken i skogbrukslandskapet. Norsk Institutt for Naturforskning Temaheft 5, Trondheim, Norway.

ANGELSTAM, P. 1984. Sexual and seasonal differences in mortality of the black grouse *Tetrao tetrix* in boreal Sweden. *Ornis Scandinavica* 15 : 123–134.

Anonymous. 1989. Goshawk breeding habitat in lowland Britain. *British Birds* 82 : 56–67.

Anonymous. 1990. Breeding biology of Goshawks in lowland Britain. *British Birds* 83: 527–540.

Anonymous. 2002. Photograph. *De Takkeling* 10 : 2.

American Ornithologists' Union. 1983. *Check-list of North American birds.* 6th ed., American Ornithological Union, D.C.

ARAUJO, J. 1974. Falconiformes del Guadarrama suroccidental. *Ardeola* 19 : 264–268.

Arbeitsgruppe Greifvögel NRW. 2002. Ergebnisse einer 30-jährigen Erfassung der Bestandsentwicklung und des Bruterfolgs beim Habicht (*Accipiter gentilis*) in Nordrhein-Westfalen von 1972–2001 (Fortschreibung 1986–2001). *Charadrius* 38 : 139–154.

ARENDONK, A.V. 1980. Goshawks at the Leeuwarden Air Base. *Journal of the North American Falconers' Association* 18–19 : 68–73.

ARKHIPOV, A.M. 2003. [Peculiarities of distribution and biology of the Goshawk in the steppes of Razdelnyanskiy district of Odessa Region]. *Strepet* 1 : 86–91 [Russian]

ARNTZ W. 1998. Voedselkeuze van de Havik in het natuurreservaat Salmorth. *De Mourik* 24(1):22–24.

AUFDERHEIDE, J. 1997. Ein habichts-Terzel 'geht fremd' – oder Zucht an den Flugdrahtanlage. *Greifvögel und Falknerei* (1996):43–44.

AUSTIN, K. 1993. Habitat use and home range size of breeding northern goshawks in the southern Cascades. MSc thesis, Oregon State University, Corvallis, Oregon, USA.

BAGYURA, J. & HARASZTHY, L. 2004. The status of birds of prey and owls in Hungary. Pp. 3–8 in CHANCELLOR, R.D. & MEYBURG, B.-U. (eds) *Raptors worldwide*. World Working Group on Birds of Prey and Owls, Berlin, Germany.

BÄHRMANN, U. 1965. Über das Variieren des Habichts in Mitteldeutschland. *Beiträge zur Vogelkunde* 6 : 186–189.

BAKKER T. 1996. Haviken in de regio Bergen op Zoom. *Veerkracht* 6(1):15–17.

BALGOOYEN, T.G. 1976. Behaviour and ecology of the American Kestrel (*Falco sparverius* L.) in the Sierra Nevada of California. *University of California Publications in Zoology* 103:1–83.

BARTELT, P.E. 1977. Management of the American Goshawk in the Black Hills National Forest. MSc thesis. University of South Dakota, Springfield, South Dakota, USA.

BASHKIROV, I.V., SHAMOVICH, D.I., KUZ'MENKO, V.V., KASHCHEEV, V.A. 2001. [On migration of raptors near Vitebsk (Belarus) in autumn 1999]. *Subbuteo* 4 : 46. [Russian, English summary]

BAUM, F. & CONRAD, B. 1978. Greifvögel als Indikatoren für Veränderung der Umweltbelastung durch chlorierte Kohlenwasserstoffe. *Tierärztliche Umschau* 33:1–19.

BECHARD, M.J., FAIRHURST, G.D. & KALTENECKER, G.S. 2006. Occupancy, Productivity, Turnover, and Dispersal of Northern Goshawks in Portions of the Northeastern Great Basin. *Studies in Avian Biology* 31:100–108.

BECKER, T.E., SMITH, D.G. & BOSAKOWSKI, T. 2006. Habitat, food habits, and productivity of Northern Goshawks nesting in Connecticut. *Studies in Avian Biology* 31:119–125.

BEDNAREK, W. 1975. Vergleichende Untersuchungen zur Populationsökologie des Habichts (*Accipiter gentilis*): Habitatbesetzung und Bestandsregulation. *Deutscher Falkenorden* (1975):47–53.

BEDNAREK, W. 1996. Zuchtbericht 1994 – Naturentfremdung behindert Greifvogelschutz. *Greifvögel und Falknerei* (1994):42–45.

BEDNAREK, W. 1997. DFO-Zuchtbericht 1995. *Greifvögel und Falknerei* (1995):33–39.

BEDNAREK, W. 1998. Zuchtbericht 1997 – Unterscheiden sich Rassen des Habichts (*Accipiter gentilis*) in ihren Verhalten? Gedanken zum Vermehren von Rassenhybriden beim Habicht zu Beizzwecken. *Greifvögel und Falknerei* (1997):69–67.

BEDNAREK, W. 1999. Zuchtbericht 1998. *Greifvögel und Falknerei* (1998):67–69.

BEDNAREK, W. in press. Seniles Alterskleid bei weiblichen Habichten (*Accipiter gentilis*). *Greifvögel und Falknerei* (2005).

BEDNAREK, W., HAUSDORF, W., JÖRISSEN, U., SCHULTE, E. & WEGENER, H. 1975. Über die Auswirkungen der chemischen Umweltbelastung auf Greifvögel in zwei Probeflächen Westfalens. *Journal für Ornithologie* 116 : 181–194.

BEDNARZ, J. C., KLEM, D., GOODRICH, L.J. & SENNER, S. E. 1990. Migration counts of raptors at Hawk Mountain, Pennsylvania, as indicators of population trends, 1934 1986. *Auk* 107 : 96 109.

BEIER, P. & DRENNAN, J.E. 1997. Forest structure and prey abundance in foraging areas of Northern Goshawks. *Ecological Applications* 7 : 564 571.

BEISHEBAEV, K.K. 1984. [Autumn counts of birds of prey in the Kungei Ala-Too]. *[Migrations of birds in Asia]* 7 : 223–228. [Russian]

BELIK, V.P. 2003. [Goshawk in steppe part of the Don River basin: distribution and ecology]. [*Proceedings of the North-Eurasian Raptor Conferences*] 4 : 15–48. [Russian]

BENGTSSON, K. 1997. Duvhöken – ständig misstrodd. *Anser* 4 : 300–303.

BENT, A.C. 1937. Life histories of North American birds of prey. *United States National Museum Bulletin* 167.

BERGER, D.D. & HAMERSTROM F. 1962. Protecting a trapping station from raptor predation. *Journal of Wildlife Management* 26 : 203–206.

BERGMANN, C. 1847. Über die Verhältnisse der Varmeökonomie der Thiere zu ihre Grösse. *Göttinger Studien* 3 : 595–608.

BERGO, G. 1996. Hønsehauken i Norge – utbreiing og bestandsforhold. Pp. 8–14 in NYGÅRD, T. & WISETH, B. (eds) Hønsehauken in skogbrukslandskapet. Norsk Institut for Naturforskning Temaheft 5, Trondheim, Norway.

BERNDT, R. 1970. Sperber (*Accipiter nisus*) und Habicht (*Accipiter gentilis*) jagen paarweise. *Vogelwelt* 91 : 31–32.

BERNERS, J. 1486. *The boke of St Albans*. Schoolmaster Printer, St. Albans, UK.

BERRY, R.B. 1972. Reproduction by artificial insemination in captive American Goshawks. *Journal of Wildlife Management* 36: 1283–1288.

BERT, E. 1619. *An approved treatise of hawks and hawking*. Richard Moore, London, UK.

BEZZEL, E., RUST, R. & KECHELE, W. 1997a. Revierbesetzung, Reproduktion und menschliche Verfolgung einer Population des Habichts, *Accipiter gentilis*. *Journal für Ornithologie* 138 : 413–441.

BEZZEL, E., RUST, R. & KECHELE, W. 1997b. Nahrungswahl südbayerischer Habichte *Accipiter gentilis* während der Brutzeit. *Ornithologischer Anzeiger* 36 : 19–30.

BILDSTEIN, K.L. 1998. Long-term counts of migrating raptors: a role for volunteers in wildlife research. *Journal of Wildlife Management* 62 : 435–445.

BIJLEVELD, M.F.I.J. 1966. Om het behoud van den Havik 1963–1964–1965. *Vogeljaar* 15 : 102–107.

BIJLSMA, R.G. 1984. Over de broedassociatie tussen Houtduiven *Columba palumbus* en Boomvalken *Falco columbarius*. *Limosa* 57 : 133–139.

BIJLSMA, R.G. 1989. Goshawk *Accipiter gentilis* and Sparrowhawk *A. nisus* in the Netherlands during the 20th century: Population trend, distribution and breeding performance. Pp. 67–89 in LUMEIJ, J.T., HUIJSKENS, W.P.F. & Croin MICHIELSEN, N. (eds). Valkerij in perspectief. Nederlands Valkeniersverbond 'Adriaan Mollen'/Stichting Behoud Valkerij, Monnickendam, The Netherlands.

BIJLSMA, R.G. 1991. Replacement of mates in a persecuted population of goshawks *Accipiter gentilis*. Pp. 155–158 in CHANCELLOR, R.D. & MEYBURG, B.U. (eds) Birds of Prey Bulletin 4. World Working Group n Birds of Prey and Owls, Berlin, Germany.

BIJLSMA, R.G. 1993. *Ecologische atlas van de Nederlandse roofvogels.* Schuyt & Co., Haarlem, The Netherlands.

BIJLSMA, R.G. 1996. Broedduur en uitkomstvolgorde van de eieren de Havik *Accipiter gentilis. Limosa* 69:67–71.

BIJLSMA, R.G. 1997. Veerafwijkingen bij nestjonge Haviken *Accipiter gentilis* veroorzaakt door 'Franse rui'? *De Takkeling* 5:40–41.

BIJLSMA, R.G. 1998–2003. Trends en broedresultaten van roofvogels in Nederland in 1997–2002. *De Takkeling* 6:4–53, 7:6–51, 8:6–51, 9:12–52, 10:7–48, 11:6–54.

BIJLSMA, R.G. 2003a. Trends en broedresultaten van roofvogels in Nederland in 2002. *De Takkeling* 11:6–54.

BIJLSMA, R.G. 2003b. Havik *Accipiter gentilis* legt superdwergei, of: leven en dood in een 30-jarig territorium op het voedselarme Planken Wambuis (Veluwe). *De Takkeling* 11:133–142.

BIJLSMA, R.G. 2004. Wat is het predatierisico voor Wespendieven *Pernis apivorus* in de Nederlandse bossen bij een afnemend voedselaanbod voor Haviken *Accipiter gentilis. De Takkeling* 12:185–197.

BIJLSMA R.G. 2005a. Speed diving is a successful hunting strategy of Goshawks *Accipiter gentilis* aiming for Racing Pigeons *Columba livia. De Takkeling* 13:112–120.

BIJLSMA, R.G. 2005b. Trends en broedresultaten van roofvogels in Nederland in 2004. *De Takkeling* 13:9–56.

BIJLSMA R.G., HUSTINGS, F. & CAMPHUYSEN, C.J. 2001. *Algemene en schaarse vogels van Nederland (Avifauna van Nederland 2).* GMB Uitgeverij/KNNV Uitgeverij, Haarlem/Utrecht, The Netherlands.

BILDSTEIN, K.L. 1998. Long-term counts of migrating raptors: a role for volunteers in wildlife research. Journal of Wildlife Management 62:435–445.

BIRKHEAD, T.R. & MØLLER, A.P. 1992. *Sperm competition in birds. Evolutionary causes and consequences.* Academic Press, London, UK.

BLOME, R. 1686. *The gentleman's recreation.* Private printing, London, UK.

BLOOM, P.H. 1987. Capturing & handling raptors. Pp. 99–123 *in* PENDLETON, B.A.G., MILLSAP, B.A., CLINE, K.W. & BIRD D.M. (eds). Raptor management techniques manual. National Wildlife Federation of scientific and technical series; no. 10.

BLOOM, P. H., STEWART, G.R., & WALTON, B.J. 1986. *The status of the northern Goshawk in California, 1981–1983.* Administrative Report 85–1. State of California, Department of Fish and Game, Wildlife Management Branch, California, USA.

BLOOM, P.H., HENCKEL, J.L., HENCKEL, E.H., SCHMUTZ, J.K., WOODBRIDGE, B., BRYAN, J.R., ANDERSON, R.L., DETRICH, P.J., MAECHTLE, T.L., MCKINLEY, J.O., MCCRARY, M.D., TITUS, K. & SCHEMPF, P.F. 1992. The dho-gaza with Great Horned Owl lure: an analysis of its effectiveness for capturing raptors. *Journal of Raptor Research* 26:167–178.

BLOXTON, T.D. 2002. Prey abundance, space use, demography, and foraging habitat of Northern Goshawks in western Washington. MSc thesis. University of Washington, Seattle, Washington, USA.

BLOXTON, T.D., ROGERS, A., INGRALDI, M.F., ROSENSTOCK, S., MARZLUFF, J.M. & FINN, S.P. 2002. Possible choking mortalities of adult northern Goshawks. *Journal of Raptor Research* 36:141–143.

BOAL, C. W. 1994a. A photographic guide to aging nestling Northern Goshawks. *Studies in Avian Biology* 16:32–40.

BOAL, C.W. 1994b. Unusual parental behaviors by male northern Goshawks. *Journal of Raptor Research* 28:120–121.

BOAL, C.W. & BACORN, J.E. 1994. Siblicide and cannibalism at Northern Goshawk nests. *Auk* 111:748–750.

BOAL, C. W. & MANNAN, R.W. 1994. Northern Goshawk diets in ponderosa pine forests on the Kaibab Plateau. *Studies in Avian Biology* 16 : 97–102.

BOAL, C.W., ANDERSEN, D.E & KENNEDY, P.L. 2003. Home range and residency status of northern goshawks breeding in Minnesota. *Condor* 105 : 811–816.

BOAL, C.W., ANDERSEN, D.E, KENNEDY, P.L. & ROBERSON, A.E. 2006. Northern Goshawk ecology in the western Great Lakes region. *Studies in Avian Biology* 33:126–134.

BOAL, C.W., ANDERSEN, D.E & KENNEDY, P.L. 2005. Productivity and mortality of Northern Goshawks in Minnesota. *Journal of Raptor Research* 39 : 222–228.

BOHNSACK, P. 1971. Eigenartige Unfall eines Habichts (*Accipiter gentilis*). *Corax* 3 : 199.

BOLD, T.J. 1850. Occurrence of the Goshawk *Falco palumbarius* in Northumberland. *Zoologist* (1850):2765.

BOND, R.M. 1942. Development of young Goshawks. *Wilson Bulletin* 54 : 81–88.

BOND, R.M. & STABLER R.M. 1941. Second-year plumage of the Goshawk. *Auk* 58 : 346–349.

BORODIN, A.I. & SOROKIN, A.G. 1986. [The Goshawk in the forests of Moscow conurbation]. [*Proceedings of the All-Union Ornithological Conferences*] 10 : 94–95. [Russian]

BORG, K., WANNTORP, H., ERNE, K. & HANKO, E. 1969. Alkyl mercury poisoning in terrestrial Swedish wildlife. *Swedish Wildlife* 6 : 302–377.

BORG, K., ERNE, K., HANKO, E. & WANNTORP, H. 1970. Experimental secondary methyl mercury poisoning in the goshawk (*Accipiter gentilis*). *Environmental Pollution* 1 : 91–104.

BOSAKOWSKI, T. 1999. *The Northern Goshawk: Ecology, Behavior, and Management in North America*. Hancock House, Blaine, Washington, USA.

BOSAKOWSKI, T. & RITHALER, J. 1997. *Goshawk and raptor inventory in the Cariboo, 1997.* Report by Beak Pacific Ltd for British Columbia Ministry of Land and Parks, Williams Lake, British Columbia, Canada. [cited in Bosakowski 1999]

BOSAKOWSKI, T. & SMITH, D.G. 1992. Comparative diets of sympatric nesting raptors in the eastern deciduous forest biome. *Canadian Journal of Zoology* 70 : 984–992.

BOSAKOWSKI, T. & SMITH, D.G. 2006. Ecology of the Northern Goshawk in the New York-New Jersey Highlands. *Studies in Avian Biology* 31:109–118.

BOSAKOWSKI, T., SMITH, D.G. & SPEISER. R. 1992. Niche overlap of two sympatric-nesting hawks Accipiter spp. in the New Jersey-New York Highlands. *Ecography* 15 : 358–372.

BOSAKOWSKI, T., MCCULLOUGH, B., LAPSANSKY, F.J. & VAUGN, M.E. 1999. Northern Goshawks nesting on a private industrial forest in western Washington. *Journal of Raptor Research* 33 : 240–244.

BOYCE, D.A., KENNEDY, P.L., BEIER, P., INGRALDI, M.F., MACVEAN, S.R., SIDERS, M.S., SQUIRES, J.R. & WOODBRIDGE, B. 2005. When are goshawks not there? Is a single visit enough to infer absence at occupied nest areas? *Journal of Raptor Research* 39 : 296–302.

BOYCE, D.A., REYNOLDS, R.T. & RUSSELL, R.T. in 2006. Goshawk status and management: what do we know, what have we done, where are we going? *Studies in Avian Biology* 31:312–325.

BRAUN, C.E., LINHART, Y.B., ENDERSON, J.H., MARTI, C.D. & FULLER, M.R. 1996. Northern Goshawk and forest management in the southwestern United States. *Wildlife Society Technical Review* 96–2.

BREE, C.R. (1850). Occurrence of the Goshawk *Falco palumbarius* near Stowmarket. *Zoologist* (1850):2649.

BREHM, H. 1969. Successful breeding experiments with goshawks. *Hawk Chalk* 8 : 20–22.

BRESINSKI, W., PIELOWSKI, Z. & SZOTT, M. 1978. Habichtszucht in Polen 1977. *Deutscher Falkenorden* (1978):19–22.

BRICHETTI, P. & FRACASSO, G. 2003. *Ornitologia italiana. 1 Gaviidae-Falconidae. Identificazione, distribuzione, consistenza e movimenti degle uccelli italiani.* Alberto Perdisa Editore, Bologna, Italy.

BRIGHT SMITH D.J. & MANNAN, R.W. 1994. Habitat use by breeding male Northern Goshawks in northern Arizona. *Studies in Avian Biology* 16 : 58 65.

BRITTEN, M.W., P.L. KENNEDY & S. AMBROSE. 1999. Performance and accuracy of small satellite transmitters. *Journal of Wildlife Management* 63 : 1349–1358.

BROO, B. 1978. Project Eagle-Owl, Southwest. Pp. 104–120 in GEER, T.A. (ed) Bird of Prey Management Techniques. British Falconers' Club, Oxford, UK.

BROOKE, A.B. 1870. Natural history of Wicklow and Kerry. *Zoologist* (1970):2281.

BROOKE, A.B. 1871. Goshawk at Galtee. *Zoologist* (1971):2524.

BROWN, L. & AMADON, D. 1968. *Birds of prey of the world.* Country Life Books, Hamlyn, Feltham, UK.

BROWN, L.H. & BROWN, B.E. 1979. The behaviour of the black sparrowhawk. Accipiter melanoleucus. *Ardea* 67 : 77–95.

BROSSET, A. 1981. Breeding the black sparrowhawk Accipiter melanoleucus in captivity. *Raptor Research* 15 : 58–63.

BRÜLL, H. 1937. *Das Leben deutscher Greifvögel – ihre Bedeutung in der Landschaft.* Jena, Germany.

BRÜLL, H. 1956. Studien über die Bedeutung des Habichts in Niederwildrevier. *Zeitschrift für Jadgwissenschaft* 2 : 165–174.

BRÜLL, H. 1964. *Das Leben deutscher Greifvögel.* Fischer, Stuttgart, Germany.

BRÜLL, H. 1984. *Das Leben europäischer Greifvögel.* Gustav Fischer Verlag, Stuttgart, Germany.

BRÜLL, H., LOOFT, V., RÜGER, A. & ZIESEMER, F. 1981. Keine Klarheit um den Habicht? *Wild und Hund* 15 : 1438–1439.

BÜHLER, U. & OGGIER, P.-A. 1987. Bestand und Bestandsentwicklung des Habichts *Accipiter gentilis* in der Schweiz. *Der Ornithologische Beobachter* 84 : 71–94.

BÜHLER, U., KLAUS, R. & SCHLOSSER, W. 1987. Brutbestand und jungenproduktion des habichts *Accipiter gentilis* in der Nordostschwiez 1979–1984. *Der Ornithologische Beobachter* 84 : 95–110.

BULL, E.L. & HOHMANN, J.H. 1994. Breeding biology of Northern Goshawks in northeastern Oregon. *Studies in Avian Biology* 16 : 103–105.

BURFIELD, I. & van BOMMEL, F. (eds) 2004. *Birds in Europe – population estimates, trends and conservation status.* BirdLife International, Cambridge, UK.

BURT, W.H. 1943. Territoriality and home range concepts as applied to mammals. *Journal of Mammalogy* 24 : 346–352.

BUSCHE, G. & LOOFT V. 2003. Zur Lage der Greifvögel im Westen Schleswig-Holsteins im Zeitraum 1980–2000. *Vogelwelt* 124 : 63–81.

BUSCHE, G., RADDATZ, H.-J. & KOSTRZEWA, A. 2004. Nistplatz-Konkurrenz und Prädation zwischen Uhu (*Bubo bubo*) und Habicht (*Accipiter gentilis*): erste Ergebnisse aus Norddeutschland. *Vogelwarte* 42 : 169–177.

BYHOLM. P. 2003. Reproduction and dispersal of goshawks in a variable environment. PhD thesis, University of Helsinki, Finland.

BYHOLM, P., BROMMER, J.E. & SAUROLA, P. 2002a. Scale and seasonal trends in northern goshawk *Accipiter gentilis* broods. *Journal of Avian Biology*. 33 : 399–406.

BYHOLM, P., RANTA, E., KAITALA, V., LINDÉN, H., SAUROLA, P., & WIKMAN M. 2002b. Resource availability and goshawk offspring ratio variation: large scale ecological phenomenon. *Journal of Animal Ecology.* 71 : 994–1001.

BYHOLM, P., SAUROLA, P., LINDÉN, H. & WIKMAN, M. 2003. Causes of dispersal in Northern Goshawks (*Accipiter gentilis*) in Finland. *Auk* 120 : 706–716.

CADE, T.J. 1960. Ecology of the peregrine and gyrfalcon populations in Alaska. *University of California Publications in Zoology* 63 : 151–290.

CADE, T.J. 2000. Progress in translocation of diurnal raptors. Pp. 343–372 in CHANCELLOR, R.D. & MEYBURG, B.-U. (eds) Raptors at risk. World Working Group on Birds of Prey and Owls, Berlin, Germany.

CADE, T.J. & BURNHAM, W. (eds) 2003. *Return of the peregrine – a North American sage of tenacity and teamwork.* The Peregrine Fund, Boise, Idaho, USA.

CADE, T.J. & TEMPLE, S.A. 1976. The Cornell University falcon programme. Pp. 245–248 in CHANCELLOR, R.D. (ed) Proceedings of the World Conference on Birds of Prey, Vienna 1975. International Council for Bird Preservation.

CADE, T.J., ENDERSON, J.H., THELANDER, C.G. & WHITE, C.M. (eds) 1988. *Peregrine falcon populations. Their management and recovery.* The Peregrine Fund, Boise, Idaho, USA.

CALLOW, E.H. 1946 Comparative studies of meat. *Journal of Agricultural Science* (Cambridge) 37 : 113–126.

CAMPBELL, J. 1780. *A treatise of modern faulconry.* Private printing, Dublin, Eire.

CAPP, P. 1993. Breeding seminar, 5th July, 1992. *The Falconer* (1992):15–22.

CARROLL, C., RODRIGUEZ, R., MCCARTHY, C. & PAULIN, K. 2006. Resource selection function models as tools for regional conservation planning for Northern Goshawk in Utah. *Studies in Avian Biology* 31:288–298.

CHAPE, S., BLYTH, S., FISH, L., FOX, P. & SPALDING M. (eds). 2003. United Nations List of Protected Areas. *IUCN, Gland, Switzerland and Cambridge, UK.*

CHILD, G. 1995. *Wildlife and people: the Zimbabwean success.* Wisdom, Harare, Zimbabwe and New York, USA.

CHUMANKIN, D.V. 2003. [On wintering Goshawks in the central part of Krasnodar Region]. [*Proceedings of the North-Eurasian Raptor Conferences*] 4 : 129–135. [Russian]

CLUTTON-BROCK, T.H., ALBON, S.D. & GUINNESS, F.E. 1985. Parental investment and sex differences in juvenile mortality in birds and mammals. *Nature* 313 : 131–133.

COCHRAN, W.W. & LORD, R.D. 1963. A radio-tracking system for wild animals. *Journal of Wildlife Management* 27:9–24.

COOK, A.S., BELL, A.A. & HAAS, M.B. 1982. *Predatory birds, pesticides and pollution.* Institute of Terrestrial Ecology, Monks Wood, Cambridge, UK.

COOPER, J.E. 1979. The history of hawk medicine. *Veterinary History* 1 : 11–18.

COOPER, J.E. 1981. A historical review of goshawk training and disease. Pages 175–184 in KENWARD, R.E. & LINDSAY, I.M. (eds), *Understanding the Goshawk.* International Association of Falconry and Conservation of Birds of Prey, Oxford, UK.

COOPER, J.E. & PETTY, J. 1988. Trichomoniasis in free-living goshawks (*Accipiter gentilis gentilis*) from Great Britain. *Journal of Wildlife Diseases* 24 : 80–87.

COOPER, M.E. 2000. Legal considerations in the international movement of diagnostic and research samples from raptors – a conference resolution. Pp. 337–343 in LUMEIJ, J.T., REMPLE, J.D., REDIG, P.T., LIERZ, M. & COOPER, J.E. (eds) Raptor Biomedicine III. Zoological Education Network, Fort Worth, USA.

CONRAD, B. 1981. Zur Situation der Pestizidbelastung bei Greifvögel und Eulen der Bundesrepublik Deutschland. *Ökologie der Vögel* 3 : 161–167.

CORDEAUX, J. 1876. Goshawk in Lincolnshire. *Zoologist* (1876):5162.
COX, H. & LASCELLES, G. 1892. *Coursing and falconry.* Longmans, Green and Co., London, UK.
CRAIGHEAD, J.J. & CRAIGHEAD, F.C. 1956. *Hawks, owls and wildlife.* Dover, New York, USA.
CRAMP, S. & SIMMONS, K.E.L. 1980. *Handbook of the birds of Europe, the Middle East and North Africa, Volume 2.* University Press, Oxford, UK.
CROCKER-BEDFORD, D.C. 1990. Goshawk reproduction and forest management. *Wildlife Society Bulletin* 18 : 262 269.
CROCKER-BEDFORD, D. C. & CHANEY, B. 1988. Characteristics of goshawk nesting stands. In GLINSKI, R.L., PENDLETON, B.G., MOSS, M.B., LEFRANC, M.N., MILLSAP, B.A., & HOFFMAN, S. A. (eds), *Proceedings of the southwest raptor management symposium and workshop.* National Wildlife Federation, Scientific Technical Series No. 11, Washington, DC, USA.
CUPPER, J. & CUPPER, I. 1981. *Hawks in focus.* Jaclin, Mildura, Australia.
CURRY-LINDAHL, K. 1950. Berguvens, *Bubo bubo*, förekomst i Sverige jämt något om dess biologi. *Vår Fågelvärld* 9 : 113–165.
CZUCHNOWSKI, R. 1993. [Birds of prey in the Niepołomicka Forest during 1987–1990.] *Notatki Ornitologiczne* 34 : 313–318. [Polish]
DANKO, Š., DIVIŠ, T., DVORSKÁ, J., DVORSKÝ, M., CHAVKO, J., KARASKA, D., KLOUBEC, B., KURKA, P., MATUŠIK, H., PEŠKE, L., SCHRÖPFER, L. & VACÍK, R. 1994. [The state of knowledge of breeding numbers of birds of prey (Falconiformes) and owls (Strigiformes) in the Czech and Slovak Republics as of 1990 and their population trends in 1970–1990.] *Buteo* 6:1–89. [Czech]
DANKO, Š., DARALOVÁ, A. & KRIŠTÍN, A. 2002. *[Bird distribution in Slovakia.]* VEDA, Bratislava, Slovakia. [Slovak]
DAW, S.K. & DESTEFANO, S. 2001. Forest characteristics of Northern Goshawk nest stands and post fledging areas in Oregon. *Journal of Wildlife Management* 65 : 59 65.
DAW, S.K., DESTEFANO, S. & STEIDL, R.J. 1998. Does survey method bias the description of Northern Goshawk nest site structure? *Journal of Wildlife Management* 62 : 1379 1384.
DECANDIDO, R. 2005. First nesting of Cooper's Hawks (*Accipiter cooperii*) in New York City since 1955. *Journal of Raptor Research* 39 : 109.
DE FRAINE, R. & VERBOVEN, R. 1997. Doorbraak van de Havik *Accipiter gentilis* als broedvogel in de Zuiderkempen (Vlaanderen, België). *Oriolus* 63 : 46–48.
DE JUANA, F. 1989. Situación actual de las rapaces diurnas (Orden Falconiformes) in España. *Ecologia* 3 : 237–292.
DEKKER, A.L. & HUT, A. 2004. Morfolgische afwijkingen bij roofvogels: beschrijving van enkele gevallen. *De Takkeling* 12 : 157–162.
DEKKER, A.L., HUT, A. & BIJLSMA, R.G. 2004. De opkomst van de Havik *Accipiter gentilis* in de stad Groningen. *De Takkeling* 12 : 205–218.
DEMANDT, C. 1927. Beobachtungen am Habichtshorst. *Beiträge zur Fortpflanzung der Vögel* 3 : 134–136.
DEMANDT, C. 1933. Neue Beobachtungen über die Flugspiele des Habichts (*Accipiter gentilis*). *Beiträge zur Fortpflanzung der Vögel* 9 : 172–175.
DEMANDT, C. 1962. Der Habicht (*Accipiter gentilis* L.) als Brutvogel des westlichen Sauerlandes. *Ornithologische Mitteilungen* 14 : 57.
DEMENTIEV, G.P. 1951. *Accipiter gentilis*. Pp. 164–181 in DEMENTIEV, G.P. & GLADKOV, N.A. (eds) *Die Vögel der Sowjetunion*, Volume 1. State Publishing House, Moscow, Russia.

DEMENTIEV, G.P. 1955. Bemerkungen über die Verbreitung der Hühnerhabichte in der Ostpaläarktis. *Vogelwelt* 76 : 161–164.

DEMENTIEV, G.P. & BÖHME, R.L. 1970. Über weisse Habichte. *Beiträge zur Vogelkunde* 16 : 67–71.

DEPPE, H.-J. 1976. Ernährungsbiologische Beobachtungen beim Habicht (*Accipiter gentilis*) in einem großstadtnahen Revier. *Ornithologische Berichte für Berlin (West)* 1 : 317–325.

DESIMONE, S.M. 1997. Occupancy rates and habitat relationships of Northern Goshawks in historic nesting areas in Oregon. MSc thesis. Oregon State University, Oregon, USA.

DESTEFANO, S. 2005. A review of the status and distribution of Northern Goshawks in New England. *Journal of Raptor Research* 39 : 342–350.

DESTEFANO, S. & CUTLER, T.L. 1998. *Diets of Northern Goshawks in eastern Oregon.* U. S. Geological Survey, Arizona Cooperative Fish and Wildlife Research Unit, Tucson, Arizona, USA.

DESTEFANO, S., DAW, S.K., DESIMONE, S.M. & MESLOW, E.C. 1994a. Density and productivity of Northern Goshawks: implications for monitoring and management. *Studies in Avian Biology* 16 : 88–91.

DESTEFANO, S., WOODBRIDGE, B. & DETRICH, P.J. 1994b. Survival of Northern Goshawks in the southern Cascades of California. *Studies in Avian Biology* 16 : 133–136.

DESTEFANO, S., MCGRATH, M.T., DESIMONE, S.T. & DAW, S.K. 2006. Ecology and habitat of breeding Northern goshawks in the inland Pacific northwest: a summary of research in the 1990s. *Studies in Avian Biology* 31:75–86.

DETRICH, P.J. & WOODBRIDGE, B. 1994. Territory fidelity, mate fidelity, and movements of color-marked Northern Goshawks in the southern Cascades of California. *Studies in Avian Biology* 16 : 130–132.

DE VOLO, S.B., REYNOLDS, R.T., TOPINKA, J.R., MAY, B. & ANTOLIN, M.F. 2005. Population genetics and genotyping for mark-recpature studies of Northern Goshawks (*Accipiter gentilis*) on the Kaibab Plateau, Arizona. *Journal of Raptor Research* 39 : 286–295.

DEWEY, S. & KENNEDY, P.L. 2001. Effects of supplemental food on parental care strategies and juvenile survival of northern goshawks. *Auk* 118 : 352–365.

DEWEY, S.R., KENNEDY, P.L. & STEPHENS, R.M. 2003. Are dawn vocalization surveys effective for monitoring goshawk nest-area occupancy? *Journal of Wildlife Management* 67 : 390–397.

DICK, T. & PLUMPTON, D. 1998. Review of information on the status of the northern goshawk (*Accipiter gentilis attricapillus*) in the Western Great Lakes Region. Unpublished report for US Fish and Wildlife Service, Fort Snelling, Minnesota, USA [cited in Kennedy 2003]

DIETRICH, J. 1982. Zur Ökologie des Habichts – *Accipiter gentilis* – in Standverband Saarbrücken. Diploma thesis, University of Saarbrücken, Germany.

DIETRICH, J. & ELLENBERG, H. 1981. Aspects of goshawk urban ecology. Pages 163–175 in KENWARD, R.E. & LINDSAY, I.M. (eds) *Understanding the Goshawk.* International Association of Falconry and Conservation of Birds of Prey, Oxford, UK.

DIETZEN, W. 1978. Der Brutbiotop des Habichts *Accipiter gentilis* in drei Gebieten Bayerns. *Anzeiger der Ornithologischen Gesellschaft in Bayern* 17 : 141–159.

DIJKSTRA, C., VUURSTEEN, L., DAAN, S. & MASMAN, D. 1982. Cluch size and laying date in the kestrel *Falcon tinnunculus*: effects of supplementary food. *Ibis* 124 : 210–213.

DIVIŠ, T. 2003. Z biologie a ekologie jestřába lesního (*Accipiter gentilis*). *Panurus* 13:3–32.

DIXON, J.B. & DIXON, R. E. 1938. Nesting of the western Goshawk in California. *Condor* 40:3–11.

DIXON, K.R. & CHAPMAN, J.A. 1980. Harmonic mean measure of animal activity measures. *Ecology* 61 : 1040–1044.

DOBLER, G. 1991. Klimatische Einflüsse auf Dichte, Brutzeit und Bruterfolg von Habicht *Accipiter gentilis* und Rotmilan *Milvus milvus*. *Vogelwelt* 112 : 152–162.

DOBLER, G. & K. SIEDLE. 1993. Fänge von Habichten (*Accipiter gentilis*) im Wurzacher Ried: Kritische Fragen zu einem behördlich genehmigten Wiedereinbürgerungsprojekt. *Journal für Ornithologie* 134 : 165–171.

DOBLER, G. & K. SIEDLE. 1994. Wurzacher Ried: Habichte illegal gefangen und getötet. *Berichte zum Vogelschutz* 32 : 61–74.

DOERR, P.D. 1968. Nesting activities and migratory status of some goshawks in Northeastern Colorado. MSc thesis, Colorado State University, Colorado, USA.

DOMASHEVSKY, S.V. 1995. [Autumnal migration of birds of prey and some waterfowl in the vicinity of the Kyiv reservoir]. *[Proceedings of Ukrainian Ornithol. Society]* 1 : 76–85. [Russian]

DOMASHEVSKY, S.V. 2001. [The birds of prey passage over Kiev city territory]. Pp 216–217 in [Problems of the study and conservation of birds of Eastern Europe and Northern Asia]. 'Matbugat jorty' Press, Kazan, Ukraine. [Russian]

DOMASHEVSKY, S.V. 2003. [The Goshawk ecology in the north of Ukraine]. *Strepet* 1 : 72–85. [Russian]

DONCASTER, C.P. & MACDONALD, D.W. 1991. Drifting territoriality in the red fox *Vulpes vulpes*. *Journal of Animal Ecology* 60: 423–439.

DÖTTLINGER, H. 1993. Breeding attempts with the goshawk (*Accipiter gentilis*). *British Falconers' Club Newsletter* 7 : 19–23.

DOUCET J. 1989. Statut évolutif d' une population d'Autour des palombes (*Accipiter gentilis*) et remarques sur les dénombrement d'animaux. *Aves* (1989), numéro spécial: 103–112.

DOYLE, F.I. 1995. Bald Eagle, *Haliaeetus leucocephalus*, and Northern Goshawk, *Accipiter gentilis*, nests apparently preyed upon by a wolverine(s) *Gulo gulo*, in the southwestern Yukon Territory. *Canadian Field-Naturalist* 109 : 115–116.

DOYLE, F.I. 2000. Timing of reproduction by Red-tailed Hawks, Northern Goshawks and Great Horned Owls in the Kluane Boreal Forest of Southwestern Yukon. MSc thesis, University of British Columbia, Canada.

DOYLE, F.I. 2006. Goshawks in Canada: population responses to harvesting and the appropriateness of using standard bird monitoring techniques to assess their status. *Studies in Avian Biology* 31:135–140.

DOYLE, F. I. & MAHON T. 2001. Inventory of the Northern Goshawk in the Kispiox Forest District, Annual Report 2000. *British Columbia Ministry of Environment, Lands and Parks. Smithers, British Columbia, Canada.*

DOYLE, F.I. & SMITH, J.M.N. 1994. Population responses of Northern Goshawks to the 10-year cycle in numbers of snowshoe hares. *Studies in Avian Biology* 16 : 122–129.

DRACHMANN, J. & NIELSEN, J.T. 2002. *Danske duehøges populationsøkologi og forvaltning.* Danish Environment Ministry report 398, Copenhagen, Denmark.

DRAULANS, D. 1984. *Dagroofvogels te Mol-Postel en omgeving.* De Wielewaal, Turnhout, Belgium.

DRAULANS, D. 1988. Timing of breeding and nesting success of raptors in a newly colonized area in north-east Belgium. *Gerfaut* 78 : 415–420.

DRAZNY, T. & ADAMSKI, A. 1996. The number, reproduction and food of the Goshawk *Accipiter gentilis* in central Silesia (SW Poland). *Populationsökologie Greifvogel- und Eulenarten* 3 : 207–219.

DRENNAN, J.E. & BEIER, P. 2003. Forest structure and prey abundance in winter habitat of Northern Goshawks. *Journal of Wildlife Management* 67 : 177–185.

DUDZIŃSKI, W. 1987. Some aspects of the effect of predators on a partridge *Perdix perdix* L. population. *Proceedings of the congress of the International Union of Game Biologists* 18 : 125–128.

DUDZIŃSKI, W. 1990. The impact of predators on a partridge population in winter. *Proceedings of the congress of the International Union of Game Biologists* 19 : 125–128.

DUMBACHER, J.P., BEEHLER, B.M., SPANDE, T.F., GARRAFFO, H.M. & DALY, J.W. 1992. Homobatrachotoxin in the Genus *Pitohui*: chemical defense in birds? *Science* 258: 799–801.

DVORAK, M., RANNER, A. & BERG, H.-M. (eds) 1993. *Atlas der Brutvögel Österreichs*. Bundesministerium für Umwelt und Familie, Wien, Austria.

EHRING, R. 2004. Bestands- und Reproduktionskontrollen am Habicht (*Accipiter gentilis*) 1970–2002 in Nordwestsachsen. Mitteilungen des Vereins Sächsischer *Ornithologen* 9: 397–405.

ELLENBERG, H. & DIETRICH, J. 1981. The goshawk as a bioindicator. Pages 69–88 in KENWARD, R.E. and LINDSAY, I.M. (eds) *Understanding the Goshawk*. International Association of Falconry and Conservation of Birds of Prey, Oxford, UK.

ELLENBERG, H. & DREIFKE, R. 1993. 'Abrition' – Der Kolkrabe als Schutzschild vor dem Habicht. *Corax* 15:2–10.

ELLENBERG, H., GAST, F. & DIETRICH, J. 1984. Elster, Krähe und Habicht: ein Beziehungsgefüge aus Territorialität, Konkurrenz und Prädation. *Verhandlungen der Gesellschaft für Ökologie* 12 : 319–330.

ELLENBERG, H., DIETRICH, J., GAST, F, HAHN, H. & MAY, R. 1986. Vögel als Biomonitoren für die Schadestoffbelastung von Landschaftsaussschnitten – Ein Überblick. *Verhandlungen der Gesellschaft für Ökologie* 14 : 403–413.

ELLIOTT, J.E. & MARTIN, P.A. 1994. Chlorinated hydrocarbons and shell thinning in eggs of (*Accipiter*) hawks in Ontario, 1989–1989. *Environmental Pollution* 86 : 1989–200.

ELTS, J., KURESOO, A., LEIBAK, E., LEITO, A., LILLELEHT, V., LUIGOJÕE, L., LÕHMUS, A., MÄGI, E. & OTS, M. 2003. Status and numbers of Estonian birds, 1998–2002. *Hirundo* 2 : 58–83.

ENALEEV, I.R. 2001. [The count of autumn migrants of diurnal birds of prey (*Falconiformes*) in central Chukotka)]. P. 230 in *[Problems of the study and conservation of birds of Eastern Europe and Northern Asia]*. 'Matbugat jorty' Press, Kazan, Ukraine. [Russian]

ENALEEV, I.R. 2003. [Autumnal migration of birds of prey at northern coast of the Sea of Okhotsk]. *[Ornithology]* 30 : 198–199. [Russian]

ENG, R. L. & GULLION, G.W. 1962. The predation of goshawks upon ruffed grouse on the Cloquet Forest Research Center, Minnesota. *Wilson Bulletin* 74 : 227–242.

ENGLEMANN, F. 1928. *Die Raubvögel Europas*. Neudamm, Germany [cited in Fischer 1980].

ERDMAN, T.C., BRINKER, D.F., JACOBS, J.P., WILDE, J. & MEYER, T.O. 1998. Productivity, population trend, and status of Northern Goshawks, *Accipiter gentilis atricapillus*, in northeastern Wisconsin. *Canadian Field-Naturalist* 112 : 17–27.

ERKENS, J. & HENDRIX, F. 1984. Prooidieren van buizerd en havik. *De Nederlandse Jager* 89 : 328–329.

ERRINGTON, P.L. 1946. Predation and vertebrate populations. *Quarterly Review of Biology* 21 : 144–177, 221–245.

ERZEPKY, R. 1977. Zur Art des Nahrungserwerbs beim Habicht (*Accipiter gentilis*). *Ornithologische Mitteilungen* 29 : 229–231.

ESTES, W. A., DEWEY S.R., & KENNEDY, P.L. 1999. Siblicide at northern Goshawk nests: Does food play a role? *Wilson Bulletin* 111 : 432–436.

FAIRCLOUGH, J. 2003. The British Falconers' Club breeding scheme. *The Falconer* (2002):253–258.

FAIRHURST, G.D. 2004. Northern Goshawk (*Accipiter gentilis*) population analysis and food habits in the Independence and Bull Run Mountains, Nevada. MSc thesis, Boise State University, Boise, Idaho, USA.

FAIRHURST, G.D. & BECHARD, M.J. 2005. Relationships between winter and spring weather and Northern Goshawk (*Accipiter gentilis*) reproduction in northern Nevada. *Journal of Raptor Research* 39:229–236.

FENZTLOFF, C. 1980. Aviary breeding goshawks. *Falconer* (1979):149–152.

FERGUSON-LEES, J. & CHRISTIE, D.A. 2005. *Raptors of the World*. Christopher Helm, London.

FERGUSON-LEES, J. & CHRISTIE, D.A. 2001. *Raptors of the World*. Christopher Helm, London.

FINN, S.P., VARLAND, D.E. & MARZLUFF, J.M. 2002. Does Northern Goshawk breeding occupancy vary with nest-stand characteristics on the Olympic Peninsula, Washington? *Journal of Raptor Research* 36:265–279.

FISCHER, D.L. 1986. Daily activity patterns and habitat use of coexisting Accipiter hawks in Utah. PhD thesis. Brigham Young University, Provo, Utah, USA.

FISCHER, W. 1979. Paarweise Jagen beim Habicht. Beiträge zur Vogelkunde 25:332.

FISCHER, W. 1980. Die Habichte. Die Neue Brehm-Bücherei, Wittenberg Lutherstadt, Germany.

FISHER, W.R. 1844. Occurrence of the Goshawk at Yarmouth. *Zoologist* (1844):491.

FLEMING, T.L. 1987. Northern Goshawk status and habitat associations in western Washington with special emphasis on the Olympic Peninsula. USDA Forest Service, Pacific Northwest Forest and Range Experiment Station, Olympia, Washington, USA.

FORSMAN, D. & SOLONEN, T. 1984. Censusing breeding raptors in southern Finland: methods and results. *Annales of. Zoologici. Fennici.* 21:317–320.

FOX, N. 1981. The hunting behaviour of trained Northern Goshawks. Pages 121–133 in KENWARD, R.E. & LINDSAY, I.M. (eds) *Understanding the Goshawk*. International Association of Falconry and Conservation of Birds of Prey, Oxford, UK.

FOX, N.C. 1995. *Understanding the bird of prey*. Hancock House, Blaine, USA.

FRANSSON T. & PETTERSSON J. 2001. *Swedish Bird Ringing Atlas. Volume 1. Divers-Raptors.* Naturhistoriska Museet, Stockholm, Sweden.

FRØSLIE, A., HOLT, G. & NORHEIM, G. 1986. Mercury and persistent chlorinated hydrocarbons in owls Strigiformes and birds of prey Falconiformes collected in Norway during the period 1965–1983. *Environmental Pollution* (Series B) 11:91–108.

FUCHS, E. 1973. Durchzug und Überwinterung des Alpenstradläufers Calidris alpina in der Camargue. *Ornithologische Beobachter* 70:113–134.

FUCHS, P. & THISSEN, J.B.M. 1981. Die Situation in den Niederlanden nach den gesetzlichen verordneten Einschränkungen im Gebrauch der chlorierten Kohlenwasserstoff-Pestizide – am Beispiel von Steinkauz und Sperber. *Ökologie der Vögel (Sonderheft)* 3:181–195.

FULLER, M.A. & MOSHER J.A. 1981. Methods of detecting and counting raptors: a review. *Studies in Avian Biology* 6:235–246.

FULLER, M.R., CHURCH, K.E., MILLSPAUGH, J.J. & KENWARD, R.E. 2005. Wildlife Radiotelemetry. Chapter 16 in BRAUN, C.L. (ed) *Manual of Wildlife Management Techniques*. The Wildlife Society, Maryland, USA.

GALBRAITH, C.A., STROUD, D.A. & THOMPSON, D.B.A. 2003. Towards resolving raptor-human conflicts. Pp. 527–535 *in* THOMPSON, D.B.A., REDPATH, S.M., FIELDING, A.H., MARQUISS, M. &

GALBRAITH, C.A. (eds). *Birds of Prey in a Changing Environment*. Scottish Natural Heritage and The Stationary Office, Edinburgh, UK.

GALUSHIN, V.M. 1970. A quantitative estimation of predatory birds' pressure upon game birds' populations in the Central Region of the European part of the USSR. *Proceedings of the International Congress of Game Biologists* 9 : 553–562.

GAMAUF, A. 1988. Der Einfluss des Waldsterbens auf die Horstbaumwahl einiger Greifvogelarten (*Accipitridae*). *Ökologie der Vögel* 10 : 79–83.

GAVIN, T.A., REYNOLDS, R.T., JOY, S.M., LESLIE, D. & MAY, B. 1998. Genetic evidence for low frequency of extra-pair fertilizations in northern Goshawks. *Condor* 100 : 556–560.

GAVIN, T.A. & MAY, B. 1996. Genetic variation and taxonomic status of Northern Goshawks in Arizona: implications for management. Unpublished Report [cited in Squires & Kennedy 2006]

GEDEON K. 1984. Daten zur Brutbiologie des Habichts, *Accipiter gentilis* (L.), im Bezirk Karl-Marx-Stadt. *Faunistische Abhandlungen Staatliches Museum für Tierkunde in Dresden* 11: 157–160.

GEER, T.A. 1978. Effects of nesting sparrowhawks on nesting tits. *Condor* 80 : 419–422.

GEER, T.A. 1979. Sparrowhawk (*Accipiter nisus*) predation on tits (*Parus* sp.). DPhil thesis, Oxford, UK.

GETZ, W.M., FORTMANN, L., CUMMING, D.H.M., DU TOIT, J., HILTY, J., MARTIN, R.B., MURPHREE, M., OWEN-SMITH, N., STARFIELD, A.M. & WESTPHAL, M.I. 1999. Sustaining natural and human capital: villagers and scientists. *Science* 283 : 1855–1856.

GLASIER, P.G. 1963. *As the falcon her bells*. Heinemann, London, UK.

GLASIER, P.G. 1978. *Falconry and hawking*. Batsford, London, UK.

GLUTZ VON BLOTZHEIM, U., BAUER, K. & BEZZEL, E. 1971. *Handbuch der Vogel Mitteleuropas*. Volume 4. Falconiformes. Akademische Verlagsgesellschaft, Frankfurt am Main, Germany.

GOOD, R.E., ANDERSON, S.H., SQUIRES, J.R. & MCDANIEL, G. 2001. Observations of northern Goshawk prey delivery behavior in southcentral Wyoming. *Intermountain Journal of Science* 7 : 34–40.

GÖRANSSON, G. 1975. Duvhökens *Accipiter gentilis* betydelse för vinterdödligheten hos fasaner *Phasianus colchicus*. *Anser* 14 : 11–22.

GÖRANSSON, G. 1982. Voljär-överwintring – ett alternative till uppfödning och utsättning av fälthöns? *Anser* 21 : 25–29.

GÖRZE, H.J. 1981. Erfolgreich Zucht von Habichten (*Accipiter gentilis*) unter Haltebedingungen. *Deutscher Falkenorden* (1980):10–16.

GOSLER, A.G., GREENWOOD, J.J.D. & PERRINS, C.M. 1995. Predation risk and the cost of being fat. *Nature* 377 : 621–623.

GOSLOW, G.E. 1971. The attack and strike of some North American raptors. *Auk* 88 : 815–827.

GOSS-CUSTARD, J.D. 1996. *The Oystercatcher: from individuals to populations*. University Press, Oxford, UK.

GOSZCZYŃSKI, J. 1997. Density and productivity of Common Buzzard *Buteo buteo* and Goshawk *Accipiter gentilis* populations in Rogów, Central Poland. *Acta Ornithologica* 32 : 149–155.

GOSZCZYŃSKI, J. 2001. The breeding performance of the Common Buzzard *Buteo buteo* and Goshawk *Accipiter gentilis* in Central Poland. *Acta Ornithologica* 36 : 105–110.

GOSZCZYŃSKI, J. & PILATOWSKI, T. 1986. Diet of common buzzards (*Buteo buteo* L.) and goshawks (*Accipiter gentilis* L.) in the nesting period. *Ekologia Polska* 34 : 655–667.

GÖTMARK F. 1994. Are bright birds distasteful? A reanalysis of HB Cott's data on the edibility of birds. *Journal of Avian Biology* 25 : 184–97.

GOTT, E. 2000. The Naxi: ancient falconry rediscovered. *British Falconers' Club Newsletter* 20 : 22–24.

GRAHAM, R.T., REYNOLDS, R.T., REISER, M.H., BASSETT, R.L. & BOYCE, D.A. 1994. Sustaining forest habitat for the northern goshawk: A question of scale. *Studies in Avian Biology* 16:12–17.

GRAHAM, R.T., RODRIGUEZ, R.L., PAULIN, K.M., PLAYER, R.L., HEAP, A.P. & WILLIAMS R. 1999. *The northern Goshawk in Utah: habitat assessment and management recommendations.* General Technical Report RMRS-GTR–22. USDA Forest Service, Rocky Mountain Research Station, Fort Collins, Colorado, USA.

GREENWALD, D.M., CROCKER-BEDFORD, D.C., BROBERG, L., SUCKLING, K.F. & TIBBITTS, T. 2005. A review of Northern Goshawk habitat selection in the home range and implications for forest management in the western United States. *Wildlife Society Bulletin* 33:120–129.

GREENWOOD, P.J., HARVEY, P.H. & PERRINS, C.M. 1978. Inbreeding and dispersal in the great tit. *Nature* 271:52–54.

GRELL, M.B. 1998. *Fuglenes Danmark.* Gads Forlag, Copenhagen, Denmark.

GROMME, O.J. 1935. The Goshawk (*Astur atricapillus atricapillus*) nesting in Wisconsin. *Auk* 52:15–20.

GRØNNESBY, S. & NYGÅRD, T. 2000. Using time-lapse video monitoring to study prey selection by breeding Goshawks *Accipiter gentilis* in Central Norway. *Ornis Fennica* 77:117–129.

GRÜNHAGEN, H. 1983. Regionale Unterschiede im Alter brütender Habichtweibchen (*Accipiter gentilis*). *Vogelwelt* 104:208–214.

GRÜNHAGEN, H. 1988. Habichte as Felsbrüter. *Greifvögel und Falknerei* (1988):79–81.

GRÜNHAGEN, H., BEDNAREK, W. & LÜTH, B. 1999. Eine neuartige Voliere für die Habichtszucht. *Greifvögel und Falknerei* (1998):63–66.

GRZYBOWSKI, J.A. & EATON, S.W. 1976. Prey items of Goshawks in southwestern New York. *Wilson Bulletin* 88:669–670.

GULLION, G.W. 1981. The impact of Goshawk predation upon ruffed grouse. *Loon* 53:82–84.

GUNN, T.E. 1870. Goshawk in Norfolk. *Zoologist* (1870):2221.

GUNN, T.E. 1888. Reported occurrence of the Goshawk in Norfolk. *Zoologist* (1888):32.

GURNEY, J.H. 1851. Occurrence of the Goshawk (Falco palumbarius) in Norfolk. *Zoologist* (1851):3027.

HADDON, B. 1981. Goshawk breeding in enclosures. Pp. 190–193 in KENWARD, R.E. & LINDSAY, I.M. (eds) *Understanding the Goshawk.* International Association of Falconry and Conservation of Birds of Prey, Oxford, UK.

HAFTORN 1971. *Norges fugler.* Universitetsforlaget, Oslo, Norway.

HAGEMEIJER, E.J.M. & BLAIR, M.J. (eds) 1997. *The EBCC atlas of European breeding birds: their distribution and abundance.* Poyser, London, UK.

HAGEN, Y. 1942. Totalgewichts-Studien bei norwegischen Vogelarten. *Archiv für NATURGESCHICHTE* N.F. 11:1–173.

HAGEN, Y. 1947. Does the merlin sometimes play a role as the protector of fieldfare colonies on the fjells? *Vår Fågelvärld* 6:137–141.

HAHN, E. 1991. *Schwermetallgehalte in Vogelfedern – ihre Ursache und der Einsatz von Federn standorttreuer Vogelarten im Rahmen von Bioindikationsverfahren.* Berichte 2493 des Forschungszentrums Jülich, Germany.

HAKKARAINEN, H. & KORPIMÄKI, E. 1991. Reversed sexual size dimorphism in Tengmalm's owl: is small male size adaptive? *Oikos* 61:337–346.

HAKKARAINEN, H. & KORPIMÄKI, E. 1993. The effect of female body size on clutch volume of Tengmalm's owls (*Aegolius funereus*) in varying food conditions. *Ornis Fennica* 70:189–195.

HALDANE, J.B.S. 1955. The calculation of mortality rates from ringing data. *Proceedings of the International Ornithological Congress* 11 : 454–458.

HALL, P.A. 1984. Characterization of nesting habitat of Goshawks (*Accipiter gentilis*) in northwestern California. MSc thesis. Humboldt State University, Arcata, California, USA.

HALLEY, D.J. 1996. Movements and mortality of Norwegian goshawks: an analysis of ringing data. Fauna Norvegica Series C, *Cinclus* 19 : 55–67.

HALLEY, D.J., NYGÅRD, T. & WISETH, B. 2000. Winter home range fidelity and summer movements of a male goshawk from fledging to first breeding. *Ornis Norvegica* 23 : 31–37.

HAMERSTROM, F. 1963. The use of Great Horned Owls in catching Marsh Hawks. *Proceedings of the International Ornithological Congress* 13 : 866–869.

HAMILTON, W.D. 1971. Geometry for the selfish herd. *Journal of Theoretical Biology* 31 : 295–311.

HANAUSKA-BROWN, L.A., BECHARD, M.J. & ROLOFF, G.J. 2003. Northern goshawk breeding ecology and nestling growth in mixed coniferous forests of west-central Idaho. *Northwest Science* 77 : 331–339.

HANTGE, E. 1980. Untersuchungen über den Jagderfolg mehrerer europäischer Greifvögel. *Journal für Ornithologie* 121 : 200–207.

HARGIS, C.D., MCCARTHY, C. & PERLOFF, R.D. 1994. Home ranges and habitats of Northern Goshawks in eastern California. *Studies in Avian Biology* 16 : 66–74.

HARGIS, C.D. & WOODBRIDGE, B. 2006. A design for monitoring Northern Goshawks (*Accipiter gentilis*) at the bioregional scale. *Studies in Avian Biology* 31:274–287.

HARRADINE, J., REYNOLDS, N. & LAWS, T. 1997. Raptors and gamebirds: a survey of game managers affected by raptors. *British Association for Shooting and Conservation, Chester, UK.*

HARTING, J.E. 1867. Goshawk in Ireland. *Zoologist* (1867):632.

HARTING, J.E. 1891. *Bibliotheca Accipitraria*. Quaritch, London.

HAUKIOJA, E. & HAUKIOJA, M. 1970. Mortality rates of Finnish and Swedish Goshawks (*Accipiter gentilis*). *Finnish Game Research* 31 : 13–20.

HAURI, R. 1963. Murmeltier als Habichtsbeute. *Ornithologische Beobachter* 60 : 143. [cited in Fischer 1980]

HAUSCH, I. 1997. Habicht *Accipiter gentilis* (Linné 1758). Chapter 8.1.11.1 in K.-H. BERCK, R. BURKHARDT, O. DIEHL, W. HEIMER, M. KORN, & W. SCHINDLER (eds) *Avifauna von Hessen*. Hessische Gesellschaft für Ornithologie und Naturschutz, Echzell, Germany.

HAWK BOARD. 1988. *Proceedings of the goshawk workshop held at Birmingham University on 13 February 1988*. Hawk Board, Gloucestershire, UK.

HAWK BOARD. 1992. The production of registered aviary-bred diurnal raptors in UK 1980–1991. *British Falconers' Club Newsletter* 4 : 18–19.

HAYWARD, G.D. & ESCAÑO, R.E. 1989. Goshawk nest-site characteristics in western Montana and northern Idaho. *Condor* 91 : 476–479.

HEIDENREICH, M. 1996. *Greifvögel: Krankheiten – Haltung – Zucht*. Blackwell, Vienna, Austria.

HEISE, G. 1986. Siedlungsdichte und Bruterfolg des Habichts (*Accipiter gentilis*) im Kreis Prenzlau, Uckermark. *Beiträge zur Vogelkunde* 32 : 113–120.

HENCKELL, T. 1997. Habichtszucht 1995 in Hamburg-Duvenstedt. *Greifvögel und Falknerei* (1995):30.32.

HENNESSY, S.P. 1978. Ecological relationships of accipiters in northern Utah with special emphasis on the effects of human disturbance. MSc thesis. Utah State University, Logan, Utah, USA.

HENNY, C.J., OLSON, R. A. & FLEMING T.L. 1985. Breeding chronology, molt, and measurements of accipiter hawks in northeastern Oregon. *Journal of Field Ornithology* 56 : 97–112.

HERRLINGER, E. 1973. Die Wiedereinburgerung des Uhus *Bubo bubo* in der Bundesrepuklik Deutschland. *Bonner Zoologischer Monographien* 4.

HERZKE, D., KALLENBORN, R. & NYGÅRD, T. 2002. Organochlorines in egg samples from Norwegian birds of prey: Congener-isomer and enantiometer specific considerations. *Science of the Total Environment* 291 : 59–71.

HICKEY, J.J. 1969. (ed.) *Peregrine Falcon populations: their biology and decline.* University of Wisconsin Press, Madison, Wisconsin, USA.

HILLERICH, K. 1978. Ergebnisse aus 20-jähriger Planberingung von Greifvögeln der Beringungsgemeinschaft Rothmann. *Luscinia* 43 : 187–205.

HINES, B. 1968. *A kestrel for a knave.* Penguin, London, UK.

HIRONS, G.J.M., HARDY, A.R. & STANLEY, P.I. 1984. Body weight, gonad development and moult in the Tawny Owl (Strix aluco). *Journal of Zoology (London)* 202 : 145–164.

HODDER, K.H. 1993. Mediated flushing: the use of avian beaters by the Brown Falcon Falco berigora. *Australian Birdwatcher* 15 : 164–165.

HODDER, K.H., KENWARD, R.E., WALLS, S.S. & CLARKE, R.T. 1998. Estimating core ranges: a comparison of techniques using the Common Buzzard (*Buteo buteo*). *Journal of Raptor Research* 32 : 82–89.

HÖGLUND, N. 1964a. Der Habichts *Accipiter gentilis* Linné) in Fennoskandia. *Swedish Wildlife Research* 2 : 195–270.

HÖGLUND, N. 1964b. Über die Ernährung des Habichts (*Accipiter gentilis* L.) in Schweden. *Swedish Wildlife Research* 2 : 271–328.

HÖGLUND, N.H. 1966. Über die Ernährung des Uhus *Bubo bubo* Lin. in Schweden. *Swedish Wildlife Research* 4 : 43–80.

HOKIICHE, H. 1822 (ed) *Youyui [Record of Breeding Hawks].* Zoku Gunsho Ruijuu – takabu [Grand Library – Hawking Division], wood press, Yedo, Japan.

HOLLING, C.S. 1959. Some characteristics of simple types of predation and parasitism. *Canadian Entomologist* 91 : 385–398.

HOLSTEIN, V. 1942. *Duehøgen Astur gentilis dubius (Sparrman).* Hirschsprung, Copenhagen, Denmark.

HOLZ, P. & NAISBITT, R. 2000. Fitness level as a determining factor in the survival of rehabilitated raptors released back into the wild – preliminary results. Pp. 321–325 in LUMEIJ, J.T., REMPLE, J.D., REDIG, P.T., LIERZ, M. & COOPER, J.E. (eds) *Raptor Biomedicine III.* Zoological Education Network, Fort Worth, USA.

HORN, G. 1846. An unusual capture. *Zoologist(1846)*:1496.

HOUSTON, D. 1977. The effect of Hooded Crows on hill sheep farming in Argyll, Scotland. *Journal of Applied Ecology* 14:1–15.

HUHTALA, K. & SULKAVA, S. 1976. Kanahaukan pesimabiologiasta. *Suomen Luonto* 15 : 299–303.

HUHTALA, K. & SULKAVA, S. 1981. Environmental influences on goshawk breeding in Finland. Pages 89–104 in KENWARD, R.E. & LINDSAY, I.M. (eds) *Understanding the Goshawk.* International Association of Falconry and Conservation of Birds of Prey, Oxford, UK.

HURRELL, L.H. 1970. The kestrel (*Falco tinnunculus*) breeds in captivity. *Captive Breeding of Diurnal Birds of Prey* 1:8–10.

HURRELL, L.H. 1977. Breeding in skylight and seclusion facilities. Pp. 30–36 in COOPER, J.E. & KENWARD, R.E. (eds) *Papers on the veterinary medicine and domestic breeding of diurnal birds of prey.* British Falconers' Club, Oxford, UK.

HUTTON, J.M. & LEADER-WILLIAMS, N. 2003. Sustainable use and incentive-driven conservation: realigning human and conservation interests. *Oryx* 37 : 215–226.

INGRALDI, M.F. 2005. A skewed sex ratio in Northern Goshawks: is it a sign of a stressed population. *Journal of Raptor Research* 39 : 247–252.

IVANOVSKY, V.V. 1991. [Breeding ecology of the Goshawk in Northern Belarus]. *[Proceedings of the All-Union Ornithological Conferences]* 10 : 238–239. [Russian]

IVANOVSKY, V.V. 1998. Current status and breeding ecology of the goshawk *Accipiter gentilis* in northern Belarus. Pp. 111–115 in MEYBURG, B.-U., CHANCELLOR, R.D. & FERRERO, J.J. (eds) *Holarctic birds of prey.* Asocación para la Defensa de la Naturaleza y los Resursos de Extramadura and World Working Group on Birds of Prey, Berlin, Germany.

IVANOVSKY, V.V., BASHKIROV, I.V. & SHAMOVICH, D.I. 2003. [Goshawk in Northern Belarus in 1995–1999]. *[Proceedings of the North-Eurasian Raptor Conferences]* 4: 80–81. [Russian]

IVERSON, G.C., HAYWARD, G.D., TITUS, K., DEGAYNER, E., LOWELL, R.E., CROCKER-BEDFORD, C.D., SCHEMPF, P.F & LINDELL, J. 1996. *Conservation assessment for the northern goshawk in Southeast Alaska.* US Department of Agriculture Forest Service, Portland, Oregon, USA.

JACK, T.A.M. 1970. Hawk sense II. The wind. *Falconer* (1970):218–228.

JACK, T.A.M. 1971. Hawk sense III. Flight. *Falconer* (1971):277–293.

JACOB, M. & WITT, K. 1986. Beutetiere des Habichts (*Accipiter gentilis*) zur Brutzeit in Berlin 1982–1986. *Ornithologische Berichte für Berlin (West)* 11 : 187–195.

JACOBS, E.A. 1996. A mechanical owl as a trapping lure for raptors. *Journal of Raptor Research* 30 : 31–32.

JAMES, J.C. 1970. Geographic size variation in birds and its relationship to climate. *Ecology* 51 : 365–390.

JAMESON, E.W. 1962. *The Hawking of Japan: The History and Development of Japanese Falconry.* Davis, California, USA.

JĘDRZEJEWSKA, B. & JĘDRZEJEWSKI, W. 1998. *Predation in Vertebrate Communities. The Białowieża Forest as a case study.* Springer Verlag, Berlin, Germany.

JEFFREYS, A.J., WILSON, V. & THEIN, S.L. 1985. Individual-specific 'finger-prints' of human DNA. *Nature* 316 : 76–79.

JENKINS, D., WATSON, A. & MILLER, G.R. 1964. Predation and red grouse populations. *Journal of Applied Ecology* 1 : 183–195.

JOHNELS, A., TYLER, G. & WESTERMARK, T. 1979. A history of mercury levels in Swedish fauna. *Ambio* 8 : 160–168.

JOHANSSON, C., HARDIN, P.J. & WHITE, C.M. 1994. Large-area goshawk habitat modelling in Dixie National Forest using vegetation and elevation data. *Studies in Avian Biology* 16 : 50–57.

JOHANSEN, R. 1957. Die Vogelfauna Westsibiriens III. *Journal für Ornithologie* 98 : 397–399.

JOIRIS, C. & DELBEKE, K. 1985. Contamination by PCBs and Organochlorine Pesticides of Belgian Birds of Prey, their eggs and their Food, 1969–1982. Pp. 403–414 in NURNBERG, H.W. (ed.) *Pollutants and their Ecotoxicological Significance.* John Wiley and Sons, London, UK.

JOLLIE, M. 1977. Phylogeny within the Falconiform groups. *Evolutionary Theory* 3 : 106–123.

JONES, C.G., HECK, W., LEWIS, R.E., MUNGROO, Y., SLADE, G. & CADE, T. 1994. The restoration of the Mauritius kestrel *Falco punctatus* population. *Ibis* 137 : 173–180.

JONES, M.D. 2003. Do we really know the imprinted goshawk? *Falconer* (2002):243–245.

JOUBERT, B. 1987. Quelques donneés sur la reproduction de l'Autour des palombes en Haute-Loire. *Le Grand-Duc* 30 : 11–15.

JOUBERT, B. 1994. Autour des palombes. Pp. 190–191 in D. Yeatman-Berthelot (ed). *Nouvel atlas des oiseaux nicheurs de France 1985–1989.* Société Ornithologique de France, Paris, France.

JOUBERT, B & MARGERIT, T. 1986. Aspects du comportement de l'autour, *Accipiter gentilis*, en Haute-Loire. *Nos Oiseaux* 38 : 209–228.

JOY, S.M., REYNOLDS, R.T. & LESLIE, D.G. 1994. Northern Goshawk broadcast surveys: hawk response variables and survey costs. *Studies in Avian Biology* 16 : 24–30.

KALABÉR, L. 1984. Note sulla biologia e lo sviluppo postembrionale dell'astore *Accipiter gentilis*, in Romania. *Riv. Ital. Orn., Milano* 45 : 179–190.

KALCHREUTER, H. 1981a. Der Habicht ist gefährlich. *Wild und Hund* 15 : 1440–1443.

KALCHREUTER, H. 1981b. The goshawk *Accipiter gentilis* in Western Europe. Pp. 18–28 in KENWARD, R.E. & LINDSAY, I.M. (eds) *Understanding the Goshawk*. International Association for Falconry and Conservation of Birds of Prey, Oxford, UK.

KARLBOM, M. 1981. Techniques for trapping goshawks. Pages 138–144 in KENWARD, R.E. & LINDSAY, I.M. (eds) *Understanding the Goshawk*. International Association for Falconry and Conservation of Birds of Prey, Oxford, UK.

KAYSER, Y. 1993. Le régime alimentaire de l'Autour des palombes, *Accipiter gentilis* (L.) en Alsace. *Ciconia* 17 : 143–166.

KEANE, J. J. 1999. Ecology of the northern goshawk in the Sierra Nevada, California. PhD Dissertation, University of California, Davis, California, USA.

KEANE, J.J. & MORRISON, M.L. 1994. Northern Goshawk ecology: effects of scale and levels of biological organization. *Studies in Avian Biology* 16:3–11.

KEANE, J.J., MORRISON, M.L. & FRY, D.M. 2006. Prey and weather factors associated with temporal variation in Northern Goshawk reproduction in the Sierra Nevada, California. *Studies in Avian Biology* 31:87–99.

KEHL & ZERNING. 1993 [cited in] ALTENKAMP, R. & S. HEROLD. 2001. Habicht (*Accipiter gentilis*). Pp. 175–179 in ABBO. *Die Vogelwelt von Brandenburg und Berlin.* Verlag Natur and Text, Rangsdorf.

KEITH, L.B. 1963. *Wildlife's ten-year cycle.* University of Wisconsin Press, Madison, Wisconsin, USA.

KEITH, L.B. & RUSCH, D.H. 1988. Predation's role in the cyclic fluctuations of ruffed grouse. *Proceedings of the International Ornithological Congress* 19 : 699–732.

KEITH, L.B. & WINDBERG, L.A. 1978. A demographic analysis of the snowshoe hare cycle. *Wildlife Monographs* 58:1–70.

KEITH, L.B., TODD, A.D., BRAND, C.J., ADAMCIK, R.S. & RUSCH, D.H. 1977. An analysis of predation during a cyclic fluctuation of snowshoe hares. *Proceedings of the International Congress of Game Biologists* 13 : 151–175.

KENDEIGH, S.C. 1970. Energy requirements for existence in relation to size of bird. *Condor* 72 : 60–65.

KENNEDY, P.L. 1988. Habitat characteristics of Cooper's hawks and northern goshawks nesting in New Mexico. *National Wildlife Federation Scientific and Technical Series* 11 : 218–227.

KENNEDY, P.L. 1991. Reproductive strategies of northern goshawks and Cooper's hawks in north-central New Mexico. PhD thesis, Utah State University. Logan, Utah, USA.

KENNEDY, P.L. 1997. The Northern Goshawk (*Accipiter gentilis atricapillus*): is there evidence of a population decline? *Journal of Raptor Research* 31 : 95–106.

KENNEDY, P.L. 1998. Evaluating northern Goshawk (*Accipiter gentilis atricapillus*) population trends: a reply to Smallwood and Crocker-Bedford. *Journal of Raptor Research* 32 : 336–342.

KENNEDY, P.L. 2003. Northern goshawk (*Accipiter gentilis atricapillus*): a technical conservation assessment. USDA Forest Service, Rocky Mountain Region. www.fs.fed.us/r2/projects/scp/assessments/northerngoshawk.pdf

KENNEDY, P.L. & GESSAMEN, J.A. 1991. Diurnal resting metabolic rates of accipiters. *Wilson Bulletin* 103 : 101–105.

KENNEDY, P.L. & STAHLECKER, D.W. 1993. Responsiveness of nesting Northern Goshawks to taped broadcasts of 3 conspecific calls. *Journal of Wildlife Management* 57 : 249–257.

KENNEDY, P.L. & WARD, J.M. 2003. Effects of experimental food supplementation on movements of juvenile northern Goshawks (*Accipiter gentilis atricapillus*). *Oecologia* 134: 284–291.

KENNTNER, N., KRONE, O., ALTENKAMP, R. & TATARUCH, F. 2003. Environmental contaminants in liver and kidney of free-ranging Northern Goshawks (*Accipiter gentilis*) from three regions of Germany. *Archives of Environmental Contamination and Toxicology* 45 : 128–135.

KENWARD, R.E. 1974. Mortality and fate of trained birds of prey. *Journal of Wildlife Management* 38 : 751–756.

KENWARD, R.E. 1976a. The effect of predation by goshawks, *Accipter gentilis*, on woodpigeon, *Columba palumbus*, populations. DPhil thesis, University of Oxford, UK.

KENWARD, R.E. 1976b. Captive breeding: a contribution of falconers to the preservation of Falconiformes. Pp. 378–381 in CHANCELLOR, R.D. (ed.) *Proceedings of the World Conference on Birds of Prey, Vienna 1975*. International Council for Bird Preservation, London, UK.

KENWARD, R.E. 1977. Predation on released pheasants (*Phasianus colchicus*) by goshawks (*Accipiter gentilis*) in central Sweden. *Swedish Game Research* 10 : 79–112.

KENWARD, R.E. 1978a. Radio transmitters tail-mounted on hawks. *Ornis Scandinavica* 9 : 220–223.

KENWARD, R.E. 1978b. Hawks and doves: factors affecting success and selection in goshawk attacks on woodpigeons. *Journal of Animal Ecology* 47 : 449–460.

KENWARD, R.E. 1978c. The influence of human and goshawk *Accipiter gentilis* activity on woodpigeons *Columba palumbus* at brassica feeding sites. *Annals of Applied Biology* 89 : 277–286.

KENWARD, R.E. 1979. Winter predation by goshawks in lowland Britain. *British Birds* 72 : 64–73.

KENWARD, R.E. 1981a. The causes of death in trained raptors. Pp. 27–29 in COOPER, J.E. & GREENWOOD, A.G. *Recent advances in the study of raptor diseases*. Chiron Publications, Keighley, UK.

KENWARD, R.E. 1981b. Goshawk re-establishment in Britain – causes and implications. *Falconer* 7 : 304–310.

KENWARD, R.E. 1982. Goshawk hunting behaviour, and range size as a function of habitat availability. *Journal of Animal Ecology* 51 : 69–80.

KENWARD, R.E. 1986. Problems of goshawk predation on pigeons and some other game. *Proceedings of the International Ornithological Congress* 18 : 666–678.

KENWARD, R.E. 1987a. *Wildlife Radio Tagging – Equipment, Field Techniques and Data Analysis*. Academic Press, London, UK.

KENWARD, R.E. 1987b. Protection versus management in raptor conservation: the role of falconry and hunting interests. Pp 1–13 in HILL, D.J. (ed.) *Breeding and Management in Birds of Prey*. Bristol University Press, Bristol, UK.

KENWARD, R.E. 1993. Modelling raptor populations: to ring or to radio tag? Pp. 157–167 in LEBRETON, J.D. & NORTH, P.M. (eds) *The use of marked individuals in the study of bird population dynamics: models, methods and software*. Birkhauser, Basle, Switzerland.

KENWARD, R.E. 1996. Goshawk adaptation to de-forestation: does Europe differ from North America. Pp. 233–243 in BIRD, D.M, VARLAND, D.E. & NEGRO J.J. (eds) *Raptors in human landscapes.* Academic Press, London, UK.

KENWARD, R.E. 2000. Socio-economic problems and solutions in raptor predation. pp. 565–570 in CHANCELLOR, R.D. & MEYBURG, B.-U. (eds.) *Raptors at Risk.* World Working Group on Birds of Prey and Owls, Berlin, Germany.

KENWARD, R.E. 2001. *A manual for wildlife radio-tagging.* Academic Press, London, UK.

KENWARD, R.E. 2002a. *Management tools for reconciling bird hunting and biodiversity.* Workpackage 4 in Reconciling Gamebird Hunting and Bioversity (REGHAB). EVK2-CT–2000–200004. www.uclm.es/irec/reghab/inicio.html

KENWARD, R.E. 2002b. Identifying the main threats to raptor populations. Pp. 15–21 in YOSEF, R., MILLER, M.L. & PEPLER, D. (eds) *Raptors in the New Millenium.* International Birding and Research Centre at Eilat, Israel.

KENWARD, R.E. 2004. Management tools for raptors. Pp. 329–339 in CHANCELLOR, R.D. & B.-U. MEYBURG (eds). *Raptors Worldwide.* World Working Group on Birds of Prey and Owls, Berlin, Germany.

KENWARD, R.E. & GAGE, M.J.G. in press. Incentive-based conservation of game birds through falconry.

KENWARD, R.E. & GARCÍA CIDAD, V. 2002. *Innovative approaches to sustainable use of biodiversity and landscape in the farmed countryside.* UNEP High-Level Conference on Agriculture and Biodiversity. http://nature.coe.int/conf_agri_2002/agri16erev.01.doc

KENWARD, R.E. & MARCSTRÖM, V. 1981. Goshawk predation on game and poultry: some problems and solutions. Pp. 152–162 in KENWARD, R.E. & LINDSAY, I.M. *Understanding the Goshawk.* International Association of Falconry and Conservation of Birds of Prey, Oxford, UK.

KENWARD, R.E. & MARCSTRÖM, V. 1988. How differential competence could sustain suppressive predation on birds. *Proceedings of the International Ornithological Congress* 19 : 733–742.

KENWARD, R.E. & SIBLY, R.M. 1977. A woodpigeon (*Columba palumbus*) feeding preference explained by a digestive bottleneck. *Journal of Applied Ecology* 14 : 815–826.

KENWARD, R.E. & WALLS, S.S. 1994. The systematic study of radio-tagged raptors: I. Survival, home-range and habitat-use. Pp. 303–315 in MEYBURG, B.-U. & CHANCELLOR, R.D. (eds) *Raptor Conservation Today.* World Working Group on Birds of Prey, Berlin, Germany.

KENWARD, R.E. & WIDÉN, P. 1989. Do goshawks need forests? Some conservation lessons from radio tracking. In B.-U. MEYBURG & R.D. CHANCELLOR (eds) *Raptors in the modern world.* World Working Group on Birds of Prey, Berlin, Germany.

KENWARD, R.E., MARCSTRÖM, V. & KARLBOM, M. 1981a. Goshawk winter ecology in Swedish pheasant habitats. *Journal of Wildlife Management* 45 : 397–408.

KENWARD, R.E., MARQUISS, M. & NEWTON, I. 1981b. What happens to goshawks trained for falconry. *Journal of Wildlife Management* 45 : 802–806.

KENWARD, R.E., HIRONS, G.J.M. & ZIESEMER, F. 1982. Devices for telemetering the behaviour of free-living birds. *Symposia of the Zoological Society of London* 49 : 129–137.

KENWARD, R.E., KARLBOM, M. & MARCSTRÖM, V. 1983. The price of success in goshawk trapping. *Raptor Research* 17 : 84–91.

KENWARD, R.E., MARCSTRÖM, V. & KARLBOM, M. 1993a. Post-nestling behaviour in goshawks, *Accipiter gentilis*: II. Sex differences in sociality and nest switching. *Animal Behaviour* 46 : 371–378.

KENWARD, R.E., MARCSTRÖM, V. & KARLBOM, M. 1993b. Post-nestling behaviour in goshawks, *Accipiter gentilis*: I. The causes of dispersal. *Animal Behaviour* 46 : 365–70.

KENWARD, R.E., MARCSTRÖM, V. & KARLBOM, M. 1999. Demographic estimates from radio-tagging: models of age-specific survival and breeding in the goshawk. *Journal of Animal Ecology* 68 : 1020–1033.

KENWARD, R.E., WALLS, S.S., HODDER, K.H., PAHKALA, M., FREEMAN, S.N. & SIMPSON, V. R. 2000. The prevalence of non-breeders in raptor populations: evidence from rings, radio-tags and transect surveys. *Oikos* 91 : 271–279.

KENWARD, R.E., PFEFFER, R.H., AL-BOWARDI, M.A., FOX, N.C., RIDDLE, K.E., BRAGIN, Y.A., LEVIN, A.S., WALLS, S.S. & HODDER, K.H. 2001a. Setting harness sizes and other marking techniques for a falcon with strong sexual dimorphism. *Journal of Field Ornithology* 72 : 244–257.

KENWARD, R.E., WALLS, S.S. & HODDER, K.H. 2001b. Life path analysis: scaling indicates priming effects of social and habitat factors on dispersal distances. *Journal of Animal Ecology* 70:1–13.

KENWARD, R.E., CLARKE, R.T., HODDER, K.H. & WALLS, S.S. 2001c. Density and linkage estimators of home range: nearest-neighbor clustering defines multi-nuclear cores. *Ecology* 82 : 1905–1920.

KENWARD, R.E., HALL, D.G., WALLS, S.S. & HODDER, K.H. 2001d. Factors affecting predation by buzzards (*Buteo buteo*) on released pheasants (*Phasianus colchicus*). *Journal of Applied Ecology* 38 : 813–822.

KENWARD, R.E., RUSHTON, S.P., PERRINS, C.M., MACDONALD, D.W, & SOUTH, A.B. 2002. From marking to modelling: dispersal study techniques for land vertebrates. Pp. 50–71 in BULLOCK, J.M., KENWARD, R.E. & HAILS, R. (eds) *Dispersal: an ecological perspective.* Symposium of the British Ecological Society, Blackwell, Oxford, UK.

KENWARD, R., KATZNER, T., WINK, M., MARCSTRÖM, V., WALLS, S., KARLBOM, M., PFEFFER, R., BRAGIN, E., HODDER, K. & LEVIN, A. (in press.) Rapid sustainability modelling for raptors with radio-tags and DNA-fingerprints. *Journal of Wildlife Management.*

KIMMEL, J.T. 1995. Spatial hierarchy of habitat use by northern goshawks in two forest regions of Pennsylvania. PhD thesis, Pennsylvania State University, University Park, Pennsylvania, USA.

KIMMEL, J.T. & YAHNER, R.H. 1990. Response of Northern Goshawks to taped conspecific and Great Horned Owl call. *Journal of Raptor Research* 24 : 107–112.

KIMMEL, J.T. & YAHNER, R.H. 1994. *The northern Goshawk in Pennsylvania: habitat use, survey protocols, and status (Final Report).* School of Forestry Resources, Pennsylvania State University, USA.

KIMSON, M. 1993. Breeding seminar, 5th July, 1992 – goshawks. *Falconer* (1992):16–18.

KIRK, D. A. & HYSLOP C. 1998. Population status and recent trends in Canadian raptors: a review. *Biological Conservation.* 83 : 91–118.

KIRKLEY, J. 1999. Foraging and roosting habitats of overwintering northern goshawks (*Accipiter gentilis*) in Southwestern Montana. Proceedings from the Raptor Research Foundation Annual Meeting. La Paz, Baja California Sur, Mexico.

KJELLEN, N. 1998. Rovfågelsträcket över Falsterbohalvön hösten 1997. *Anser* 37 : 19–35.

KLAAS, C. 1967. Lebenstätte und Beuteauswahl dreier Habichtspaare. *Natur und Volk* 97 : 347–353.

KLAWES, M. 1956. Seltener Todesfall eines Habichts. *Falke* 3 : 70.

KLEINSCHMIDT, O. 1922–23. *Berajah, Zoografia infinita. Die Realgattung Habicht, Falco columbarius (Kl.).* Gebauer-Schwetschke, Halle, Germany.

KLEINSCHMIDT, O. 1934. *Die Raubvögel der Heimat.* Quelle & Meyer, Leipzig, Germany.

KLUTH, S. 1984. Untersuchungen zur Beutewahl des Habichts (*Accipiter gentilis* L.): test der Telemetrie und Kritik bisher angewandter Methoden. Diploma thesis, Ludwig-Maximillians University, Munich, Germany.

KNÜWER, H. 1981. Ergebnisse einer fünfjährigen Greifvogelbestandsaufnahme im Münsterland. *Charadrius* 17 : 131–143.

KOEHLER, A. 1970. Breeding birds of prey in captivity. *Captive breeding of Diurnal Birds of Prey* 1:2–6.

KOEHLER, B. & BAUMGART, W. 1972. Toxi-Infektionen durch Clostridium perfringens Typ A bei Greifvögeln. *Monatsheft Veterinärischen Medizin* 25 : 348–352.

KOEMAN, J.H. & VAN GENDEREN H. 1975. Some preliminary notes on residues of chlorinated hydrocarbon insecticides in birds and mammals in the Netherlands. *Mededelingen Landbouwhogeschool Gent* 30 : 1879–1887.

KOEMAN, J.H., ENSKIN, H.J.A., FUCHS, P., HOSKAM, E.G., MÖRZER-BRUYNS, M.F. & DE VOS, R.H. 1968. Vogelsterfte door landbouwvergiften. *Landbouwker Tijdschrift* 80 : 206–214.

KOEMAN, J.H., VINK, J.A.J. & de GOEIJ, J.J.M. 1969. On the causes of mortality in predatory birds in the Netherlands in the winter of 1968/1969. *Ardea* 57 : 67–76.

KOEMAN J.H., VAN BEUSEKOM, C.F. & DE GOEIJ, J.J.M. 1972. Eggshell and population changes in the Sparrow-hawk (*Accipiter nisus*). *TNO nieuws* 1972 : 542–550.

KOLLINGER, D. 1962. Bemerkenswerte Habichtsbeobachtungen. *Deutscher Falkenorden* (1962): 52–56.

KOLLINGER, D. 1964. Weitere Beobachtungen zur Biologie des Habichts. *Deutscher Falkenorden* (1964):63–67.

KOLLINGER, D. 1975. Erkenntnisse über den Habicht (*Accipiter gentilis*) und seinen heutigen Stand. *Deutscher Falkenorden* (1974):9–18.

KOMDEUR, J. & PEN, I. 2002. Adaptive sex allocation in birds: the complexities of linking theory and practise. *Philosophical Transactions of the Royal Society of London B,* 357 : 373–380.

KOMDEUR, J., DAAN, S., TINBERGEN, J. & MATEMAN, A.C. 1997. Extreme adaptive modification of sex ratio in the Seychelles warbler's eggs. *Nature* 385 : 522–526.

KONING 2000. [cited in RUTZ *et al.* 2006].

KORELOV, M.N. & PFANDER, P.V. 1983. [On new southern region of the Goshawk breeding]. Pp. 65–66 in *[Ecology of Birds of Prey]*. Nauka, Moscow, Russia [Russian]

KORPIMÄKI, E. 1986. Reversed size dimorphism in birds of prey, especially in Tengmalm's owl *Aegolius funereus*: a test of the 'starvation hypothesis'. *Ornis Scandinavica* 17 : 326–332.

KORPIMÄKI, E., MAY, C.A., PARKIN, D.T., WETTON, J.H. & WIEHN, J. 2000. Environmental and condition-related variation in sex ratio of kestrel broods. *Journal of Avian Biology* 31: 128–134.

KOS, R. 1980. Der Habicht in der Bundesrepublik Deutschland. *Vogelwelt* 101 : 161–175.

KOSTRZEWA, A. 1987. Quantitative Untersuchungen zur Habitattrennung von Mäusebussarde (*Buteo buteo*), Habicht (*Accipiter gentilis*) und Wespenbussard (*Pernis apivorus*). *Journal fur Ornithologie* 128 : 209–229.

KOSTRZEWA, A. 1991. Interspecific interference competition in three European raptor species. *Ethology, Ecology, and Evolution* 3 : 127–143.

KOSTRZEWA, A. 1996. A comparative study of nest site occupancy and breeding performance as indicators for nesting-habitat quality in three European raptor species. *Ethology, Ecology, and Evolution* 8: 1–18.

KOSTRZEWA, R. & KOSTRZEWA, A. 1990. The relationship of spring and summer weather with density and breeding performance of the buzzard *Buteo buteo*, goshawk *Accipiter gentilis*, and kestrel *Falco tinnunculus*. *Ibis* 132 : 550–559.

KOSTRZEWA, A. & SPEER, G. (eds). 2001. *Greifvögel in Deutschland*: Bestand, Situation, Schutz. 2., vollst. neu bearb. und erw. Auflage. AULA-Verlag, Wiebelsheim, Germany.

KOSTRZEWA, A., SPEER, R., VON DEWITZ, W. & H. WEISER. 2000. Zur Populationsökologie des Habichts (*Accipiter gentilis*) in der Niederrheinischen Bucht (1981–1998). *Charadrius* 36: 80–93.

KRAMER, K. 1973. *Habicht und Sperber.* Neue Brehm Bücherei, Wittenberg-Lutherstadt, Germany.

KRASNOSHTANOVA, M.N. 2001. [Autumn passage of Falconiformes at South Baikal in 1995–1998, 2000]. Pp. 110–118 in *[The modern problems of the Lake Baikal area].* Irkutsk University Press, Irkutsk, Russia. [Russian]

KRASNOSHTANOVA, M.N. 2003. [Details of Falconiformes monitoring in South-Baikalian migration corridor]. *[Proceedings of the North-Eurasian Raptor Conferences]* 4 : 207–209. [Russian]

KREBS, C.J., BOUTIN, S. & BOONSTRA, R. 2001. *Ecosystem Dynamics of the Boreal Forest: The Kluane Project.* Oxford University Press, UK.

KRECHMAR, A.V. & PROBST, R. 2003. Der weisse Habicht *Accipiter gentilis albidus* in Nordost-Sibirien – Portrat eines Mythos. *Limicola* 17 : 289–305.

KRÓL, W. 1985. Breeding density of diurnal raptors in the neighbourhood of Susz (Ilawa Lakeland, Poland) in the years 1977–79. *Acta Ornithologica* 21 : 95–114.

KRONE, O., PRIEMER, J., STREICH, J., SÖMMER, P., LANGGEMACH, T. & LESSOW, O. 2001. Haemosporidia of birds of prey and owls from Germany. *Acta Protozoologica* 40 : 281–289.

KRONE, O., ALTENKAMP, R. & KENNTNER, N. 2005. Prevalence of Trichomonas gallinae in Northern Goshawks from the Berlin area of northeastern Germany. *Journal of Wildlife Disease* 41 : 305–309.

KRÜGER, O. 2002a. Interactions between common buzzard *Buteo buteo* and Goshawk *Accipiter gentilis*: tradeoffs revealed by a field experiment. *Oikos* 96 : 441–452.

KRÜGER, O. 2002b. Analysis of nest occupancy and nest reproduction in two sympatric raptors: common buzzard *Buteo buteo* and goshawk *Accipiter gentilis*. *Ecography* 25 : 523–532.

KRÜGER, O. 2005. Age at first breeding and fitness in goshawk *Accipiter gentilis*. *Journal of Animal Ecology* 74 : 266–273.

KRÜGER, O & LINDSTRÖM, J. 2001. Habitat heterogeneity affects population growth in Goshawk *Accipiter gentilis*. *Journal of Animal Ecology* 70 : 173–181.

KRÜGER, O. & STEFENER, U. 1996. Nahrungsökolgie und Populationsdynamik des Habichts *Accipiter gentilis* in östlichen Westfalen. *Vogelwelt* 117:1–8.

KRÜGER, O. & STEFENER, U. 2000. Populationsfluktuation und die Rolle der Reproduktion in einer Population des Habichts *Accipiter gentilis*. *Populationsökologie Greifvogel- und Eulenarten* 4 : 263–271.

KÜHNAPFEL, O. & BRUNE, J. 1995. Die Mauserfeder als Hilfsmittel zur Altersbestimmung und Individualerkennung von Habichten (*Accipiter gentilis*). *Charadrius* 31 : 120–125.

KUTSCHER, S. 1981. Greifvögel und Taubenhaltung – Erfahrungen eines Brieftaubenzüchters. *Vogelschutz* 2 : 16–19.

LACK, D. 1954. *The natural regulation of animal numbers.* University Press, Oxford, UK.

LACK, D. 1966. *Population studies of birds.* University Press, Oxford, UK.

LAPINSKI, N., BOWERMAN, W. & SJOGREN, S. 2000. Factors affecting the Northern Goshawk in the Upper Peninsula of Michigan. Pp. 182–191 in YOSEF, R., MILLER, M.L. & PEPLER, D. (eds) *Raptors in the new millenium.* International Birding and Research Center, Eilat, Israel.

LA SORTE, F.A., MANNAN, R.W., REYNOLDS, R.T. & GRUBB, T.G. 2004. Habitat associations of sympatric Red-tailed Hawks and Northern Goshawks on the Kaibab Plateau. *Journal of Wildlife Management* 68 : 307–317.

LAUGHLIN, K.F. 1975. The bioenergetics of the Tufted Duck *Aythya fuligula* (L.). PhD thesis, University of Stirling, UK.

LEE, J.A. 1981a. Comparative breeding behavior of the Goshawk and the Cooper's hawk. MSc thesis, Brigham Young University, Provo, Utah, USA.

LEE, J.A. 1981b. Habituation to human disturbance in nesting accipiters. *Journal of Raptor Research* 15: 48–52.

LELOV E. 1991. Breeding raptors and owls at Halinga, SW Estonia, in 1978–1989. *Ornis Fennica* 68: 119–122.

LE MUNYAN, C.D., WHITE, W., NYBERT, E. & CHRISTIAN, J.J. 1959. Design of a miniature radio transmitter for use in animal studies. *Journal of Wildlife Management* 23: 107–110.

LENSINK, R. 1997. Range expansion of raptors in Britain and the Netherlands since the 1960s: testing an individual-based diffusion model. *Journal of Animal Ecology* 66: 811–826.

LEOPOLD, A. 1933. *Game management.* Charles Scribners Sons, New York, USA.

LESTER, S. 1999. The return of the goshawk. *Shooting and Country Magazine* (March 1999): 35–36.

LEWIS, S.B. 2001. Breeding season diet of goshawks in Northeast Alaska with a comparison of techniques used to examine raptor diet. MSc thesis, Boise State University, Idaho, USA.

LEWIS, S.B., M. R. FULLER & K. TITUS. 2004. A comparison of 3 methods for assessing raptor diet during the breeding season. *Wildlife Society Bulletin* 32: 373–385.

LINDÉN, H. & WIKMAN, M. 1983. Goshawk predation on tetraonids: availability of prey and diet of the predator in the breeding season. *Journal of Animal Ecology* 52: 953–968.

LINDNER, K. 1973. *Beiträge zur Voglefang und Falknerei im Altertum.* Walter de Gruyter, Berlin, Germany.

LINDSAY, I.M. 1981. The uses of trained Accipiters in research. Pages 134–138 in KENWARD, R.E. & LINDSAY, I.M. (eds) *Understanding the Goshawk.* International Association of Falconry and Conservation of Birds of Prey, Oxford, UK.

LINK, H. 1977. Beiträge zur Bestandssituation, Ökologie, Brutbiologie und Beutewahl einer nordbayerischer Population des Habichts (*Accipiter gentilis*). Diploma thesis, University of Erlangen, Germany.

LINK, H. 1986. Untersuchungen am Habicht (*Accipiter gentilis*). PhD thesis, Friedrich-Alexander Universität, Erlangen-Nürnberg, Germany.

LINKOLA, P. 1956. Kanahaukkakannan romahdus vuonna 1956. *Luonnon Tutkija* 61: 49–58.

LÕHMUS, A. 1993. Kanakulli (*Accipiter gentilis*) toitumisest Eestis aastatel 1987–92. *Hirundo* 2:3–14.

LÕHMUS, A. 2004. [Monitoring of raptors and owls in Estonia, 1999–2003: decline of the Goshawk and the clockwork of vole-cycles.] *Hirundo* 17:3–18. [Estonian]

LOOFT, V. 1984. Die Entwicklung des Habichtbestandes (*Accipiter gentilis*) in Schleswig-Holstein 1968–1984. *Corax* 10: 395–400.

LOOFT, V. 2000. The ups and downs of a northern goshawk population over a 30 year period – natural dynamics or an artefact. Pp 499–506 in CHANCELLOR, R.D. & MEYBURG, B.-U. (eds) *Raptors at risk.* World Working Group on Birds of Prey and Owls, Berlin, Germany.

LOOFT, V. & BIESTERFELD, G. 1981. Habicht – *Accipiter gentilis.* Pp. 101–115 in LOOFT, V. & BUSCHE, G. (eds) *Vogelwelt Schleswig-Holsteins, Band 2: Greifvögel.* Karl Wachholtz Verlag, Neumünster, Germany.

LÜDERS, O. 1938. *Beringungs-Ergebnisse beim Habicht.* Deitscher Falkenorden (1938):38–48.

LUNDBERG, A. 1986. Adaptive advantages of reversed sexual selection in European owls. *Ornis Scandinavica* 17: 133–140.

MACDONALD, H. 2005. Hawks, history, heritage: observations on falconry in Europe. Presentation at the Symposium 'Falconry: a world heritage', Abu Dhabi, 13–15 September 2005.

MACLEOD, R., GOSLER, A.G. & CRESSWELL, W. 2005. Diurnal mass gain strategies and perceived predation risk in the great tit *Parus major. Journal of Animal Ecology* 74: 956–964.

MACPHERSON, H.A. 1883. Goshawk near Oxford. *Zoologist* (1883):31.

MADSEN, J. 1988. Duehg *Accipiter gentilis* forstyrrer og draeber knortegase Branta bernicla ved specialiseret jadgteknik. *Dansk Ornitologiska Forenings Tidsskrift* 82 : 57 : 58.

MAHON, T. & DOYLE, F.I. 2005. Effects of timber harvesting near nest sites on the reproductive success of Northern Goshawks (Accipiter gentilis). *Journal of Raptor Research* 39 : 335–341.

MAHON, T., DOYLE, F.I. & NELLIGAN M. 2003. *Effect of forest development on the reproductive success of Northern Goshawks (Accipiter gentilis) in the Prince Rupert Forest Region.* Unpublished report for Babine Forest Products, Houston Forest Products and Skeena Cellulose Inc., Canada.

MAÑOSA, S. 1991. Biologia trofica, us de l'habitat I biologia de la reproduccio de l'Astor *Accipiter gentilis* (Linnaeus 1758) a la Segarra. PhD thesis, University of Barcelona, Spain.

MAÑOSA, S. 1993. Selección de hábitat de nidificación en el Azor (*Accipiter gentilis*): recomendaciónes para su gestión. *Alytes* 6 : 125–136.

MAÑOSA, S. 1994. Goshawk diet in a Mediterranean area of north-eastern Spain. *Journal of Raptor Research* 28 : 84–92.

MAÑOSA, S., REAL, J. & SANCHEZ E. 1990. Comparació de l'ecologia de dues poblacions d'astor *Accipiter gentilis* a Catalunya: el Vallès-Moianès I la Segarra. *El Medi Natural del Vallès* 2 : 204–212.

MAÑOSA, S., R. MATEO, C. FREIXA, & R. GUITART. 2003. Persistent organochlorine contaminants in eggs of northern goshawk and Eurasian buzzard from northeastern Spain: temporal trends related to changes in the diet. *Environmental Pollution* 122 : 351–359.

MARCSTRÖM, V. & KENWARD, R.E. 1981a. Sexual and seasonal variation in condition and survival of Swedish goshawks (*Accipiter gentilis*). *Ibis* 123 : 311–327.

MARCSTRÖM, V. & KENWARD, R.E. 1981b. Movements of wintering goshawks in Sweden. *Swedish Game Research* 12:1–35.

MARCSTRÖM, V. & WIDÉN, P. 1977. Hur skulle det gå för duvhöken om inte Fasanen fanns? *Svensk Jakt* 115 : 98–101.

MARCSTRÖM, V., KENWARD, R.E. & ENGREN, E. 1988. The impact of predation on boreal tetraonids during vole cycles: an experimental study. *Journal of Animal Ecology* 57 : 859–872.

MARCSTRÖM, V., KENWARD, R.E. & KARLBOM, M. 1990. *Düvhöken och dess plats i naturen.* Trycksaker, Norrköping, Sweden.

MARQUISS, M. 1981. The goshawk in Britain – its provenance and current status. Pages 43–55 in KENWARD, R.E. & LINDSAY, I.M. (eds) *Understanding the Goshawk.* International Association of Falconry and Conservation of Birds of Prey, Oxford, UK.

MARQUISS, M. & NEWTON, I. 1982a. A radio-tracking study of the ranging behaviour and dispersion of European Sparrowhawks *Accipiter nisus. Journal of Animal Ecology* 51 : 111–133.

MARQUISS, M. & NEWTON, I. 1982b. The Goshawk in Britain. *British Birds* 75 : 243–260.

MARQUISS, M., PETTY, S.J., ANDERSON D.I.K. & LEGGE, G. 2003. Contrasting population trends of the Northern Goshawk (*Accipiter gentilis*) in the Scottish / English Borders and North-East Scotland. Pp. 143–148 *in* THOMPSON, D.B.A, REDPATH, S.M., FIELDING, A.H., MARQUISS, M. & GALBRAITH, C.A. (eds) *Birds of Prey in a Changing Environment.* Scottish Natural Heritage and The Stationary Office, Edinburgh, UK.

MARTÍ, R. & DEL MORAL, J.C. (eds) 2003. *Atlas de las Aves de España.* Dirección General de Conservación de la Naturaleza–Sociedad Española de Ornitología. Madrid, Spain.

MAURER, J.R. 2000. Nesting habitat and prey relations of the northern goshawk in Yosemite National Park, California. MSc thesis, University of California, Davis, USA.

MAVROGORDATO, J.G. 1937. Nesting activities of a female goshawk in 1936. *Falconer* 1:9–11.

MAVROGORDATO, J.G. 1960. *A hawk for the bush.* Witherby, London, UK.

MAYO, L. 2002. Birds and the hand of power: a political geography of avian life in the Gansu Corridor, ninth to tenth centuries. *East Asian History* 24:1–66.

MCCLAREN, E.L., P.L. KENNEDY & S.R. DEWEY. 2002. Do some northern goshawk nest areas consistently fledge more young than others? *Condor* 104:343–352.

MCCLAREN, E.L., KENNEDY, P.L. & CHAPMAN, P.L. 2003. Efficacy of male goshawk food delivery calls in broadcast surveys on Vancouver Island. *Journal of Raptor Research* 37:198–208.

MCCLAREN, E.L., KENNEDY, P.L. & DOYLE, D.D. 2005. Northern Goshawk (Accipiter gentilis laingi) post-fledging areas on Vancouver Island, British Columbia. *Journal of Raptor Research* 39:253–263.

MCCLEERY, R.H. & PERRINS, C.M. 1991. Effects of predation on the numbers of great tits *Parus major*. Pp. 129–147 in PERRINS, C.M., LEBRETON, J.-D. & HIRONS, G.J.M. (eds) *Bird Population Studies*. University Press, Oxford, UK.

MCCLOSKEY, J.T. & S.R. DEWEY. 1999. Improving the success of a mounted great horned owl lure for trapping northern goshawks. *Journal of Raptor Research* 33:168–169.

MCCOY, R.H. 1999. Effects of prey delivery on the fledgling success of northern goshawks. MSc thesis. Humboldt State University, Arcata, California, USA.

MCELROY, H. 1977. *Desert Hawking II.* Privately printed, USA.

MCGOWAN, J.D. 1975. *Distribution, density and productivity of Goshawks in interior Alaska. Projects W–17–3, W–17–4, W–17–5, and W–17–6.* Alaska Department of Fish and Game, Juneau, Alaska, USA.

MCGRADY, M.J. 1991. The ecology and breeding behaviours of urban sparrowhawks *Accipiter nisus* in Edinburgh, Scotland. PhD thesis, University of Edinburgh, UK.

MCGRATH, M.T., DESTEFANO, S., RIGGS, R.A., IRWIN, L.L. & ROLOFF, G.J. 2003. Spatially explicit influences on Northern Goshawk nesting habitat in the interior Pacific Northwest. *Wildlife Monographs* 154.

MEARNS, R. & NEWTON, I. 1984. Turnover and dispersal in a Peregrine *Falco peregrinus* population. *Ibis* 126:347–355.

MEIJER, T. 1988. Reproductive decisions in the Kestrel *Falco tinnunculus*. A study in physiological ecology. PhD thesis, University of Groningen, The Netherlands.

MEINERTZHAGEN, R. 1950. The goshawk in Britain. *Bulletin of the British Ornithologists' Club* 70:46–49.

MEINERTZHAGEN, R. 1959. *Pirates and predators.* Oliver & Boyd, London, UK.

MELCHIOR, E., MENTGEN, E., PELTZER, R., SCHMITT, R. & WEISS, J. (eds). 1987. *Atlas der Brutvögel Luxemburgs.* Lëtzebuerger Natur- a Vulleschutzliga, Luxembourg.

MELNIKOV, V.N. & BUSLAEV, S.V. 2003. [Goshawk in the Ivanovo Region]. *[Proceedings of the North-Eurasian Raptor Conferences]* 4:84–90. [Russian]

MENDELSOHN, J.M. 1986. Sexual size dimorphism and roles in raptors – fat females, agile males. *Durban Museum Novitates* 13:321–336.

MENG, H.K. 1959. Food habits of nesting Cooper's Hawks and goshawks in New York and Pennsylvania. *Wilson Bull.* 71:169 174.

MENG, H.K. 1971. The Swedish goshawk trap. *Journal of Wildlife Management* 55:832–835.

MENZBIR, M.A. 1882. Der Habicht (*Astur palumbarius*, L.). Pp. 438–444 in Ornithogeographie des Europäischen Russlands. Translated in *Greifvögel und Falknerei* (1998):88–91.

MENZBIR, M.A. 1895. *Die Vogelwelt Russlands.* State Publishing House, Moscow, Russia.

MIKKELSEN, J.D. 1984 Effekt af duehøge og andre rovfugle, ved Fasanudsaetningssteder. MSc thesis, University of Copenhagen, Denmark.

MIKKOLA, H. 1983. *Owls of Europe.* Poyser, Calton, UK.

MILLSPAUGH, J.J. & MARZLUFF J.M. (eds) 2001. *Radio tracking and animal populations.* Academic Press, San Diego, California, USA.

MILONOFF, M. 1994. An overlooked connection between goshawk and tetraonids – corvids! *Suomen Riista* 40: 91–97.

MITROFANOV, O.B. 2003. [On the Goshawk biology in Altai Nature Reserve]. *[Proceedings of the International Ornithological Conferences of Russia]* 2: 142–144. [Russian]

MÖCKEL, R. & GÜNTHER, D. 1987. Die Reproduktion des Habichts *Accipiter gentilis* (L.) im Westerzgebirge in den Jahren 1974–1983. *Populationsökologie Greifvogel- and Eulenarten* 1: 217–232.

MOILANEN, P. 1976. Kanahaukkantapot ja fasaani. *Suomen Luonto* 35: 315–318.

MØLLER, A.P. 1987. Copulatory behaviour in the Goshawk, *Accipiter gentilis. Animal Behaviour* 35: 755–763.

MONNERET, R.J. 1978. Project peregrine. Pp. 56–61 in GEER, T.A. (ed.) *Bird of prey management techniques.* British Falconers' Club, Oxford, UK.

MOORE, K.R. & HENNY, C.J. 1983. Nest site characteristics of three coexisting Accipiter hawks in northeastern Oregon. *Journal of Raptor Research* 17: 65–76.

MORAN, D. 1995. Observations on wild goshawks. *The British Falconers' Club Newsletter* 10: 25–27.

MORANT, G.F. 1875. *Game preservers and bird preservers.* Longmans, Green and Co., London, UK.

MORILLO, C. & LALANDA, J. 1972. Primeros datos sobre la ecología de las Falconiformes en los montes de Toledo. *Boletín de la Estación Central de Ecología* 2: 57–70.

MORIMOTO, T. 2005. Japanese falconry: history and cultural aspects. Presentation at the Symposium *'Falconry: a world heritage',* Abu Dhabi, 13–15 September 2005.

MOSHER, J.A. & MATRAY, P.F. 1974. Size dimorphism: a factor in energy savings for Broad-winged Hawks. *Auk* 91: 325–341.

MUELLER, H.C. 1986. The evolution of reversed sexual dimorphism in owls: an empirical analysis of possible selective factors. *Wilson Bulletin* 98: 387–406.

MUELLER, H. C. & BERGER D.D. 1967. Some observations and comments on the periodic invasions of goshawks. *Auk* 84: 183–191.

MUELLER, H. C. & BERGER D.D. 1968. Sex ratios and measurements of migrant Goshawks. *Auk* 85: 431–436.

MUELLER, H. C. & MEYER, K. 1985. The evolution of reversed sexual dimorphism in size. A comparative analysis of the Falconiformes of the Western Palaearctic. Pp. 65–101 in JOHNSTON, R.F. (ed.) *Current Ornithology,* Vol.2. Plenum Press, New York, USA.

MUELLER, H. C., BERGER, D.D. & ALLEZ, G. 1977. The periodic invasions of Goshawks. *Auk* 94: 652–663.

MUELLER, H. C., BERGER, D.D. & ALLEZ, G. 1979. Age and sex differences in size of Sharp-shinned Hawks. *Bird Banding* 50: 34–44.

MUELLER, H. C., BERGER, D.D. & ALLEZ, G. 1981. Age, sex and seasonal differences in size of Cooper's Hawks. *Journal of Field Ornithology* 52: 112–126.

MÜLLER, K. 2000. Dieter Rockenbauch: Der Wanderfalke in Deutschland und umliegenden Gebieten. *Greifvögel und Falknerei* (1999):209–212.

MUNTHE-KAAS LUND, H. 1950. Hønsehauk. *Jakt, Fiske og Friluftsliv* 79 : 100–103.
MURTON, R.K. 1974. The impact of agriculture on birds. *Annals of Applied Biology* 76 : 358–365.
MURTON, R.K., ISAACSON, A.J. & WESTWOOD, N.J. 1963. The feeding ecology of the woodpigeon. *British Birds* 56 : 345–375.
MURTON, R.K., WESTWOOD, N.J. & ISAACSON, A.J. 1964. A preliminary investigation of the factors regulating population size in the Woodpigeon. *Ibis* 106 : 482–507.
MURTON, R.K., ISAACSON, A.J. & WESTWOOD, N.J. 1966. The relationships between Woodpigeons and their clover food supply and the mechanisms of population control. *Journal of Animal Ecology* 3 : 55–96.
MUSCHIOL, W. 1964. Erfolgreicher Brut- und Aufzuchtversuch mit einem Beizhabicht. *Deutcher Falkenorden* (1964):67–69.
MUSGROVE, A. 1996. Peregrines and pigeons: investigations into a raptor-human conflict. PhD thesis, University of Bristol, UK.
MYRBERGET, S. 1989. Diet of goshawks during the breeding season in northern coastal Norway. Fauna Norvegica. Series C. *Cinclus* 12 : 100–102.
NECHAEV, V.A. 1969. [The birds of the South Kurile Islands]. Nauka Press, St Petersberg, Russia. [Russian]
NEEDHAM, J., COOPER, J.E. & KENWARD, R.E. 1979. A survey of the bacterial flora of the feet of free-living goshawks (*Accipiter gentilis*). *Avian Pathology* 8 : 285–288.
NEIDEMAN, C. & SCHÖNBECK, E. 1990. Erfarenheter från 10 års ringmärkning av fångade duvhökar. *Anser* 29 : 245–260.
NELSON, M.W. 1978. Preventing electrocution deaths and the use of nesting platforms on power lines. Pp. 42–54 in GEER, T.A. (ed) *Bird of prey management techniques*. The British Falconers' Club, Oxford, UK.
NEWTON, I. 1968. The temperatures, weights and body composition of molting Bullfinches. *Condor* 70 : 323–332.
NEWTON, I. 1972. *Finches*. Collins, London, UK.
NEWTON, I. 1979. *Population ecology of raptors*. Poyser, Berkhamsted, UK.
NEWTON, I. 1986. *The sparrowhawk*. Poyser, Calton, UK.
NEWTON, I. 1991. Population limitation in birds: a comparative approach. In PERRINS, C.M., PEBERTON, J.D. & HIRONS, G.M. (eds). *Bird population studies*. University Press, Oxford, UK.
NEWTON, I. 1998. *Population limitation in birds*. Academic Press, San Diego, California, USA.
NEWTON, I. 2004. The recent declines of farmland bird populations in Britain: an appraisal of causal factors and conservation actions. *Ibis* 146 : 579–600.
NEWTON, I. & MARQUISS, M. 1982. Fidelity to breeding area and mate in Sparrowhawks *Accipiter nisus*. *Journal of Animal Ecology* 51 : 327–341.
NEWTON, I., BOGAN, J. & MARQUISS, M. 1981. Organochlorine contamination and age in sparrowhawks. *Environmental Pollution* (Series A) 25 : 155–160.
NEWTON, I., MARQUISS, M. & VILLAGE, A. 1983. Weights, breeding and survival in European Sparrowhawks. *Auk* 100 : 344–354.
NICHOLLS, M.K., LOVE, O.P. & BIRD, D.M. 2000. An evaluation of methyl anthranilate, aminoacetophenone and unfamiliar coloration as feeding repellents to American Kestrels. *Journal of Raptor Research* 34 : 311–318.
NIELSEN, J.T. 1986. Duehøgen (*Accipiter gentilis*) i Vendsyssel 1977–85. *Accipiter* 3 : 133–174.

NIELSEN, J.T. 1998. Duehøgens (*Accipiter gentilis*) prædation på brevduer i Vendsyssel. *Dansk Ornitologisk Forenings Tidsskrift* 92 : 327–332.

NIELSEN, J.T. 2003a. Duehøgens (*Accipiter gentilis*) byttevalg uden for yngletiden. *Dansk Ornitologisk Forenings Tidsskrift* 97 : 193–198.

NIELSEN, J.T. 2003b. Lav duehøgebestand en følge af ulovlig bekæmpelse ved fasanudsætninger. *Dansk Ornitologisk Forenings Tidsskrift* 97 : 173–174.

NIELSEN, J.T. & DRACHMANN, J. 1999a. Dispersal of Danish Goshawks *Accipiter gentilis* as revealed by ringing recoveries. *Dansk Ornithologisk Forenings Tidsskrift* 93 : 235–240.

NIELSEN, J.T. & DRACHMANN, J. 1999b. Prey selection of Goshawks *Accipiter gentilis* during the breeding season in Vendsyssel, Denmark. *Dansk Ornithologisk Forenings Tidsskrift* 93 : 85–90.

NIELSEN, J.T. & DRACHMANN, J. 1999c. Development and productivity in a Danish Goshawk *Accipiter gentilis* population. *Dansk Ornithologisk Forenings Tidsskrift* 93 : 153–161.

NIELSEN, J.T. & DRACHMANN, J. 2003. Age-dependent reproductive performance in Northern Goshawks *Accipiter gentilis*. *Ibis* 145:1–8.

NILSSON, I.N., NILSSON, S.G. & SYLVÉN, M. 1982. Diet choice, resource depression and the regular nest spacing of birds of prey. *Biological Journal of the Linnean Society* 18:1–9.

NISBET, J.C.T. & SMART, T.C. 1959. Unusual goshawks in Cambridgeshire and Norfolk. *British Birds* 50 : 164–166.

NOER, H. & SECHER, H. 1990. Effects of legislative protection on survival rates and status improvements of birds of prey in Denmark. *Danish Review of Game Biology* 14(2):1–63.

NORE, T. 1979. Rapaces diurnes communs en Limousin pendant la période de nidification. II: autour, épervier et faucon crécerelle. *Alauda* 47 : 259–269.

NOVAK, V.A. 1998. [On study of the Goshawk feeding]. *[Proceedings of the Conference on Birds of Prey of Eastern Europe and Northern Asia]* 3 : 92. [Russian]

NYGÅRD, T. 1991. *Rovfugl som indikatorer på forurensning i Norge.* NINA Utredning 21:1–34.

NYGÅRD, T. 1997. Temporal and spatial trends of pollutants in birds in Norway: birds of prey and Willow Grouse used as biomonitors. PhD thesis, Norwegian Technical University of Natural Sciences, Trondheim, Norway.

OAKLEAF, R.J. 1975. *Population surveys, species distribution and key habitats of selected non-game species.* Federal Aid in Wildlife Restoration Project W–53-R. Nevada Department of Fish and Game, Nevada, USA.

OAKS, J.L., GILBERT, M., VIRANI, M.Z., WATSON, R.T., METEYER, C.U., RIDEOUT, B.A., SHIVAPRASAD, H.L., AHMED, S., CHAUDHRY, M.J.I., ARSHAD, M., MAHMOOD, S., ALI, A. & KHAN, A.A. 2004. Diclofenac residues as the cause of vulture population decline in Pakistan. *Nature* 427 : 630–633.

O'BRIEN, W.J., EVANS, B.I. & BROWMAN, H.I. 1989. Flexible search strategies and efficient foraging in salutatory searching animals. *Oecologia* 80 : 100–110.

ODUM, E.P. 1960. Lipid deposition in nocturnal migrant birds. *Proceedings of the International Ornithological Congress* 14 : 563–576.

OELKE, H. 1981. Greifvogel-Monitoruntersuchung 1977–1980 im Landkreis Peine (Hannover-Braunschweig, Niedersachsen). *Beiträge zur Naturkunde Niedersachsens* 34 : 12–50.

OGGIER, P.-A. 1981. Dichte und Verteilung des Habichts (*Accipiter gentilis*) in der Schweiz. *Vorläufige Ergebnisse Nationalpark Berchtesgaden, Forschungsbericht* 3 : 25–31.

OGGIER, P.-A. & BÜHLER, U. 1998. Habicht. Pp. 196–197 in SCHMID, H., LUDER, R., NAEF-DAENZER, B., GRAF, R. & ZBINDEN, N. (eds). *Schweizer Brutvogelatlas. Verbreitung der Brutvögel in der Schweiz und im Fürstentum Liechtenstein 1993–1996.* Schweizerische Vogelwarte, Sempach, Switzerland.

OLECH, B. 1997. Diet of the Goshawk *Accipiter gentilis* in Kampinoski National Park (Central Poland) in 1982–1993. *Acta Ornithologica* 32 : 191–200.

OLECH, B. 1998. Population dynamics and breeding performance of the goshawk *Accipiter gentilis* in Central Poland. Pp. 101–110 in MEYBURG, B.-U., CHANCELLOR, R.D. & FERRERO, J.J. (eds) *Holarctic birds of prey.* Asociación para la Defensa de la Naturaleza y los Recursos de Extremadura and World Working Group on Birds of Prey, Berlin, Germany.

OLSEN, P. & OLSEN, J. 1987. Sexual size dimorphism in raptors: intrasexual competition in the larger sex for a scarce breeding resource, the smaller sex. *Emu* 87 : 59–62.

OPDAM, P. 1975. Inter- and intraspecific differentiation with respect to feeding ecology in two sympatric species of the genus *Accipiter. Ardea* 63 : 30–54.

OPDAM, P. & MÜSKENS, G. 1976. Use of shed feathers in population studies of Accipiter hawks. *Beaufortia* 24 : 55–62.

OPDAM, P., THISSEN, J., VERSCHUREN, P. & MÜSKENS, G. 1977. Feeding ecology of a population of Goshawk (*Accipiter gentilis*). *Journal für Ornithologie* 118 : 35–51.

ORTLIEB, R. 1978. Pestizidschädungen auch beim Habicht. *Falke* 25 : 78–87.

ORTLIEB, R. 1981. Habitatsbeobachtungen in Südharz. *Beiträge zur Vogelkunde* 27 : 78–87.

ORTLIEB, R. 1998. Langjährig besetzte Brutreviere des Habichts. *Greifvögel und Falknerei* (1997):96–99.

PADIAL, J.M., BAREA, J.M., CONTRERAS, F.J., AVILA, E. & PÉREZ, J. 1998. Dieta del Azor Común (*Accipiter gentilis*) en las Sierras Béticas de Granada durante el periodo de reproducción. *Ardeola* 45 : 55–62.

PALMA, L., ONOFRE, N. & POMBAL, E. 1999. Revised distribution and status of diurnal birds of prey in Portugal. *Avocetta* 23:3–18.

PARKIN, D.T. 1987. The value of genetic fingerprinting to the breeding and conservation of birds of prey. Pp. 81–86 in HILL, D.J. (ed.) *Breeding and management in Birds of Prey.* University Press, Bristol, UK.

PATLA, S. M. 1997. Nesting ecology and habitat of the northern Goshawk in undisturbed and timber harvest areas on the Targhee National Forest, Greater Yellowstone ecosystem. MSc thesis, Idaho State University, Pocatello, Idaho, USA.

PATLA, S. M. 2005. Monitoring results of Northern Goshawk nesting areas in the Greater Yellowstone ecosystem: is decline in occupancy related to habitat change? *Journal of Raptor Research* 39 : 324–334.

PECK, G.K. & JAMES, R.D. 1983. *Breeding birds of Ontario: ecology and distribution. Vol. I, Non-passerines.* Royal Ontario Museum of Life Sciences, Toronto, Ontario, Canada.

PENNYCUICK, C.J. 1972. *Animal flight.* Arnold, London, UK.

PENNYCUICK, C.J. 1989. *Bird flight performance: a practical calculation manual.* University Press, Oxford, UK

PENTERIANI, V. 1997. Long-term study of a Goshawk breeding population on a Mediterranean mountain (Abruzzi Apennines, central Italy): density, breeding performance and diet. *Journal of Raptor Research* 31 : 308–312.

PENTERIANI, V. 1999. Dawn and morning goshawk courtship vocalizations as a method for detecting nest sites. *Journal of Wildlife Management* 63 : 511–516.

PENTERIANI, V. 2001. The annual and diel cycles of Goshawk vocalizations at nest sites. *Journal of Raptor Research* 35 : 24–30.

PENTERIANI, V. 2002. Goshawk nesting habitat in Europe and North America: a review. *Ornis Fennica* 79:149–163.

PENTERIANI, V. & FAIVRE, B. 1997. Breeding density and nest site selection in a Goshawk *Accipiter gentilis* population of the Central Apennines (Abruzzo Italy). *Bird Study* 44:136–145.

PENTERIANI, V. & FAIVRE, B. 2001. Effects of harvesting timber stands on Goshawk nesting in two European areas. *Biological Conservation* 101:211–216.

PENTERIANI, V. & KENWARD, R.E. MS in review. Environmental differences generate variation in foraging and selection pressures on body size in Goshawks.

PENTERIANI, V., FAIVRE, B. & FROCHOT, B. 2001. An approach to identify factors and levels of nesting habitat selection: a cross-scale analysis of Goshawk preferences. *Ornis Fennica* 78:159–167.

PENTERIANI, V., MATHIAUT, M. & BOISSON, G. 2002. Immediate responses to catastrophic natural disturbances: windthrow effects on density, productivity, nesting stand choice, and fidelity in Northern Goshawks (*Accipiter gentilis*). *Auk* 119:1132–1137.

PENTERIANI, V., FAIVRE, B., MAZUC, J. & CEZILLY, F. 2003. Pre-laying vocal activity as a signal of mate and nest stand quality in Goshawks. *Ethology, Ecology and Evolution* 14:9–17.

PERCO, F. & BENUSSI, E. 1981. *Nidificazione e distribuzione territoriale dell' Astore (Accipiter gentilis gentilis L.) sul Carso Triestino*. Atti Primo Convegno Ecologia Territori Carsici, La Grafica, Gradisca d'Isonzo:207–216.

PERDECK, A.C. 1960. Observations on the reproductive behaviour of the great skua or bonxie, Stercorarius skua skua (Brunn), in Shetland. *Ardea* 48:111–136.

PERRINS, C.M. 1979. *British tits*. Collins, London.

PERRINS, C.M. & GEER, T.A. 1980. The effect of Sparrowhawks on tit populations. *Ardea* 68:133–142.

PETRONILHO, J.M.S. & VINGADA, J.V. 2002. First data on feeding ecology of Goshawk *Accipiter gentilis* during the breeding season in the Natura 2000 site Dunas de Mira, Gândara e Gafanhas (Beira Litoral, Portugal). *Airo* 12:11–16.

PETROV, V.S. & GUSEV, V.M. 1995. [On hawk feeding in the Caucasia]. *[Proceedings of Teberda Nature Reserve]* 14:170–186. [Russian]

PETTY, S.J. 1996a. *Reducing disturbance to goshawks during the breeding season*. Forestry Commission Research Information Note 267. Forestry Commission, Edinburgh, UK.

PETTY, S.J. 1996b. History of the northern goshawk *Accipiter gentilis* in Britain. Pp. 95–102 in HOLMES, J.S. & SIMONS, J.R. (eds) *The introduction and naturalisation of birds*. HM Stationery Office, London, UK.

PETTY, S.J. 2002. Northern Goshawk (Goshawk) *Accipiter gentilis*. pp232–234 in WERNHAM, C.V., TOMS, M.P., MARCHANT, J.H., CLARK, J.A., SIRIWARDENA, G.M. & BAILLIE, S.R. (eds) *The Migration Atlas: movements of the birds of Britain and Ireland*. Poyser, London, UK.

PETTY, S. J., ANDERSON, D.I.K., Davidson. M., LITTLE, B., SHERRATT, T.N., THOMAS, C.J. & LAMBIN, X. 2003a. The decline of Common Kestrels *Falco tinnunculus* in a forested area of northern England: the role of predation by Northern Goshawks *Accipiter gentilis*. *Ibis* 145:472–483.

PETTY, S.J., LURZ, P.W.W. & RUSHTON, S.P. 2003b. Predation of red squirrels by northern Goshawks in a conifer forest in northern England: can this limit squirrel numbers and create a conservation dilemma? *Biological Conservation* 111:105–114.

PEUS, F. 1954. Zur Kentnis der Brutvögel Griechenlands. *Bonner Zoologische Beiträge (Sonderband)*:1.50.

PEYTON, R.B., VORRO, J., GRISE, L., TOBIN, R. & EBERHARDT, R. 1995. A profile of falconers in the United States: falconry practises, attitudes and conservation behaviours. *Transactions of the North American Wildlife and Natural Resources Conference* 60 : 181–192.

PIELOWSKI, Z. 1961. Über den Unifikationseinfluss der selektiven Nahrungswahl des Habichts, Accipiter gentilis L., auf Haustauben. *Ekologia Polska* A 9 : 183–194.

PIELOWSKI, Z. 1968. Studien über die Bestandsverhältnissen einer Habichtspopulation in Zentralpolen. *Beiträge zur angewandten Vogelkunde* 5 : 125–136.

PIENKOWSKI, M.W., LLOYD, C.S. & MINTON, C.D. 1979. Seasonal and migrational weight changes in Dunlins. *Bird Study* 26 : 134–148.

POLIS, G.A., MYERS, C.A. & HOLT, R.D. 1989. The ecology and evolution of intraguild predation: potential competitors that eat each other. *Annual Review of Ecology and Systematics* 20 : 297–330.

PÖPPELMANN, B. 1994. Der 'Hühnerhabicht'. *Greifvögel und Falknerei (1993)*:90–91.

PÖPPELMANN, B. 1997. Gedanken zur Habichtszucht. *Greifvögel und Falknerei (1995)*:28–32.

PÖPPELMANN, B. 2000. Gedanken zur neuartigen Voliere für die Habichtszucht. *Tinnunculus* 11: 42–43.

PORTER, T.W. & WILCOX, H.H. 1941. Goshawk nesting in Michigan. *Wilson Bulletin* 53 : 43–44.

POTTER, S. & SARGENT, L. 1973. *Pedigree: words from nature.* Collins, London.

PRITCHARD, M. 1970. Goshawk breeding project. *Captive Breeding of Diurnal Birds of Prey* 1 : 16.

PRILL, H. 1959. Habicht als Aasfresser. *Falke* 6 : 33.

PRINS, H.H.T., GROOTENHUIS, J.G. & DOAN, T.T. (eds) 2000. *Wildlife conservation by sustainable use.* Kluwer, Dordrecht, The Netherlands.

PUGACEWICZ, E. 1996. Lęgowe ptaki drapieżne Polskiej części puszczy Białowieskiej. *Notatki Ornitologiczne* 37 : 173–224.

PULLIAM, H.R. & DANIELSON, B.J. 1991. Sources sinks and habitat selection: a landscape perspective on population dynamics. *American Naturalist* 137 : 50–66.

QUINN, J.L. & CRESSWELL, W. 2004. Predator hunting behaviour and prey vulnerability. *Journal of Animal Ecology* 73 : 143–154.

QUINN, J.L., PROP, J., KOKOREV, Y. & BLACK, J.M. 2002. Predator protection or similar habitat selection in red-breasted goose nesting associations: extremes along a continuum. *Animal Behaviour* 65 : 297–307.

RADDATZ, H.-J. 1997. Greifvogelbestände im Kreis Pinneberg (Schleswig-Holstein) von 1985 bis 1997. *Hamburger avifaunistische Beiträge* 29 : 137–158.

RAINES, R.J. 1946. Goshawks in Nottinghamshire. *British Birds* 39 : 155–156.

RAINES, R.J. 1956. Goshawk in Cheshire. *British Birds* 49 : 280–281.

RAND, A.L. 1952. Secondary sexual characters and ecological competition. *Fieldiana Zoology* 34 : 65–70.

RANSON, J. 1863. The Goshawk (*Falco palumbarius*) nesting in Yorkshire. *Zoologist* (1863):8679.

RANTA, E., BYHOLM, P., KAITALA, V., SAUROLA, P. & LINDÉN, H. 2003. Spatial dynamics of a predator: the connection to prey availability. *Oikos* 102 : 391–396.

RASSMUSSEN, L.U. & STORGÅRD, K. 1989. Ynglende rovfugle i Sydøstjylland 1973–1987. *Dansk Ornitologisk Forenings Tidsskrift* 83 : 23–34.

RATCLIFFE, D.A. 1980. *The Peregrine Falcon.* Poyser, Berkhamsted, UK.

READING, C.J. 1990. Molt pattern and duration in a female northern goshawk (*Accipiter gentilis*). *Journal of Raptor Research* 24 : 91–97.

REDIG, P.T., FULLER, M.R. & EVANS, D.L. 1980. Prevalence of *Aspergillus fumigatus* in free-living Goshawks (*Accipiter gentilis atricapillus*). *Journal of Wildlife Diseases* 16 : 169–174.

REDPATH, S.M. & THIRGOOD, S.J. 1997. *Birds of Prey and Red Grouse.* HM Stationery Office, London, UK.

REDPATH, S.M. & THIRGOOD, S.J. 1999. Numerical and functional responses in generalist predators: hen harriers and peregrines on Scottish grouse moors. *Journal of Animal Ecology* 68 : 879–892.

REDPATH, S.M. & THIRGOOD, S.J. 2003. The impact of Hen Harrier (*Circus cyaneus*) predation on Red Grouse (*Lagopus l. scoticus*) populations: linking models with field data. Pp. 499–511 in THOMPSON, D.B.A., REDPATH, S.M., FIELDING, A., MARQUISS, M. and GALBRAITH, C.A. (eds) *Birds of Prey in a Changing Environment.* The Stationery Office, London, UK.

REDPATH, S.M., THIRGOOD, S. & LECKIE, F. 2001. Does supplementary feeding reduce harrier predation on red grouse? *Journal of Applied Ecology* 38 : 1157–1168.

REYNOLDS, R.T. 1972. Sexual dimorphism in accipiter hawks: a new hypothesis. *Condor* 74 : 191–197.

REYNOLDS, R.T. 1975. Distribution, density and productivity of three species of *Accipiter* hawks in Oregon. MSc thesis, Oregon State University, Corvallis, Oregon, USA.

REYNOLDS, R.T. 1978. Food and habitat partitioning in two groups of coexisting accipiters. PhD thesis. Oregon State University, Corvallis, Oregon, USA.

REYNOLDS, R.T. & JOY, S.M. 1998. *Distribution, territory occupancy, dispersal, and demography of Northern Goshawks on the Kaibab Plateau, Arizona.* Arizona Game and Fish, Heritage Project Report No. I94045.

REYNOLDS, R.T. & JOY, S.M. 2006. Demography of Northern Goshawks in Northern Arizona, 1991–1996. *Studies in Avian Biology* 33:63–74.

REYNOLDS, R.T. & MESLOW E.C. 1984. Partitioning of food and niche characteristics of coexisting *Accipiter* during breeding. *Auk* 101 : 761–779.

REYNOLDS, R.T. & WIGHT, H.M. 1978. Distribution, density and productivity of *Accipiter* hawks breeding in Oregon. *Wilson Bulletin* 90 : 182–196.

REYNOLDS, R.T., MESLOW, E.C. & WIGHT, H. M. 1982. Nesting habitat of coexisting *Accipiter* in Oregon. *Journal of Wildlife Management* 46 : 124 138.

REYNOLDS, R.T., GRAHAM, R.T., REISER, M.H., BASSETT, R.L., KENNEDY, P.L., BOYCE, D. A., GOODWIN, G., SMITH, R. & FISHER, E.L. 1992. *Management recommendations for the northern goshawk in the southwestern United States.* U.S. Forest Service, General Technical Report RM 217, Rocky Mountain Forest and Range Experiment Station, Fort Collins, Colorado.

REYNOLDS, R.T., JOY, S.M. & LESLIE, D.G. 1994. Nest productivity, fidelity, and spacing of Northern Goshawks in Arizona. *Studies in Avian Biology* 16 : 106–113.

REYNOLDS, R.T., WHITE, G.C., JOY, S.M. & MANNAN, R.W. 2004. Effects of radiotransmitters on Northern Goshawks: do tailmounts lower survival of breeding males? *The Journal of Wildlife Management* 68 : 25–32.

REYNOLDS, R.T., WIENS, D.A., JOY, S.M. & SALAFSKY, S.R. 2005. Sampling considerations for demographic and habitat studies of Northern Goshawks. *Journal of Raptor Research* 39 : 274–285.

REYNOLDS, R.T., BOYCE, D.A. & GRAHAM, R.T. 2006 a. An ecosystem-based conservation strategy for the Northern Goshawk. *Studies in Avian Biology* 31:299–311.

REYNOLDS, R.T., WIENS, D.A. & SALAFSKY, S.R. 2006 b. A review and evaluation of factors limiting Northern Goshawk populations. *Studies in Avian Biology* 31: 260–273.

RICHMOND, W.K. 1959. British birds of prey. *Lutterworth Press, London, UK.*

RIPLEY, S.D. 1975. Foreword. Pp. 7–9 in ZIMMERMAN, D.R. (ed.) *To save a bird in peril*. Coward, McCann & Geoghegan, New York, USA.

RISCH, M., LOOFT, V. & ZIESEMER, F. 2004. Alter und Reproduktion weiblicher Habichte (*Accipiter gentilis*) in Schleswig-Holstein – ist Seneszenz nachwiesbar? *Corax* 19 : 323–329.

RITO, A. 1808. Topography of the Province Kawatsi (1801–1808). [cited in Glasier 1978]

RISEBOROUGH, R.W. 2004. Population collapses of three species of Gyps vultures in the Indian sub-continent. Pp. 197–214 in CHANCELLOR, R.D. & MEYBURG, B.-U. (eds) *Raptors Worldwide*. World Working Group on Birds of Prey and Owls. Berlin, Germany.

ROBERTS, C.M., BOHNSACK, J.A., GELL, F., HAWKINS, J.P. & GOODRIDGE, R. 2002. Marine reserves and fisheries management. *Science* 295 : 1234–1235.

ROBERTSON, P.A., N.J. AEBISCHER, R.E. KENWARD, I.P. HANSKI & N.P. WILLIAMS. 1998. Simulation and jack-knifing assessment of home-range indices based on underlying trajectories. *Journal of Applied Ecology* 35 : 928–940.

ROCKENBAUCH, D. 1998. *Der Wanderfalke in Deutschland und umliegenden Gebieten*. Ludwigsberg, Germany.

ROGERS, A.S., INGRALDI, M.F. & DESTEFANO, S. 2006. Diet, prey delivery rates, and prey biomass of Northern Goshawks in east-central Arizona. *Studies in Avian Biology* 31: 219–227.

ROHNER, C. & DOYLE, F.I. 1992. Food-stressed great horned owl kills adult goshawk: exceptional observation or community process. *Journal of Raptor Research* 26 : 261–263.

ROOS, G. 1974. Sträckräckningar vid Falsterbo hösten 1973. *Vår Fågelvärld* 33 : 270–285.

ROOT, M. & DESIMONE, P. 1980. *A progress report on the status and breeding ecology of the Goshawk, Red-shouldered Hawk and Barred Owl in northwest Connecticut*. Privately published, USA. [cited in Rutz *et al*. in press]

ROOT, M. & B. ROOT. 1978. A nesting census of the uncommon raptors in Northwest Connecticut '77. *Hawk Mountain News* 35:5–13. [cited in Bosakowski *et al*. 2005]

RÖRIG, G. 1909. Magen- und Gewöllenuntersuchungen heimischer Raubvögel. *Arbeite der Biologisches Anstalt für Land- und Forstwissenschaft* 7 : 484–484.

ROSENDAAL, C.W.C. 1990. Haviken in Zuid-Twente I: voedselonderzoek 1984–1988. *Vogeljaar* 38 : 198–207.

ROSENFIELD, R.N., BIELEFELDT, J., AFFELDT, J.L. & BECKMANN, D.J. 1995. Nesting density, nest area reoccupancy, and monitoring implications for Cooper's hawks in Wisconsin. *Journal of Raptor Research* 29:1–4.

ROSENFIELD, R.N., BIELEFELDT, J., TREXEL, D.R. & DOOLITTLE, T.C.J. 1998. Breeding distribution and nest-site habitat of Northern Goshawks in Wisconsin. *Journal of Raptor Research* 32 : 189–194.

ROSENFIELD, R.N., BIELEFELDT, J., SONSTHAGEN, S.A. & BOOMS, T.L. 2000. Comparable reproductive success at conifer plantation and non-plantation nest sites for Cooper's Hawks in Wisconsin. *Wilson Bulletin* 112 : 417–421.

RSPB 2005. Web-site of the Royal Society for the Protection of Birds, Sandy, UK.

RUDNICK, J.A., KATZNER, T.E., BRAGIN, E.A., RHODES, O.E. & DEWOODY, J.A. in press. Using naturally shed feathers for individual identification, genetic parentage analyses, and population monitoring in an endangered Eastern imperial eagle (*Aquila heliaca*) population from Kazakhstan. *Molecular Ecology*.

RUMMEL, E. 1962. Diet specialisation among Goshawks inhabiting the village of Puka. *Loodusuurijate Seltsi Aastaraamat* 54 : 223–226.

RÜPPELL, W. 1948. Neue Ergebnisse über Heimfinden beim Habicht. *Vogelzug* 11 : 57–64.

RUST, R. 1977. Zur Populationsdynamik und Ernährung das Habichts (*Accipiter gentilis*) in Südbayern. *Garmischer Vogelkundlicher Berichte* 2:1–9.

RUST. R. & KECHELE, W. 1996. Altersbestimmung von Habichten *Accipiter gentilis*: Langfristige Vergeliche gemauserter Handschwingen. *Ornithologischer Anzeiger* 35 : 75–83.

RUST, R. & MISCHLER, T. 2001. Auswirkungen legaler und illegaler Verfolgung auf Habichtpopulationen in Südbayern. *Ornithologischer Anzeiger* 40 : 113–136.

RUTHKE, P. 1929. Beobachtungen an einem Habichtshorst. *Beiträge zur Fortpflanzung der Vögel* 5 : 201–204.

RUTZ, C. 2001. Raum-zeitliche Habitatnutzung des Habichts *Accipiter gentilis* in einem urbanen Lebensraum. Diploma thesis, University of Hamburg, Germany.

RUTZ, C. 2003. Assessing the breeding season diet of Goshawks *Accipiter gentilis*: biases of plucking analysis quantified by means of continuous radio-monitoring. *Journal of Zoology* 259 : 209–217.

RUTZ, C. 2004. Breeding season diet of Northern Goshawks *Accipiter gentilis* in the city of Hamburg. *Corax* 19 : 311–322.

RUTZ, C. 2005a. Extra-pair copulation and intraspecific nest intrusions in the Northern Goshawk *Accipiter gentilis*. *Ibis* 146 : 831–835.

RUTZ, C. 2005b. The Northern Goshawk – population dynamics and behavioural ecology. DPhil thesis, University of Oxford, UK.

RUTZ, C. 2005c. The dynamics of wildlife synurbanisation. MS in RUTZ, C. 2005b.

RUTZ, C. 2005d. Selection of odd prey had direct fitness benefits. MS in RUTZ, C. 2005b.

RUTZ, C. & BIJLSMA, R.G. 2006. Food limitation in a generalist predator. Proceedings of the Royal Society B 273: 2069–2076.

RUTZ, C. & BIJLSMA, R.G. 2005. Collapse of breeder and non-breeder segments in a food-limited raptor population. MS in RUTZ, C. 2005b.

RUTZ, C., ZINKE, A., BARTELS, T. & WOHLSEIN, P. 2004. Congenital neuropathy and dilution of feather melanin in nestlings of urban-breeding northern goshawks (*Accipiter gentilis*). *Journal of Zoo and Wildlife Medicine* 35 : 97–103.

RUTZ, C., MARQUISS, M., BIJLSMA, R.G., KENWARD, R.E. & NEWTON, I. 2005. Continental-scale abundance profile in an avian top predator. MS in RUTZ, C. 2005b

RUTZ, C., BIJLSMA, R.G., MARQUISS, M. & KENWARD, R.E. 2006b. Population limitation in the Northern Goshawk in Europe. *Studies in Avian Biology* 31: 158–197.

RUTZ, C., WHITTINGHAM, M.J. & NEWTON, I. 2006b. Age-dependent diet choice in an avian top predator. Proceedings of the Royal Society B 273: 579–568.

RYABTSEV, V.V., Durnev, Yu.A., FEFELOV, I.V. 2001. [Autumn passage of Falconiformes in the south-western shore of the Lake Baikal]. *[Russian Journal of Ornithology]* 130 : 63–68 [Russian]

RYTTMAN, H. 1999. Duvhöken – näringskonkurrent till människan? *Fauna och Flora* 94 : 113–119.

SAAR, C. 1988. Reintroduction of the peregrine falcon in Germany. Pp. 629–635 in CADE, T.J., ENDERSON, J.H., THELANDER, C.G. & C.M. White (eds). *Peregrine falcon populations, their management and recovery.* The Peregrine Fund Inc., Boise, Idaho, USA.

SAAR, C. 2000. Wanderfalken-Auswilderungbericht 1999. *Greifvögel und Falknerei (1999)*:38–38.

SAAR, C., HENCKELL, T. & TRUMF, J. 1999. Bekämpfung einer Kaninchenplage mit Beizhabichten. *Greifvögel und Falknerei (1998)*:160–166.

SALAFSKY, S.R., REYNOLDS, R.T. & NOON, B.R. 2005. Patterns of temporal variation in goshawk reproduction and prey resources. *Journal of Raptor Research* 39 : 237–246.

SARKANEN 1971. [cited in RUTZ *et al.* 2006]

SATHEESAN, S.M. 2005. *Vulture soup for truth-seeking Gypsophil souls.* Message of 1 May 2005 to Vulture Conservation discussion group of the World Working Group on Birds of Prey and Owls, Berlin, Germany.

SAUNDERS, L.B. 1982. Essential nesting habitat of the Goshawk (*Accipiter gentilis*) on the Shasta-Trinity National Forest, McCloud District. MSc thesis. California State University, Chico, California, USA.

SAUROLA, P. 1976. Kanahaukkan kuolevuus ja kuolinsyyt. *Suomen Luonto* 35 : 310–314.

SAUROLA, P. 1978. Artificial nest construction in Europe. Pp. 72–80 in GEER, T.A. (ed) *Bird of prey management techniques.* British Falconers' Club, Oxford, UK.

SAUROLA, P. 2001. Bird ringing in Finland 2000. *Linnut vuosikirja 2000*: 91–101.

SAVINICH, I.B. 1999. Ubijstvo serymi voronami *Corvus cornix* yastreba-teterevyatnika *Accipiter gentilis*. *Russian Journal of Ornithology* 69:8–9.

SCHÄFER, E. 1938. Ornithologische Ergebnisse zweier Forschungsreisen nach Tibet. *Journal für Ornithologie* (Sonderheft) 83 : 151–152.

SCHAFFER, W.W. 1998. Northern Goshawk (*Accipiter gentilis*) habitat characterization in central Alberta. MSc thesis, University of Alberta, Edmonton.

SCHARENBERG, W. & LOOFT, V. in press. Reduction of organochlorine residues in goshawk eggs (*Accipiter gentilis*) from northern Germany (1971–2002). *Ambio*.

SCHIERMANN, G. 1925. Wanderfalken und Hühnerhabicht in den Mark Brandenberg. *Journal für Ornithologie* 73 : 277–283.

SCHIØLER, E.L. 1931. *Danmarks Fugle III*, Copenhagen, Denmark. [cited in Fischer 1980]

SCHMIDT-BEY, W. 1913. Neckereien der Raubvögel nebst Gedanken über die Entstehung ihrer sekundären Geschlechtsunterschiede. *Ornithologische Monatsschrift* 38: 400–416.

SCHNELL, J.H. 1958. Nesting behavior and food habits of Goshawks in the Sierra Nevada of California. *Condor* 60: 377–403.

SCHNURRE, O. 1934. Zwei Habichts-Bruten in gegensätzlichen Landschaftsformationen. *Mitteilungen des Vereins Sächsischer Ornithologen* 4 : 99–109.

SCHNURRE, O. 1956. Ernährungsbiologische Studien an Raubvögel und Eulen der Darβhalbinsel (Mecklenburg). *Beiträge zur Vogelkunde* 4 : 211–245.

SCHNURRE, O. 1963. Lebensbilder märkischer Habichte. *Milu* 1 : 221–238.

SCHNURRE, O. 1965. Zur beuteauswahl beim Habicht. *Zeitschrift für Jagdwissenschaft* 11 : 121–135.

SCHNURRE, O. 1973. Ernährungsbiologische Studien an Greifvögel der Insel Rügen (Mecklenburg). *Beiträge zur Vogelkunde* 19:1–16.

SCHÖNBRODT, R. & H. TAUCHNITZ. 1987. Ergebnisse 10-jähriger Planberingung von jungen Greifvögeln in den Kreisen Halle, Halle-Neustadt und Saalkreis. *Populationsökologie Greifvogel- und Eulenarten* 1 : 67–84.

SCHÖNBRODT, R. & H. TAUCHNITZ. 1991. Greifvogelhorstkontrollen der Jahre 1986 bis 1990 bei Halle. *Populationsökologie Greifvogel- und Eulenarten* 2 : 61–74.

SCHULZ, R. 1981. Remarks on artificial insemination on goshawks. Pp. 189–190 in KENWARD, R.E. & LINDSAY, I.M. (eds) *Understanding the Goshawk.* International Association of Falconry and Conservation of Birds of Prey, Oxford, UK.

SCHWEIGMANN, H.F.A. 1941. Havik, *Accipiter gentilis* gallinarum (Brehm) nestelend op den grond. *Ardea* 30 : 269.

SELANDER, R.K. 1966. Sexual dimorphism and differential niche utilization in birds. *Condor* 68 : 113–151.
SELÅS, V. 1997a. Nest-site selection by four sympatric forest raptors in Southern Norway. *Journal of Raptor Research* 31 : 16–25.
SELÅS, V. 1997b. Influence of prey availability on re-establishment of Goshawk *Accipiter gentilis* nesting territories. *Ornis Fennica* 74 : 113–120.
SELÅS, V. 1998a. Does food competition from red fox (Vulpes vulpes) influence the breeding density of goshawk (*Accipiter gentilis*)? Evidence from a natural experiment. *Journal of Zoology (London)* 246 : 325–335.
SELÅS, V. 1998b. Hønsehaukbestanden I tilbakegang – også I Aust-Agder. *Vår fuglefauna* 21: 149–154.
SELÅS, V. & STEEL, C. 1998. Large brood sizes of pied flycatcher, sparrowhawk and goshawk in peak microtine years: support for the mast depression hypothesis. *Oecologia* 116: 449–455.
SELLECK, D. & GLADING, B. 1943. Food habits of nesting Barn Owls and Marsh Hawks at Dune Lakes, California, as determined by the 'cage-nest' method. *California Fish and Game* 29 : 122–131.
SERGIO, F. & NEWTON, I. 2003. Occupancy as a measure of territory quality. *Journal of Animal Ecology* 72 : 857–865.
SERGIO, F., MARCHESI, L. & PEDRINI, P. 2003. Spatial refugia and the coexistence of a diurnal raptor with its intraguild owl predator. *Journal of Animal Ecology* 72 : 232–245.
SHEPEL, A.I. 1992. *[Birds of Prey and Owls of the Perm Cis-Kama River area]*. Irkutsk University Press, Irkutsk, Russia. [Russian]
SHERROD, S. K., HEINRICH, W.R., BURNHAM, W.A., BARCLAY, J.H. & CADE, T.J. 1981. *Hacking: a method for releasing Peregrine Falcons and other birds of prey.* The Peregrine Fund, Cornell, New York, USA.
SHUSTER, W. C. 1976. Northern Goshawk nesting densities in montane Colorado. *Western Birds* 7 : 108–110.
SHUSTER, W. C. 1980. Northern Goshawk nest site requirements in the Colorado Rockies. *Western Birds* 11 : 89–96.
SIBLEY, C. & MONROE, B. 1990. *Distribution and taxonomy of birds of the world.* Yale University Press, New Haven, USA.
SIDERS, M.S. 1995. Nesting habitat of sympatric accipiters in north central New Mexico, MSc thesis, Colorado State University, Fort Collins, Colorado, USA.
SIDERS, M.S. & KENNEDY, P.L. 1994. Nesting of Accipiter hawks: is body size a consistent predictor of nest habitat characteristics? *Studies in Avian Biology* 16 : 92–96.
SIDERS, M. S., & P. L. KENNEDY. 1996. Forest structural characteristics of accipiter nesting habitat: Is there an allometric relationship? *Condor* 98 : 123–132.
SIEWERT, H. 1933. Die Brutbiologie des Hühnerhabichts. *Journal für Ornithologie* 81 : 44–94.
SIMMONS, R.E., AVERY, D.M. and AVERY, G. 1991. Biases in diets determined from pellets and remains: correction factors for a mammal and bird-eating raptor. *Journal of Raptor Research* 25: 63–67.
SLÁDEK, J. 1963. Beitrag zur Nährungsökologie des Hühnerhabichts. *Journal für Ornithologie* 81 : 44–94.
SLAGSVOLD, T. 1978. Hønsehauk som kråkespesialist. *Vår Fuglefauna* 1 : 126.
SLIJPER, H.J. 1978. Een 18e eeuws schilderij van een Havik met veertien staartpennen. *Falknerij in de Nederlanden (1977)*:24–26.
SLIJPER, H.J. 1980. Flight speeds of goshawks. Falknerij in de Nederlanden (1979).

SMALLWOOD, K. S. 1998. On the evidence needed for listing northern Goshawks (*Accipiter gentilis*) under the Endangered Species Act: a reply to Kennedy. *Journal of Raptor Research* 32 : 323–329.
SMITH, K.D. 1965. On the birds of Morocco. *Ibis* 107 : 493–526.
SMITH, R.N, CAIN, S.L., ANDERSON, S.H., DUNK, J.R. & WILLIAMS, W.S. 1998. Blackfly-induced mortality of nestling Red-tailed Hawks. *Auk* 115 : 368–375.
SMITH, S. 1982. Raptor 'reverse' dimorphism revised: a new hypothesis. *Oikos* 9 : 118–122.
SMITHERS, B.L. 2003. Northern goshawk food habits in Minnesota: an analysis using time-lapse video recording systems. MSc thesis, Texas Technical University, Lubbock, Texas, USA. [cited in Boal *et al.* 2006]
SMITHERS, B.L., BOAL, C.W. & ANDERSEN, D.A. 2005. Northern Goshawk diet in Minnesota: an analysis using video cameras. *Journal of Raptor Research* 39 : 264–273.
SNYDER, N.F.R. & WILEY, J.W. 1976. Sexual size dimorphism in hawks and owls of North America. *Ornithological Monographs* 20:1–96.
SNYDER, N.F.R., SNYDER, H.A., LINCER, J.L. & REYNOLDS, R.T. 1973. Organochlorines, heavy metals, and the biology of North American accipiters. *BioScience* 23: 300–305.
SOLLIEN, A. 1978. Vandringar hos norsk hønsehauk. *Vår Fuglefauna* 1 : 52–59.
SOLLIEN, A. 1979. Bestandsutvikling hos hønsehauk *Accipiter gentilis* in Norge de siste 100 år. *Vår Fuglefauna* 2 : 95–106.
SONERUD, G.A. 1992. Search tactics of a pausetravel predator: adaptive adjustments of perching times and move distances by Hawk Owls (*Surnia ulula*). *Behavioural Ecology and Sociobiology* 30 : 207–217.
SONSTHAGEN, S. 2002. Year-round habitat, movement, and gene flow of northern Goshawks breeding in Utah. MSc thesis, Brigham Young University, Provo, Utah, USA.
SONSTHAGEN, S.A., RODRIGUEZ, R. & WHITE, C.M. 2006a. Satellite telemetry of Northern Goshawks (*Accipiter gentilis*) breeding in Utah. I. Annual movements. *Studies in Avian Biology* 31: 239–251.
SONSTHAGEN, S.A., RODRIGUEZ, R. & WHITE, C.M. 2006b. Satellite telemetry of Northern Goshawks (*Accipiter gentilis*) breeding in Utah. II. Annual habitats. *Studies in Avian Biology* 31: 252–259.
ŠOTNÁR, K. 2000. A contribution to the breeding biology and feeding ecology of the Goshawk (*Accipiter gentilis*) in the Horné Ponitrie Region, Slovakia. *Buteo* 11 : 43–50.
SOUTH, A.B., RUSHTON, S.P., KENWARD, R.E., & MACDONALD, D.W 2002. Modelling vertebrate dispersal and demography in real landscapes: how does uncertainty regarding dispersal behaviour influence predictions of spatial population dynamics? Pp. 327–349 in BULLOCK, J.M., KENWARD, R.E. & HAILS, R. (eds*) Dispersal: an ecological perspective.* Symposium of the British Ecological Society, Blackwell, Oxford, UK.
SOUTHERN, W.E. 1964. Additional observations on winter bald eagle populations: including remarks on biotelemetry techniques and immature plumages. *Wilson Bulletin* 76 : 222–237.
SPEISER, R. 1992. Notes on the natural history of the Northern Goshawk. *Kingbird* 42 : 133–137.
SPEISER, R. & BOSAKOWSKI, T. 1984. History, status, and future management of Goshawk nesting in New Jersey. *Records of New Jersey Birds* 10 : 29–33.
SPEISER, R. & BOSAKOWSKI, T. 1987. Nest site selection by Northern Goshawks in northern New Jersey and southeastern New York. *Condor* 89 : 387–394.
SPEISER, R. & BOSAKOWSKI, T. 1989. Nest trees selected by Northern Goshawks along the New York New Jersey border. *Kingbird* 39 : 132–141.

SPEISER, R. & BOSAKOWSKI, T. 1991. Nesting phenology, site fidelity, and defense behavior of Northern Goshawks in New York and New Jersey. *Journal of Raptor Research* 25 : 132–135.

SQUIRES, J.R. 1995. Carrion use by northern Goshawks. *Journal of Raptor Research* 29 : 189–190.

SQUIRES, J.R. 2000. Food habits of northern Goshawks nesting in south central Wyoming. *Wilson Bulletin* 112 : 536–539.

SQUIRES, J.R. & KENNEDY, P.L. 2006. Northern Goshawk ecology: an assessment of current knowledge and information needs for conservation and management. *Studies in Avian Biology* 31: 8–62.

SQUIRES, J.R. & RUGGIERO, L.F. 1995. Winter movements of adult northern Goshawks that nested in southcentral Wyoming. Journal of Raptor Research *29:5–8.*

SQUIRES, J.R. & RUGGIERO, L.F. 1996. Nest-site preference of Northern Goshawks in southcentral Wyoming. *Journal of Wildlife Management* 60 : 170–177.

STAUDE J. 1987. Ergebnisse mehrjäriger Brutbestandsaufnahmen von Greifvögeln im Wesenbergerland. *Vogelkundliche Berichte aus Niedersachsen* 19 : 37–45.

STEEN, O.F. 2004. Hønsehauken i Buskerud – tetthet, bestand och hekkesuksess. *Vår Fuglefauna* 27 : 18–24.

STEENHOF, K. & KOCHERT, M.N. 1982. An evaluation of techniques used to estimate raptor nesting success. *Journal of Wildlife Management* 46 : 885–893.

STEINER, H. 1999. Der Mäusebussard (*Buteo buteo*) als Indikator für Struktur und Bodennutzung des ländlichen Raumes: Produktivität im heterogenen Habitat, Einfluβ von Nahrung und Witterung und Vergleiche zum Habicht (*Accipiter gentilis*). *Stapfia* 62:1–74.

STEPHENS, R.M. 2001. Migration, habitat use, and diet of northern Goshawks (*Accipiter gentilis*) that winter in the Uinta Mountains, Utah. MSc thesis. University of Wyoming, Laramie, Wyoming, USA.

STEVENSON, H. 1859a. Occurrence of the Goshawk in Norfolk. *Zoologist (1859)*:6325.

STEVENSON, H. 1859b. Occurrence of the Goshawk in Suffolk. *Zoologist (1859)*:6443.

STORER, R.W. 1966. Sexual dimorphism and food habits in three North American accipiters. *Auk* 83 : 423–436.

STORGÅRD, K. & BIRKHOLM-CLAUSEN, F. 1983. En status over Duehøgen i Sydjylland. *Proceedings of the Third Nordic Congress of Ornithology (1981)*:59–64.

STRAZDS, M., PRIEDNIEKS, J. & VĀVERIŅŠ, G. 1994. Latvija putnu skaits. *Putni dāba* 4:3–18.

STROUD, D.A. 2003. The status and legislative protection of birds of prey and their habitats in Europe. Pp. 51–84 *in* THOMPSON, D.B.A, REDPATH, S.M., FIELDING, A.H., MARQUISS, M. & GALBRAITH, C.A. (eds). *Birds of Prey in a Changing Environment.* Scottish Natural Heritage and The Stationary Office, Edinburgh, UK.

STUBBE, M., H. ZÖRNER, H. MATTHES, & W. BÖHM. 1991. Reproduktionsrate und gegenwärtiges Nahrungsspektrum einiger Greifvogelarten im nördlichen Harzvorland. *Populationsökologie Greifvogel- und Eulenarten* 2 : 39–60.

STÜLCKEN, K. 1943. Aufzeichnungen am Habichtshorst in einem Südholsteinischen Waldgebiet. *Deutscher Falkenorden* (1943):31–48.

SUNDE, P. 2002. Starvation mortality and body condition of Goshawks *Accipiter gentilis* along a latitudinal gradient in Norway. *Ibis* 144 : 301–310.

SULKAVA, S. 1964. Zur Nährungsbiologie des Habichts, *Accipiter gentilis* L. *Aquilo Seria Zoologica* 3:1–103.

SULKAVA, P. & SULKAVA, S. 1981. Petolintujen syksyisestä pesärakentamista. *Lintumies* 16 : 77–80.

SULKAVA, S., HUHTALA, K. & TORNBERG, R. 1994. Regulation of goshawks *Accipiter gentilis* breeding in western Finland over last 30 years. Pp. 67–76 in MEYBURG, B.-U. & R. D. CHANCELLOR (eds), *Raptor conservation today*. World Working Group on Birds of Prey, Berlin, Germany.

SUNDE, P. 2002. Starvation mortality and body condition of Goshawks *Accipiter gentilis* along a latitudinal gradient in Norway. *Ibis* 144 : 301–310.

SUTHERLAND, W.J. 1996. *From individual behaviour to population ecology*. University Press, Oxford, UK.

SUTTON, G.M. 1931. The status of the Goshawk in Pennsylvania. *Wilson Bulletin* 43: 108–113.

SVENSSON, S. 2002. Duvhökens *Accipiter gentilis* beståndsutveckling I Sverige sedan 1975. *Ornis Svecica*. 12 : 147–156.

SWEM, T. & ADAMS, M. 1992. A northern Goshawk nest in the tundra biome. *Journal of Raptor Research* 26 : 102.

TABERLET, P. & LUIKHART, F. 1999. Non-invasive genetic sampling and individual identification. *Biological Journal of the Linnean Society* 68 : 41–55.

TAPPER, S. (ed) 1999. *A question of balance – game animals and their role in the British countryside*. The Game Conservancy Trust, Fordingbridge, UK.

TAVERNER, P.A. 1940. Variation in the American Goshawk. *Condor* 42 : 157–160.

TELLA, J.L. & MAÑOSA, S. 1993. Eagle owl predation on Egyptian vulture and northern Goshawk: possible effect of a decrease in European rabbit availability. *Journal of Raptor Research* 27 : 111–112.

THIOLLAY, J.-M. 1967. Ecologie d'une population de rapaces diurnes en Lorraine. *La Terre et la Vie – Revue D'Écologie Appliquée* 21 : 116–183.

THIOLLAY, J.-M., BRETAGNOLLE, V. & SÉRIOT, J. 2003. *Inventaire des rapaces de France 2000–2003*. Rapport LPO/MED. Rochefort, France.

THIRGOOD, S. & REDPATH, S. 1997. Red grouse and their predators. *Nature* 390 : 547.

THIRGOOD, S.J., REDPATH, S.M., HAYDON, D.T., ROTHERY, P., NEWTON, I. & HUDSON, P.J. 2000a. Habitat loss and raptor predation: disentangling long- and short-term causes of red grouse declines. *Proceedings of the Royal Society of London Series B* 267 : 651–656.

THIRGOOD, S.J., REDPATH, S.M., NEWTON, I. & HUDSON, P. 2000b. Raptors and Red Grouse: Conservation Conflicts and Management Solutions. *Conservation Biology* 14 : 95–104.

THIRGOOD, S.J., REDPATH, S.M., ROTHERY, P. & AEBISCHER, N.J. 2000c. Raptor predation and population limitation in red grouse. *Journal of Animal Ecology* 69 : 504–516.

THIRGOOD, S.J., REDPATH, S.M., CAMPBELL, S. & SMITH, A.A. 2002. Do habitat characteristics influence predation on red grouse? *Journal of Applied Ecology* 39 : 217–225.

THISSEN, J., MÜSKENS, G. & OPDAM, P. 1981. Trends in the Dutch Goshawk *Accipiter gentilis* population and their causes. Pages 28–43 in KENWARD, R.E. & LINDSAY, I.M. (eds). *Understanding the Goshawk*. International Association for Falconry and Conservation of Birds of Prey, Oxford, UK.

THRAILKILL, J.A., L.S. ANDREWS & R.M. CLAREMONT. 2000. Diet of breeding northern goshawks in the Coast Range of Oregon. *Journal of Raptor Research* 34 : 339–340.

TINBERGEN, L. 1936. Gegevens over het voedsel von Nederlandse Haviken (*Accipiter gentilis gallinarum* (Brehm)). *Ardea* 25 : 195–200.

TITUS, K., C. FLATTEN & R. LOWELL. 1997. *Goshawk ecology and habitat relationships on the Tongass National Forest, 1996 field progress report and preliminary stable isotope analysis*. Alaska Department of Fish and Game, Division of Wildlife Conservation, Juneau, Alaska, USA.

TOMIAŁOJĆ, L. & STAWARCZYK, T. 2003. *Awifauna Polski*. Rosmieszczenie, liczebność i zmiany. PTPP 'pro Natura', Wroclaw, Poland.

TØMMERAAS, P.J. 1977. Høge som ådelsaedere. *Naturen Verden* 1 : 37–40.
TØMMERAAS, P.J. 1980. Hvor åtselet er, der skal ørnene samles. *Trondheims Turistforenings Årbok* 93 : 25–35.
TØMMERAAS, P.J. 1986. Prosjekt hønsehauk I Trøndelag 1985. *Trøndersk Natur* 13 : 97–100.
TORDOFF, H.B., MARTELL, M.S. & REDIG, P.T. 1998. Effect of fledge site on choice of nest site by Midwestern Peregrine Falcons. *Loon* 70 : 127–129.
TORNBERG, R. 1997. Prey selection of the goshawk *Accipiter gentilis* during the breeding season: the role of prey profitability and vulnerability. *Ornis Fennica* 74 : 15–28.
TORNBERG, R. 2000. Effect of changing landscape structure on the predator-prey interaction between goshawk and grouse. PhD thesis. University of Oulu, Oulu, Finland.
TORNBERG, R. 2001. Pattern of Goshawk *Accipiter gentilis* predation on four forest grouse species in northern Finland. *Wildlife Biology* 7 : 245–256.
TORNBERG, R. & COLPAERT, A. 2001. Survival, ranging, habitat choice and diet of the Northern Goshawk *Accipiter gentilis* during winter in northern Finland. *Ibis* 143 : 41–50.
TORNBERG, R. & SULKAVA, S. 1991. The effect of changing tetraonid populations on the nutrition and breeding success of the Goshawk (*Accipiter gentilis* L.) in northern Finland. *Aquilo Seria Zoologica* 28 : 23–33.
TORNBERG, R. & VIRTANEN, V. 1997. When and why do Goshawks die? *Linnut* 32 : 10–13.
TORNBERG, R., KORPIMÄKI, E. & BYHOLM, P. 2006. Ecology of the Northern Goshawk in Fennoscandia. *Studies in Avian Biology* 31: 141–157.
TORNBERG, R., MÖNKKÖNEN, M. & PAHKALA, M. 1999. Changes in diet and morphology of Finnish Goshawks from 1960s to 1990s. Oecologia *121 : 369–376.*
TORNBERG, R., KORPIMÄKI, E., REIF, V., JUNGELL, S. & Mykrä, S. 2005. Delayed numerical response of goshawks to population fluctuations of forest grouse. *Oikos* 111 : 408–415.
TOYNE, E.P. 1997a. Notes on Northern Goshawks nesting in an abandoned heronry in Wales. *Journal of Raptor Research* 31 : 89.
TOYNE, E.P. 1997b. Nesting chronology of northern goshawks (*Accipiter gentilis*) in Wales: implications for forest management. *Forestry* 70 : 121–127.
TOYNE, E. P. 1998. Breeding season diet of the Goshawk *Accipiter gentilis* in Wales. *Ibis* 140 : 569–579.
TOYNE, E.P. & ASHFORD, R.W. 1997. Blood parasites of nestling goshawks. *Journal of Raptor Research* 31 : 81–83.
TRIVERS, R.L. 1974. Parent-offspring conflict. *American Zoologist* 14 : 249–264.
TROFIMENKO, V.V. 2002. [Winter ecology of hawks in the south of Rostov Region: Goshawk *Accipiter gentilis* (L.)]. *[Proceedings of Teberda State Biosphere Nature Reserve]* 31 : 165–167. [Russian]
TROMMER, G. 1981. The goshawk as a foster parent for peregrines. Pages 185–187 in KENWARD, R.E. & LINDSAY, I.M. (eds) *Understanding the Goshawk.* International Association of Falconry and Conservation of Birds of Prey, Oxford, UK.
TROMMER, G. 1996. Die Adoption von jungen Wanderfalken beim Habicht – eine gute Auswilderungsmethode oder ein Risiko? *Greifvögel und Falknerei* (1994):58–61.
TROMMER, G., SIELICKI, S. & WIELAND, P. 2000. Der Wanderfalke – nun auch wieder Brutvogel in Polen. *Greifvögel und Falknerei (1999)*: 48–56.
TUCK, J.G. 1877. Goshawk in Norfolk. *Zoologist* (1877):179.

TUFTS, R.W. 1961. *Birds of Nova Scotia*. Nova Scotia Museum, Halifax, Nova Scotia, Canada. [cited in Squires & Kennedy 2006]

TYACK, A.J., KENWARD, R.E. & WALLS, S.S. 1998. Behaviour in the post-nestling dependence period of radio-tagged Common Buzzards *Buteo buteo*. *Ibis* 140: 58–63.

TYE, A. 1989. A model of search behavior for the Northern Wheatear *Oenanthe oenanthe* (*Aves, Turdidae*) and other pausetravel predators. *Ethology* 83:1–18.

UNDERWOOD, J., WHITE, C. & RODRIGUEZ, R. 2006. Winter movement and habitat use of Northern Goshawks (*Accipiter gentilis*) breeding in Utah. *Studies in Avian Biology* 31: 228–238.

UPTON, R.C. 1980. *A bird in the hand – celebrated falconers of the past*. Debretts Peerage, London, UK.

UTTENDÖRFER, O. 1939. *Die Ernährung der deutschen Raubvögel und Eulen und ihre Bedeutung in der heimischen Natur*. Neumann-Neudamm, Melsungen, Germany.

UTTENDÖRFER, O. 1952. *Neue Ergebnisse uber die Ernahrung der Greifvogel und Eulen*. Eugen Ulmer, Stuttgart, Germany.

VÄISÄNEN, R.A., KOSKIMIES, P. & LAMMI, E. 1998. *Distribution, numbers and population changes of Finnish breeding birds*. Keuruu, Otava, Finland.

VALKAMA, J. KORPIMÄKI, E., ARROYO, B., BEJA, P., BRETAGNOLLE, V., BRO, E., KENWARD, R., MAÑOSA, S., REDPATH, S.M, THIRGOOD, S. & VIÑUELA, J. 2005. Birds of prey as limiting factors of gamebird populations in Europe: a review. *Biological Review* 80: 171 203.

VAN BEUSEKOM, C.F. 1972. Ecological isolation with respect to food between sparrowhawk and goshawk. *Ardea* 60 : 72–96.

VAN HAAFF, G. 2001. Postduifringen als indicator voor vroegere nestbezetting door *Accipiter gentilis*. Over de archeologie van Treekse havikhorsten. *De Takkeling* 9 : 137–149.

VAN LENT, T. 2004. De Havik *Accipiter gentilis* op de Utrechtse Heuvelrug van 1965–70: broedresultaten, prooikeus en ruiveren. *De Takkeling* 12 : 118–144.

VAN ROSSEM, A.J. 1938. A Mexican race of the Goshawk (*Accipiter gentilis*) [Linnaeus]. *Proceedings Biological Society Washington* 51 : 99–100. Describes apache

VAURIE, C. 1965. *The birds of the Palaearctic fauna: non-Passeriformes*. Witherby, London, UK.

VEDDER, O. 2000. Veerafwijking bij nestjonge Havik Accipiter gentilis, en mogelijk oorzaak. *De Takkeling* 8 : 221–222.

VEDDER, O. & DEKKER, A.L. 2004. Kan een sperwervrouw *Accipiter nisus* haar nest tegen predatie door een Havik *Accipiter gentilis* verdedigen? *De Takkeling* 12 : 150–155.

VEIGA, J.P. 1982. Ecologia de las rapaces de un ecosistema de montaña. Aproximación a su estructura comunitaria. PhD thesis, University of Madrid, Spain. [cited in Manosa 1991]

VEIT, R.R. & PETERSEN, W.R. 1993. *Birds of Massachusetts*. Massachusetts Audubon Society, Lincoln, Massachusetts, USA.

VERDEJO, J. 1994. Datos sobre la reproducción y alimentación del Azor (*Accipiter gentilis*) en un área mediterranea. *Ardeola* 41 : 37–43.

VILLAGE, A. 1983. *The kestrel*. Poyser, Calton, UK.

VINCENT, J. & WORMALD, H. 1943. Goshawk in Norfolk. *British Birds* 36 : 181.

VIÑUELA, J. 2002. *Reconciling Gamebird Hunting and Biodiversity* (REGHAB). EVK2-CT–2000–200004. www.uclm.es/irec/reghab/initio.html

VITOVICH, O.A. 1985. [The Goshawk in Teberda Nature Reserve]. Pp. 129–139 in *[Birds of North-western Caucasia]*. State Publishing House, Moscow, Russia. [Russian]

VLUGT, D. 2002. De postduif Columa livia als pooi van de Havik Accipiter gentilis in de duinen van Noord-Holland. *De Takkeling* 10 : 135–149.
VOOUS, K.H. & WATTLE, J. 1972. 'Tropische' Varietät eines deutschen Habichts (*Accipiter gentilis*). *Journal für Ornithologie* 113 : 214–218.
VON BITTERA, J. 1916. Über die Nahrung des Habichts und Sperbers. *Aquila* 22 : 216–217.
VON SCHWEPPENBURG, G. 1938. Frit *Accipiter gentilis* Aas? *Ornithologischer Mitteilungen* 46 : 183–184.
VORONIN, R.N. 1995. [Goshawk] *Accipiter gentilis buteoides* (Linnaeus, 1758). Pp. 73–75 in *[Birds. Vol.1. Nonpasseriformes]*. Nauka, St-Petersburg, Russia [Russian]
WAARDENBURG, P. 1977a. Die Auswirkungen einiger menschlicher Storungsfaktorern auf die Siedlungsdichte des Habichts (*Accipiter gentilis*). *Deutscher Falkenorden* (1976/77):46–49.
WAARDENBURG, P. 1977b. Freilandbruten von einjährigen Habichten und von (entflogenen) Beizhabichten. *Deutscher Falkenorden* (1976/77):50–51.
WALLACE, M.P. 2001. Recovery Efforts for the California Condor. P. 197 in *Abstracts of the 4th Eurasian Congress on Raptors*. Seville-Spain 25–29 September 2001.
WALLS, S.S. 2005. How sociality, weather and other factors affect the leaving, transition and settling phases of dispersal in the buzzard *Buteo buteo*. PhD thesis, University of Reading, UK.
WALLS, S.S. & KENWARD, R.E. 2001. Spatial consequences of relatedness and age in buzzards. *Animal Behaviour* 61 : 1069–1078.
WALLS, S.S., KENWARD, R.E. & HOLLOWAY, G.J. 2005. Weather to disperse? Evidence that climatic conditions influence vertebrate dispersal. *Journal of Animal Ecology* 74: 190–197.
WALTER, H. 1979. *Eleonora's falcon, adaptations to prey and habitat in a social raptor*. University Press, Chicago, USA.
WARD, J.M. & KENNEDY, P.L. 1994. Approaches to investigating food limitation hypotheses in raptor populations: an example using the northern goshawk. *Studies in Avian Biology* 16 : 114–118.
WARD, J.M. & KENNEDY, P.L. 1996. Effects of supplemental food on growth and survival of juvenile northern goshawks. *Auk* 113 : 200–208.
WARNCKE, K. 1961. Beitrag zur Brutbiologie von Habicht und Sperber. *Vogelwelt* 82:6–12.
WATSON, J.W., HAYS, D.W. & FINN, S.P. 1998. Prey of breeding Northern Goshawks in Washington. *Journal of Raptor Research* 32 : 297–305.
WATSON, J.W., HAYS, D.W. & PIERCE, D.J. 1999. Efficacy of Northern Goshawk broadcast surveys in Washington State. *Journal of Wildlife Management* 63 : 98–106.
WATTEL, J. 1973. Geographical differentiation in the genus Accipiter. *Bulletin of the Nuttall Ornithological Club* 13, Cambridge, Massachusetts, USA.
WATTEL, J. 1981. The goshawk and its relatives. Some remarks on systematics and evolution. Pages 6–14 in KENWARD, R.E. & LINDSAY, I.M. (eds) *Understanding the Goshawk*. International Association of Falconry and Conservation of Birds of Prey, Oxford, UK.
WEBER, M. 2001. Untersuchungen zu Greifvogelbestand, Habitatstruktur und Habitatveränderung in ausgewählten Gebieten von Sachsen-Anhalt und Mecklenburg-Vorpommern. PhD thesis. University of Halle, Germany.
WELANDER, E. 1924. Duvhökens, *Astur gentilis* L., ruvningstid. *Fauna och Flora* 19 : 185–187.
WERNERY, U., KINNE, J., SHARMA, A., BOEHNEL, H. & SAMOUR, J.H. 2000. Clostridium enterotoxaemia in Falconiformes in the United Arab Emirates. Pp. 35–42 in LUMEIJ, J.T., REMPLE, J.D., REDIG, P.T., LIERZ, M. & COOPER, J.E. (eds) *Raptor Biomedicine III*. Zoological Education Network, Fort Worth, USA.

WERNHAM, C.V., TOMS, M.P., MARCHANT, J.H., CLARK, J.A., SIRIWARDENA, G.M. & BAILLIE, S.R. (eds) 2004. *The migration atlas: movements of the birds of Britain and Ireland.* Poyser, London, UK.

WESTCOTT, P.W. 1964. Unusual feeding behaviour of a goshawk. *Condor* 66 : 163.

WHALEY, W.S. & WHITE, C.M. 1994. Trends in geographic variation of Cooper's hawk and northern Goshawk: a multivariate analysis. *Western Foundation of Vertebrate Zoology* 5 : 161–209.

WHITE, C. & KIFF, L. 1998. Language use and misapplied selective science; their roles in swaying public opinion and policy as shown with two North American raptors. Pp. 547–560 in MEYBURG, B.-U., CHANCELLOR, R.D. & FERRERO, J.J. (eds) *Holarctic birds of prey.* Asociación para la Defensa de la Naturaleza y los Recursos de Extremadura and World Working Group on Birds of Prey, Berlin, Germany.

WHITE, C.M., LLOYD, G.D. & RICHARDS, G.L. 1965. Goshawk nesting in the Upper Sonoran in Colorado and Utah. *Condor* 67 : 269.

WHITE, G.C. & GARROTT, R.A. 1990. *Analysis of wildlife radio tracking data.* Academic Press, San Diego, USA.

WHITE, T.H. 1951. *The goshawk.* Jonathan Cape, London, UK.

WIDÉN, P. 1981. Activity pattern of goshawks in Swedish boreal forests. Pages 114–120 in KENWARD, R.E. & LINDSAY, I.M. (eds) *Understanding the Goshawk.* International Association of Falconry and Conservation of Birds of Prey, Oxford, UK.

WIDÉN, P. 1982. Radio monitoring the activity of goshawks. *Symposia of the Zoological Society of London* 49 : 153–160.

WIDÉN, P. 1984a. Reversed sexual size dimorphism in birds of prey: revival of an old hypothesis. *Oikos* 43 : 259–263.

WIDÉN, P. 1984b. Activity patterns and time-budget in the Goshawk *Accipiter gentilis* in a boreal forest area in Sweden. *Ornis Fennica* 61 : 109–112.

WIDÉN, P. 1985. Breeding and movements of Goshawks in boreal forests in Sweden. *Holarctic Ecology* 8 : 273–279.

WIDÉN, P. 1987. Goshawk predation during winter, spring and summer in a boreal forest area of Sweden. *Holarctic Ecology* 10 : 104–109.

WIDÉN, P. 1989. The hunting habitats of Goshawks *Accipiter gentilis* in boreal forests of central Sweden. *Ibis* 131 : 205–231.

WIDÉN, P. 1997. How, and why, is the Goshawk (*Accipiter gentilis*) affected by modern forest management in Fennoscandia? *Journal of Raptor Research* 31 : 107–113.

WIDÉN, P., ANDRÉN, H., ANGELSTAM, P. & LINDSTRÖM, E. 1987. The effect of prey vulnerability: goshawk predation and population fluctuations of small game. *Oikos* 49 : 233–235.

WIEHN, J. & KORPIMÄKI, E. 1998. Resource levels, reproduction and resistance to haematozoan infections. *Proceedings of the Royal Society of London* B 265 : 1197–1201.

WIELICZKO, A., T. PIASECKI, G. M. DORRESTEIN, A. ADAMSKI, & M. MAZURKIEWICZ. 2003. Evaluation of the health status of goshawk chicks (*Accipiter gentilis*) nesting in Wroclaw vicinity. *Bulletin of the Veterinary Institute of Pulawy* 47 : 247–257.

WIENS, J.D. 2004. Post-fledging survival and natal dispersal of juvenile northern goshawks in Arizona. MSc thesis, Colorado State University, Fort Collins, Colorado, USA.

WIENS, J.D. & REYNOLDS, R.T. 2005. Is fledging success a reliable index of fitness in Northern Goshawks? *Journal of Raptor Research* 39 : 210–221.

WIGGINS, J. 1971. An attempt to breed goshawks. *Captive Breeding of Diurnal Birds of Prey* 2 : 11–12.

WIKLUND, C.G. 1982. Fieldfare (*Turdus pilaris*) breeding success in relation to colony size, nest position and association with merlins (*Falco columbarius*). *Behavioural Ecology and Sociobiology* 11 : 165–172.

WIKMAN, M. 1976. Sex ratio of goshawk nestlings. *Suomen Luonto* 35 : 307–309.

WIKMAN, M. & LINDÉN, H. 1981. The influence of food supply on goshawk population size. Pp. 105–113 in KENWARD, R.E. & LINDSAY, I.M. (eds) *Understanding the Goshawk*. International Association of Falconry and Conservation of Birds of Prey, Oxford, UK.

WIKMAN, M. & TARSA, V. 1980. Kanahaukan pesimäaikaisesta ravinnosta Länsi-Uudellamaalla 1969–77. *Suomen Riista* 28 : 86–96.

WILLIAMS, N.P. & EVANS, J. 2000. The application of DNA technology to enforce raptor conservation legislation within Great Britain. Pp. 859–867 in CHANCELLOR, R.D. & MEYBURG, B.-U. (eds) *Raptors at risk*. World Working Group on Birds of Prey and Owls, Berlin, Germany.

WILLIAMS, W.J., HARTERT, E. & WITHERBY, J. 1919. American goshawk in Ireland. *British Birds* 13 : 31–32.

WILLOUGHBY, E.J. & CADE, T.J. 1964. Breeding behaviour of the American Kestrel (Sparrowhawk). *Living Bird* 3 : 75–96.

WILLOUGHBY, S. 1843. Rare birds of Lincolnshire. *Zoologist*(1843):247.

WINK, M. 1998. Application of DNA markers to study the ecology and evolution of raptors. Pp. 49–71 in MEYBURG, B.-U., CHANCELLOR, R.D. & FERRERO, J.J. (eds) *Holarctic birds of prey*. Asociación para la Defensa de la Naturaleza y los Recursos de Extremadura and World Working Group on Birds of Prey, Berlin.

WINK, M. & SAUER-GÜRTH, H. 2004. Phylogenetic relationships in diurnal raptors based on nucleotide sequences of mitochondrial and nuclear marker genes. Pp. 483–498 in CHANCELLOR, R.D. & MEYBURG, B.-U. (eds) *Raptors worldwide*. World Working Group on Birds of Prey and Owls, Berlin, Germany.

WINK, M., SEIBOLD, I., LOTFIKHAH, F. & BEDNAREK, W. 1998. Molecular systematics of Holarctic raptors (Order Falconiformes). Pp. 29–48 in MEYBURG, B.-U., CHANCELLOR, R.D. & FERRERO, J.J. (eds) *Holarctic birds of prey*. Asociación para la Defensa de la Naturaleza y los Recursos de Extremadura and World Working Group on Birds of Prey, Berlin, Germany.

WOETS, D. 1998. De Havik *Accipiter gentilis* als broedvogel in De Weerribben: 1980–1997 (deel I). *De Noordwesthoek* 25: 51–58.

WOOD, M. 1938. Food and measurements of Goshawks. *Auk* 55 : 123–124.

WOODBRIDGE, B. & DETRICH, P.J. 1994. Territory occupancy, and habitat patch size of Northern Goshawks in the southern Cascades of California. *Studies in Avian Biology* 16 : 83–87.

WOODBRIDGE, B., DETRICH, P. & BLOOM, P.H. 1988. *Territory fidelity and habitat use by nesting northern goshawks: implications for management*. Unpubl. report, USDA Forest Service, Klamath National Forest, Macdoel, California, USA.

WOODFORD, M.H. 1960. *A manual of falconry*. A&C Black, London, UK.

WORTELAERS, F. 1951. *Accipiter gentilis* L. *Gerfaut* 41 : 168.

WORTON, B.J. 1989. Kernel methods for estimating the utilisation distribution in home range studies. *Ecology* 70 : 164–168.

WÜRFELS, M. 1994. Entwicklung einer städtischer Population des Habichts (*Accipiter gentilis*) und die Rolle der Elster (*Pica pica*) im Nahrungsspektrum des Habichts. *Charadrius* 30 : 82–93.

WÜRFELS, M. 1999. Ergebnisse weiterer Beobachtungen zur Populationsentwicklung des Habichts (*Accipiter gentilis*) im Stadtgebiet von Köln 1993–1998 und zur Rolle der Elster (*Pica pica*) im Nahrungsspektrum des Habichts. *Charadrius* 35 : 20–32.

WYRWOLL, T. 1977. Die Jadgbereitschaft des Habichts (*Accipiter gentilis*) in Beziehung zum Horstort. *Journal für Ornithologie* 118 : 21.34.

YALDEN, D.W. 1987. The natural history of Domesday Cheshire. *Naturalist* 112 : 125–131.

YAZDANI, A. 2005. *Falconry: a cultural heritage in Iran.* Presentation at the Symposium 'Falconry: a world heritage', Abu Dhabi, 13–15 September 2005.

YE, XAODIE. 2005. *Falconry in the minorities of China.* Presentation at the Symposium 'Falconry: a world heritage', Abu Dhabi, 13–15 September 2005.

YOM-TOV, Y. & YOM-TOV, S. in review. Decrease in wing and beak length of juvenile Danish goshawks during the 20th century.

YOUNK, J.V. & BECHARD, M.J. 1994. Breeding ecology of the Northern Goshawk in high-elevation aspen forests of northern Nevada. *Studies in Avian Biology* 16 : 119–121.

ZACHEL, C.R. 1985. Food habits, hunting activity, and post-fledging behavior of northern Goshawks (*Accipiter gentilis*) in interior Alaska. MSc thesis, University of Alaska, Fairbanks, Alaska, USA.

ZAHAVI, A. & ZAHAVI, A. 1997. *The handicap principle: a missing piece of Darwin's puzzle.* Oxford University Press, New York, USA.

ZAWADZKA, D. & ZAWADZKI, J. 1998. The Goshawk *Accipiter gentilis* in Wigry National Park (NE Poland) – numbers, breeding results, diet composition and prey selection. *Acta Ornithologica* 33 : 182–190.

ZIESEMER, F. 1981a. Habichte verlieren Ringe. *Corax* 8 : 211–212.

ZIESEMER, F. 1981b. Methods of assessing goshawk predation. Pp. 144–151 in KENWARD, R.E. & LINDSAY, I.M. (eds) *Understanding the goshawk.* International Association for Falconry and Conservation of Birds of Prey, Oxford, UK.

ZIESEMER, F. 1982. Eine Stempelfarbe zur dauerhaften Markierung von Vögeln. *Vogelwarte* 31 : 465–466.

ZIESEMER, F. 1983. *Untersuchungen zum Einfluβ des Habichts (Accipiter gentilis) auf Populationen seiner Beutetiere.* Beiträge zur Wildlbiologie 2. Hartmann, Kronshagen, Germany.

ZINN, L. J. & TIBBITTS, T.J. 1990. *Goshawk nesting survey, 1990. North Kaibab Ranger District, Kaibab National Forest, Arizona.* CCSA # 07–90–02, Nongame and Endangered Wildlife Program, Arizona Game and Fish Department, Phoenix, Arizona, USA.

ZIRRER, F. 1947. The Goshawk. *Passenger Pigeon* 9 : 79–94.

Index

Accipiter gentilis 19, 28, 29–30
    Accipiter gentilis albidus 27, 29, 30, 142, 262
    apache 26, 29, 34
    arrigoni 29
    atricapillus 26, 28, 29
    buteoides 27, 29, 30, 142, 262
    fujiyamae 29
    gentilis 28, 29, 30
    laingi 28, 29
    marginatus 28, 29
    schvedowi 29, 30
Accipitridae 30, 31, 32
age 23
    of dispersal 127
    effect on clutch size 83
    effect on diet 175–7
    effect on productivity 228
    of first breeding 248–9
    of fledging 123
aggression 77, 79, 103
attacks 159–60, 187–90, 216
    see also kill rates
Australian Brown Goshawk 31, 35

bands *see* rings
bathing 159–60
beak 21
behaviour 153–5
    see also courtship behaviour; flight behaviours
bells 263
'Bergmann's Rule' 54
Black Goshawk 31, 32
body mass 39, 40, 42–6, 55
    of pigeons 190
    seasonal variations 46–51
    variation with latitude 51–2
Brassica crops 185, 189–91, 197–8, 261–2
breeding 24, 107
    domestic 266–9
    habitat for 60, 84
    rates of 220, 229–33
    risks of 112
    weather effect on 82, 83, 222, 249
    see also copulation; courtship behaviour; egg-laying; nests
broadcast surveys 86
brooding 59, 94, 100, 112, 114
    adoptive broods 129
    failures of 220–4
    switching of broods 128–9
    see also hatching; incubation; parenting
Buzzard 31, 32, 123, 204, 214

caching 95
cainism 106, 107
calling 73–4, 78, 88, 99, 143
calling off 260–1

carrion 30, 177
claws 21
Clostridia 61
cluster analysis polygons 168, 169
clutch size 80, 82–4, 89, 220
    average 245
    effect on incubation time 93
    factors affecting 249, 251
    failure to hatch 94
clutches, repeat 84
colonisation 145
    see also recolonisation
colour 26, 28, 34
    see also plumage
competition
    nest site 287–8
    prey 52
    sperm 78
    territory 53
conservation 34, 278, 296–302
conservation and management implications
    death and demography 248–50
    diet and foraging 180–2
    falconry & management 273–6
    incubation & rearing 112–13
    markers and movements 145–6
    names, races & relatives 33–4
    nesting and laying 84
    prey selection & predation 210–15
    weights and measures 54–5
consumption figures 195
'continuous selection' hypothesis 30
conventions 296–9
Cooper's Hawk 31–3, 35, 204
copulation 77–9
corvids 173, 179, 193–4, 202–4, 212
courtship behaviour 59, 70–1, 88–9, 150
    calling 73–4, 143
    flight behaviour 71–3
    provisioning 76–7
    territoriality 75
crows *see* corvids

day length 34
death, causes of 100–2, 106, 107, 233, 234–7, 265
density, dynamics and 243–4, 251
Diclofenac 282, 283
diet
    comparing between races 177–80
    divergence of 53
    requirements 195–6
    studying at nests 149–51
    variation between sexes 175–7
    variation with altitude and latitude 170–1, 182
    variation with time 171–5
    in winter 151–3
    see also feeding; food supplies; prey

Index

dimorphism *see* sexual dimorphism
disease 223, 234–5, 236–7
dispersal 125–8, 131–7, 146–7
    see also movements
displays 71–3, 75
distribution 19, 25
DNA studies 31, 119, 146, 234, 249
dominance hypotheses 52
dyes 120, 146
dynamics 243–4, 248

eagles 31
egg size 80
egg-laying 220
    date of 80–1, 81, 82, 249
    leading up to 59, 71
    loss of mass during 50
    moulting 107
    timing of 34
egg-white protein analysis 31
electronic identification 121–3
    see also radio tags
emigration 139, 218, 219, 234, 241, 249
entering 261
evolution, dynamics and 248
extinction 285
eyasses 257–8, 260
eye colour 22–3, 27

Falconiformes 30, 31
falconry 252–3
    hacking 270, 276
    history of 253–7, 276
    location aids 263–4
    training Goshawks 257–63
fat 43, 44, 47, 55
fault bars 112
feathers 19–23, 26
    content of 43, 55
    growth of 107–12, 114
    identification by 119
    see also plumage
feeding 92, 95–8, 159–60
    see also food supplies; foraging; prey
fighting 75
    see also aggression
fixed kernel contours 168, 169
fledging 123–5, 146
flight behaviours 71–3, 75, 157, 160
flying 44–5, 123
food supplies 105–6, 249, 288
    effect on clutch size 82, 83
    effect on dispersal 134–5
    influence on range size 170
    variation in 145
foraging
    behaviour when 153–4
    in Fennoscandia 160–3
    interference in 53
    in lowland Britain 155–9
    in North America 165–6
    in towns 163–5
'Freie-folge' technique 262, 263

genetics 31, 33, 35
Grey Goshawk 31
Grey-bellied Goshawk 32
grouse 134, 171–3, 182, 202, 206, 226, 290

habitat 30, 33, 59, 293–4
    choice of 162–3
    destruction of 299
    nest 60
    variation in 145
hacking 270, 276
haggards 257–8, 260
harriers 31, 32, 278
hatching 93–5
Henst's Goshawk 31, 32, 35
home-ranges 166–70
homing 137–8
human interference 223–4, 225, 237, 278
humans, death due to 251, 284–7, 302
hunting habits 33, 127–8

identification 119, 120
illegal killing 145, 234, 284–7, 290, 291, 302, 303
immigration 218–19, 249, 274, 287
imprinting 258
incubation 91–2, 93, 112, 113, 150
interbreeding 18
irruptions 140–2, 147, 241

jesses 259
juveniles
    body mass of 46–7, 48
    killing of 291–3
    mortality rates of 237–9
    movements of 146
    plumage of 19–20, 22, 34, 119
    wing-loading of 50

Kestrels 204, 264
kill rates 187, 188, 195, 196, 216
kites 31, 32

Latin names 304–6
laying *see* egg-laying
loafing 159–60
location aids 263–4
    see also radio tags

magpies *see* corvids
management *see* conservation and management
manning 259–60
mantling 58, 98, 103
markers/marking 112–13, 116, 119–23, 146
mass *see* body mass
mating 77–9
measurements 36–56
Merlins 264
Meyer's Goshawk 31, 35
migration 138–40, 140, 142–3
minimum convex polygon (MCP) 166–8, 169, 182
mortality rates 220, 233–4, 237–40, 249, 251
moulting 49, 107–12, 110, 111
movements
    choosing 142–3
    dispersal 125–8, 131–7, 146–7
    first-winter 130–1
    juveniles 146
    post-fledging 123–5
    pre-nuptial 143–5

names 18–19, 34
Nearctic races 24, 33, 39
    see also specific race
nest area/site 61–70, 86–8

nestlings 108
   death of 100–2, 106, 107
   development of 100–4
   feeding of 95–8
   plumage 19–20
nests 60–1
   building of 71, 75–6
   counting 218
   defence of 53
   destruction of 222
   habitat for 60, 62–5, 85, 88
   height of 62–3, 65, 88
   leaving 123–5, 146
   productivity of 245
   spacing of 246
   studying diet at 149–51

occupancy 68–70, 229–33
owls 204, 205, 287–8

pair formation 143–4
Palaearctic races 24, 33
   see also specific race
parenting 97–100
passagers 257–8, 260
'past isolation' hypothesis 29, 30, 34, 35
pellets 149, 150, 151, 178
Peregrine 264, 266
persecution 284–7, 300
pesticides 219, 279–84, 302
Pheasant 151, 160–1, 168–70, 191–3, 199–201, 206–8
plucking 151–2
plumage 19–23, 26–7, 32, 37
   concealment 29–30
   juveniles 119
   selective pressure 34
pollution 234, 302
population 33–4, 302
   decline of 290–1
   density of 295–6
   dynamics of 218–19
   models of 241–2
   regulation of 244–8
post-fledging 123–5, 128–9, 270
predation 152, 153–4, 181, 187, 210
   as a cause of death 237, 284, 287–9
   concealment from 29
   effect on brood size 223
   estimating the impacts of 196–8
   management of 270–3
   relates to prey density 206–10
   risk of 100–1
prey
   analysis of 149–51
   choice of 33
   deficits in 288–93
   density of 141–2, 186–7, 206–10
   encounter rate of 187
   remains collection/analysis of 149–50, 151, 174, 182
   selection of 190–4, 216
   staple 170–1, 183
   vulnerability 185–7, 190, 215
productivity 224–9, 245, 250–1
promiscuity 79
protection 234, 297
protein 43, 47, 55
provisioning 76–7, 99–100, 113, 247

races 17–18, 24–30, 34–5
   see also specific race
radio tags 116, 121–3, 146, 147, 263
radio-tracking
   diet analysis 149, 151, 152, 180, 182
   dispersal 131, 132, 145
   irruptions 142
   migration 139
recolonisation 145, 285, 296
recovery 130–4
references 312–57
relatives, Goshawk 30–2
relocation 273, 276, 278
reproductive success 240, 249
reversed size dimorphism 52–3
   see also sexual dimorphism
rings and ringing 120, 146, 147, 233, 251
runting 101–2

sea-eagles 31, 32
seasonal variations 46–51, 55
sex ratios 104–7, 114, 142
sexing 42, 100
sexual dimorphism 29, 55
   body composition 42–6
   origins of 52–3
   seasonal variations 46–8
   size 37–42, 39, 45
Sharp-shinned Hawk 31, 32, 33, 204
shedding *see* moulting
short-stay perched hunting (SSPH) 160, 165, 182
siblicide 106, 107
site fidelity 143–5
size 24–6, 27
   rules of 54
   selective pressure 34
   variation in sexes 37–42
sky-diving 72
Snowshoe Hare 141, 229
soaring 160
Sparrowhawk 31, 32, 33, 35, 204, 214, 264
specialisation 194–5
starvation 100–1, 234–6, 251
stomach content analysis 151, 174
survival rates 233–4

taste 193
territoriality 75, 166–70, 247
tracking 116, 154–5, 180–1
   see also radio-tracking
trained hawks 188, 276
training 257–63
traps & trapping 116–18, 272, 300

United States Endangered Species Act 84

vultures 31

warbling 103
weights and measures 36–56
whitewash 90, 91, 100, 112, 113, 114
wingbeats 23, 71–2
wing-loading 45, 50
wings 26, 36, 38–41, 55, 104
Woodpigeon 192, 197–9, 203, 204, 216, 228, 289–90

yarak 261–2